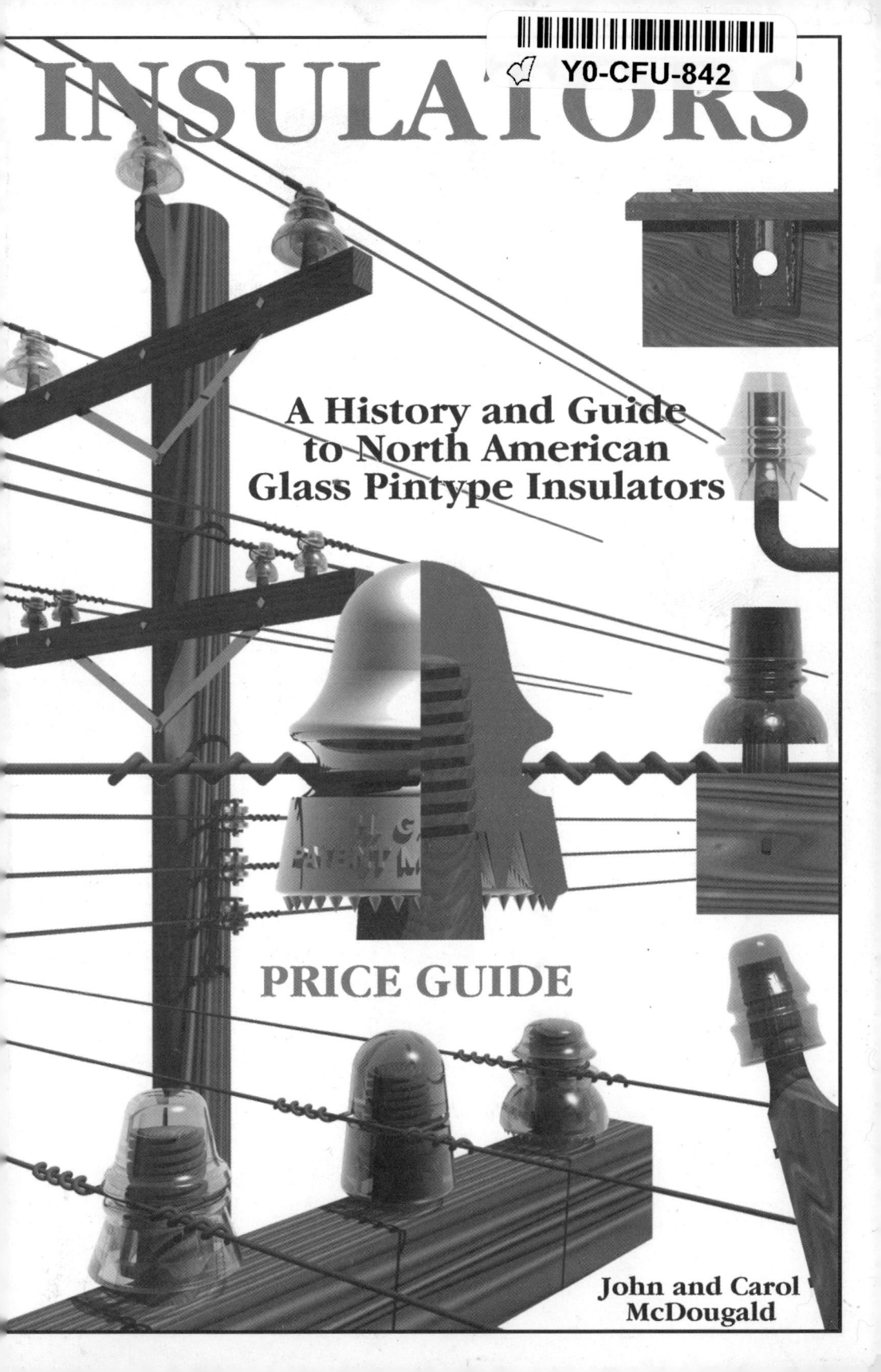

INSULATORS

A History and Guide to North American Glass Pintype Insulators

PRICE GUIDE

John and Carol McDougald

Additional copies of the
PRICE GUIDE
for
INSULATORS -
A History and Guide to North American Glass Pintype Insulators
may be ordered from:

The McDougalds
Box 1003
St. Charles, Illinois 60174-7003
or from our website: **www.crownjewelsofthewire.com**

@ $29.95 each postpaid

Library of Congress Catalog Card Number: 99-93172
ISBN 1-928701-00-0

Copyright © 1999 by John and Carol McDougald

All Rights Reserved. No part of this book may be reproduced, transmitted or stored in any form or by any means, electronic or mechanical, without prior written permission from the publisher.

Published by The McDougalds
St. Charles, Illinois

TABLE OF CONTENTS

How To Use This Price Guide .... i-x
    New Features in the 1999 Price Guide i
    General Rules of Order i
    How Do I Find My Insulator? ii
        CD Number ii
        Using the Appendices to Find Your Insulator iii
        Primary Embossing iii
        Insulator Parts iv
        Specific Embossing iv
        Base v
    Other Information About Embossings v
        Brackets ( ) [ ] { } v
        Slanted Line / v
        Punctuation vi
        NO., No., NO, No, Nº, Nº, etc. vi
        [Letter or number], [Number], and [Numbers and dots] vi
        No Name Categories vi
        Summary vii
    Color vii
    Pricing viii
    Catalog Number viii
    Acknowledgments x
Threaded Glass Pintypes -- CD 100 - CD 342 1
Threadless Glass Pintypes -- CD 700 - CD 796 175
Threadless Glass Blocks -- CD 1000 - CD 1016 189
Miscellaneous Glass -- CD 1030, CD 1038, CD 1040 191
Glass Spools and Nail Knobs -- CD 1049 - CD 1110 192
Guy Wire Strains -- CD 1130 - CD 1142 195
Battery Rests -- CD 10 - CD 61 196
Commemoratives 199
Private Issue Commemoratives 201
Salesman Miniature Insulators 202
Private Issue Miniature Insulators 203
Appendix I
    Cross Reference of Manufacturer's Style Number
    to CD Number 205
Appendix II
    Cross Reference of Primary Embossing to CD Number 211
Appendix III
    Cross Reference of Patent Dates to CD Number 217
Appendix IV
    Abbreviated Primary Embossing Cross Reference --
        Primary Embossing to Abbreviation 221
        Abbreviation to Primary Embossing 223
Insulator Reference Material 225

HOW TO USE THIS PRICE GUIDE

This Price Guide is designed to be used in conjunction with
INSULATORS - A History and Guide to North American Glass Pintype Insulators
by John and Carol McDougald

The primary objective of this price guide is to provide readers with a logical listing of each of the confirmed styles and embossings of North American glass pintype insulators, along with an estimate of the fair market value for each listing. Significant changes have been made since this price guide was last issued in 1995. Nearly 500 additional embossing variations have been included and, in total, nearly 2,000 new embossing/color combinations have been identified.

New Features in the 1999 Price Guide

One very important new feature has been included in this price guide. Mr. N. R. Woodward, the originator of the CD numbering system for threaded glass insulators (CD 100 - CD 375), was kind enough to allow the inclusion of his scale engineering drawings of those styles as part of the price guide. CD numbers outside the threaded glass pintype range were not part of Mr. Woodward's original design. These other CD numbers have gained general acceptance throughout the insulator hobby over the past 30 years.

Where scale drawings were not available, photographs have been provided so that each style has either a scale drawing or photograph with width (1st) and height (2nd) measurements in inches. We also want to acknowledge Elton Gish for his scale drawings of the battery rests and Gus Stafford for his scale drawings of the spools and nail knobs.

A few CD number changes have been made to more closely align with Mr. Woodward's assignments. The Fluid Insulator, previously shown as CD 180.1 is now listed as CD 180.5. The CD 162 American has been reassigned as CD 160.4, the CD 163 Whitall Tatum [020] has been reassigned to CD 163.4, and CD 166.2 E.L.Co. is now CD 166. Mr. Woodward has reassigned all of the CD 134 Americans to CD 133. However, due to the number of related insulators (Oakman and several No Names) and the possible confusion this might cause, these styles have been left in CD 134 for this edition of the price guide. You may want to make the appropriate adjustments in your records to coincide with Mr. Woodward's assignment.

A few CD numbers/listings that appeared in the 1995 edition have been deleted from the current edition. We have been unable to verify the existence of CD 185.2 B.E.L.Co. Therefore, this listing has been dropped. The spool embossed B.E.L.Co. now appears as CD 1110. CD 782 M.T.Co. has now been verified as a battery rest, and so this piece has been renumbered in the battery rest section as CD 61. Several listings (e.g. CD 190/191 Brookfield) had been reported by collectors and included in previous editions, but we have been unable to confirm their existence, and they have been dropped from this edition.

Two insulators embossed VMR NAPOLI (CD 122 and CD 122.6) were included in the 1995 price guide. Since that time, research has confirmed that manufacture of these insulators took place in Italy, not in North America. Even though the CD 122 was manufactured for use in the U.S. (1" pinhole) and has been found in service in great numbers in the U.S., we do not believe that it should be included in this listing since it is of non-North American manufacture. The CD 122.6 does not have a standard 1" pin, has not been found in service in the U.S., and therefore should be listed as a foreign glass insulator.

General Rules of Order

In order for you to use this price guide most effectively, you need to understand the conventions (rules) that were used to develop the listing. These few pages will

i

take a while to digest, but hopefully they will assist you when it is time for you to look up a specific listing. Pricing is discussed on page viii. Listings are sequenced by CD Number (the Consolidated Design numbering system developed by Mr. N. R. Woodward to designate style and general use of the insulator), Primary Embossing (Brookfield, Hemingray, Whitall Tatum, etc.), Specific Embossing (detailed below) and finally Base.

Within the "Specific Embossing" category, listings are sequenced based on the location of the embossing. The following priority for embossing location was used in determining the listing sequence: Dome, Front Crown (F-Crown), Front Skirt (F-Skirt), Rear Crown (R-Crown), Rear Skirt (R-Skirt), Base, Other (specified). That means that all listings within a given CD number and primary embossing that have dome embossing will appear before any listing without dome embossing. The same logic applies throughout the listing sequence. If you are hopelessly confused at this point, don't give up. The next section should help you understand the process a little better.

How Do I Find My Insulator?

CD Number

If you know the CD number of your insulator, skip to the paragraph on "Primary Embossing". If you are totally unfamiliar with the CD numbering system, the following information along with the appendices at the back of this volume will be somewhat helpful. However, you will probably need a reference book that depicts CDs for positive identification, such as the two-volume reference book *INSULATORS - A History and Guide to North American Glass Pintype Insulators,* by John and Carol McDougald. (See advertisement on page 228.)

CD numbers are sequenced based on insulator usage. Generally speaking, the larger the insulator, the larger the CD number, although there are many notable exceptions to the latter rule. In describing uses, a few terms may need clarification. "Wire grooves" are the indentations that normally go all the way around the insulator that permit the conducting wire to be fastened securely to the insulator. There are both "side" wire grooves and "top" (or "saddle") wire grooves. "Petticoat" is a term used to describe the number of skirts an insulator has, including the outside surface of the insulator. This can be determined by looking up through the pinhole of the insulator from the bottom. The number of "rings" formed by the glass will be the number of petticoats. Don't forget to count the outside skirt. Using these rules, the following chart describes the groupings for the CD numbering system:

The first grouping of CDs (CD 100 - CD 375) is for insulators with a threaded pinhole which allows the insulator to be secured in place by screwing it onto a similarly threaded pin. These are referred to as "threaded insulators".

CD 100 - CD 144: Side wire groove, single petticoat. (Some of these will have two side wire grooves.)
CD 145 - CD 184: Side wire groove, double petticoat.
CD 185 - CD 189: Pinhole goes all the way through the insulator. (Many of these are referred to as "mine insulators".)
CD 190 - CD 205: Transposition. (These will all have multiple side wire grooves, may have multiple petticoats, and may consist of two insulators mounted on a single pin.)
CD 206 - CD 249: Saddle wire groove, double petticoat.
CD 250 - CD 279: Cable, single or double petticoat. (These all have saddle wire grooves and are formed so that the wire appears to sit between two "ears".)
CD 280 - CD 289: Side wire groove, triple petticoat.
CD 290 - CD 309: Saddle wire groove, triple petticoat.
CD 310 - CD 314: Sleeves. (These covered the pins and provided additional electrical resistance for the insulator. They are not "stand alone" insulators.)
CD 315 - CD 334: One-piece, large power styles.
CD 335 - CD 375: Cemented multipiece, large power styles.

Insulators in the following CD number series (CD 700 - CD 799) are referred to as "threadless insulators" because they have no internal threading in the pinhole. Various means were used to attach these insulators to the pins.

CD 700 - CD 700.1: "Egg", Pinhole goes all the way through the insulator.
CD 701 - CD 718: "Egg", Pintypes.
CD 721 - CD 725: Pintype, designed to be used with a wood cover. (These insulators do not have wire grooves.)
CD 726 - CD 728.8: Side wire groove signal, straight skirt.
CD 729 - CD 732.2: Side wire groove signal, concave skirt.
CD 734 - CD 735.5: Side wire groove "Pilgrim Hat", one wire ridge.
CD 735.6 - CD 739.5: Side wire groove "Pilgrim Hat", two wire ridges.
CD 740 - CD 742.3: Side wire groove "Pilgrim Hat", rounded dome.
CD 743.1 - CD 743.3: Side wire groove "Beehive".
CD 780: "Bureau Knob".
CD 784 - CD 796: Cross top; Slash top; Teapot; Miscellaneous styles.

The following CD number series includes nonpintype insulators that have become inseparable from this aspect of the hobby and are, therefore, included here for completeness. Most of these listings are new additions to the current edition and are included because of the growing interest in collecting several of these specialty areas.

CD 10 - CD 99: Battery rests (found in the price guide after CD 1142).
CD 1000 - CD 1016: Threadless glass blocks.
CD 1030 - CD 1040: Miscellaneous glass
CD 1049 - CD 1110: Glass spools and nail knobs.
CD 1130 - CD 1142: Guy wire strains

Using all of the CD number groupings shown above, you should be able to turn to the appropriate section of the price guide and compare the shape and size of your insulator to the drawings and pictures that are included with the listings.

Using the Appendices to Find Your Insulator

There are three appendices at the back of the price guide to help you identify your insulator. Appendix I cross references the manufacturer's style number with a CD number. If your insulator is embossed "HEMINGRAY N⁰ 8", a reference to Appendix I would tell you that the CD number for that style is CD 112.4.

Unfortunately, many insulators were manufactured without style numbers. Appendix II cross references Primary Embossing (discussed below) to CD number. This provides a listing of all of the CDs which contain a specific primary embossing. For example, if your insulator is embossed "CAL ELEC WORKS", a reference to Appendix II will limit your search to CDs 130 and 130.1.

Appendix III cross references patent dates to CD numbers. In some cases, where only patent dates appear on the insulator, this will be the easiest way to identify your insulator. In other cases, where primary embossings also exist, this appendix should help you "lock in" on the right CD number.

Primary Embossing

Once you have identified the CD number of your insulator, you want to identify the primary embossing. This generally refers to a name that is embossed on the insulator. It is usually either the company that manufactured the insulator or the company for whom it was made. Look the insulator over and pick out a name. See if you can find it in the bold listings that contain information on your CD. If absolutely no markings appear on the insulator, look through the NO EMBOSSING categories. If you are unable to find a primary embossing category that works, you may have either selected the wrong CD number, or you may have an insulator that hasn't been cataloged yet. The latter condition would be unusual, but not impossible.

Insulator Parts

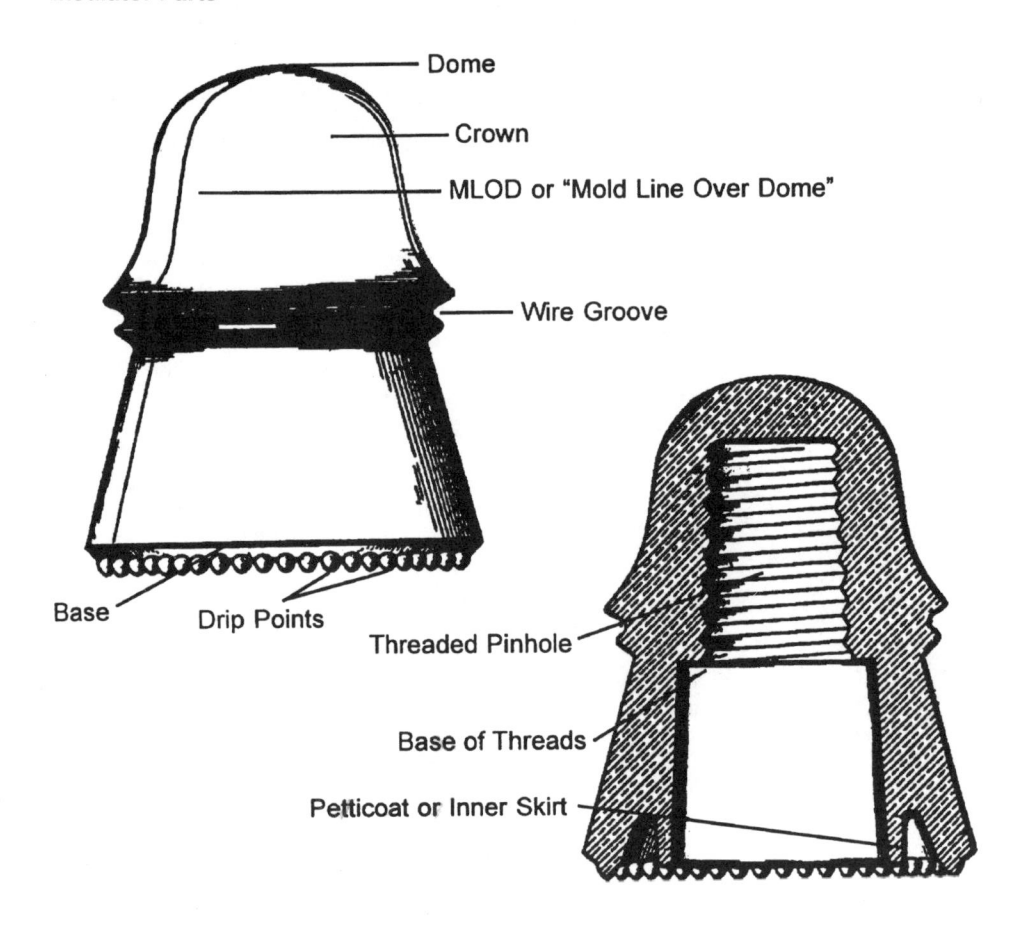

- Dome
- Crown
- MLOD or "Mold Line Over Dome"
- Wire Groove
- Base
- Drip Points
- Threaded Pinhole
- Base of Threads
- Petticoat or Inner Skirt

Specific Embossing

Once you have identified the CD number and the primary embossing, your problems are almost over, unless you happen to have an insulator with a lot of embossing variations. (CD 126 Brookfields are my personal favorite when it comes to tough embossings.) Use the accompanying drawing and look at specific locations on the insulator for embossing. Start with the dome. Insulators formed with two-piece molds (referred to as "MLOD" or "Mold Line Over Dome") cannot be dome-embossed. Therefore, if there is a seam running through the dome, embossing found near the top of the insulator is considered to be either front crown or rear crown, not dome embossing.

If there is no MLOD and there is dome embossing, confine your search to the first few listings that start "(Dome)". They will all be together.

If there is no dome embossing, look on the front crown. If only one side of the crown has significant embossing, it is considered to be the front. If embossing appears on the front crown, confine your search to listings that begin "(F-Crown)" and that have the exact embossing on the front crown that your insulator has. By following this procedure and the sequencing of embossing locations discussed earlier, you should be able to limit your search to 8-10 listings, even on the most complex listings. Be careful though, because when you get close to your embossing, there are some very subtle differences in listings that need to be reviewed to ensure you have selected the right one.

Base

Once you have gotten this far, you may find several identical listings, differing only in the last component, which describes the base of the insulator, including the presence or absence of drip points. The following abbreviations have been used to describe various bases:

SB - Smooth base.
RDP - Round drip points.
SDP - Sharp drip points.
FDP - Flat drip points {often found on Maydwell and McLaughlin insulators}.
RB - Rounded base {found on some Canadian CD 102, CD 134, CD 143, CD 162 [Hamilton], and CD 162.4 insulators}.
CD - Continuous drip {describes a rounded projection that goes all of the way around the base and forms a ledge on the inside and outside of the base; found on some Pennycuick styles and the skirt -embossed CD 112 SBT&T}.
CB - Corrugated base {found on newer Hemingray and Kimble products}
WDP - Wedge drip points {found on some CD 102 and CD 162.3 Star insulators}.
GB - Grooved base {found on some CD 143 Dwight, Great Northwestern and Canadian threadless insulators}.
BE - Base-embossed {found on American, Oakman and several threadless styles}.

All listings will contain one of these abbreviations. Hopefully, by this time you have found your insulator and are in the process of adding some pricing information to the listing of your collection.

No attempt has been made to provide individual listings based on the number of drip points on an insulator, which can vary widely for any given embossing (CD 160 H.G.CO. is probably the best example). Counting and cataloging drip points has been left to the discretion of the individual collector. Generally, the number of drip points does not affect the value of the insulator.

OTHER INFORMATION ABOUT EMBOSSINGS

Below is listed some additional information to help you interpret the symbols that are used in the embossing listings. Some other data is included to help explain what rules were used to decide whether an embossing should be listed separately or not.

Brackets ( ) [ ] { }

Three types of brackets have been used to help clarify the listings:

( ) - These serve two functions: (1) to describe the location of specific embossing - (F-Skirt); (2) to indicate embossing that is not in a straight line - (Arc) or (Circle).

[ ] - These are used to describe embossing that cannot be reproduced in the book with precision. For example, [Glass Dot] means that there is a glass dot at that particular location on the insulator. It does not mean that the words "Glass Dot" appear on the insulator. They are also used in the front of each listing to enclose a three- digit embossing index number. See the section on Catalog Numbers for more information.

{ } - These are used to describe unusual features of the embossing - {Note backwards letter}, or unusual features of the insulator itself - {Oakman style}. This bracket is particularly useful in distinguishing between NO EMBOSSING styles for a specific CD number. In addition, these brackets have been used to distinguish the embossing on the two parts of the two-piece transposition insulators - {191 Top}.

Slanted Line /

The slanted line should be read - "over". However, it has been used in two different ways. When no blotted out em-

v

bossing exists, the various lines of embossing appear sequentially down the insulator, with each individual line of embossing separated by the slanted line. If blotted out embossing appears in the listing, the blotout may appear directly under some other embossing. The important thing to know is that a blotout appears somewhere on that portion of the surface. CD 121's have been found with 'HEMINGRAY' appearing above, below, or directly on top of ['AM.TEL.& TEL. Co.' blotted out]. All three are included in the listing:
(F-Skirt)HEMINGRAY/['AM.TEL.& TEL.Co.' blotted out]

The slanted line is also used on base-embossed insulators. If no slanted line appears, the base embossing can be read continuously in a "clockwise" direction around the base. The slanted line is used to designate a change in direction of the embossing. Typically, this would indicate that the top half of the base embossing would be read in a "clockwise" direction (a slanted line would appear in the listing here), and the bottom half would be read "counterclockwise".

Punctuation

Punctuation presented the most difficult problem in deciding what to list as a separate embossing. The presence or absence of punctuation, periods versus commas, location of dashes, etc., can all be a function of the engraver, the age of the mold and/or the quality of the glass. To cite one example, there are 16 possible variations on the embossing "(F-Skirt) AM.TEL.& TEL.Co.", depending on which periods are there. All 16 variations exist. To avoid confusion, all of these possible variations have been included under the one listing shown above. In general, minor punctuation differences have not been included as separate listings and typically don't affect value. A few listings have been included where punctuation has affected price, and recognition is given to these listings with comments in the { } brackets. Otherwise, a common, or generic, listing has been included to cover all of the possible variations. Further detailing is left to the specialty collector.

NO., No., NO, No, N⁰, N⁰, etc.

There are lots of different ways to spell "number", at least as it applies to style numbers on insulators. Some embossings are absolutely consistent. Others have great variety. Seven different ways were used to say "N°20" on CD 164 McLaughlins. Only two appear in the book, and there would have only been one, except one particular embossing seemed to always appear with one drip point style so it was included. Again, the most common style was identified and included to represent all of the possibilities. Once again, the specialty collector has a great opportunity.

[Letter or number], [Number], and [Numbers and dots]

[Letter or number] and [Number] generally refer to mold information. The presence or absence of mold information and the exact location of that mold information has been recorded with some care. This information seems to be of interest to collectors and in many cases, separate listings have been included because of differences in mold number locations.

[Numbers and dots] is almost exclusively a Hemingray and related company marking. It contains mold and date information. Since dots were added to signify additional years of production, [Numbers and dots] may represent an embossing that, in fact, has no dots because it is the first year of use of the mold. In other words, some liberties may have to be taken with the use of this description.

No Name Categories

No Name insulators, those without a specific company name embossed on them, have always been a problem because they represent such a large portion of the total insulator population. Several No Name categories are included in this edition based on identifiable embossing to further delineate and identify insulators that display some form of embossing. One example of this is an insulator that is embossed with just a patent date, which is now listed under the category "PATENT - OTHER". See Appendix III for a cross reference chart of patent dates and CD numbers. Insulators that are completely unembossed (no markings of any kind) are in a primary embossing category called "NO EMBOSSING".

In addition, a number of insulators that either have blotted-out embossing that is legible (e.g. CD 102.3 blotted-out K.C.G.W.) or known company symbols (e.g. Ⓐ and £) have been included in the appropriate primary embossing category.

Summary

This information has been provided to help you understand the rules that were used to select separate embossing listings for this price guide. You may prefer to be either more or less exacting in your own record keeping. At least, you will have a set of rules to follow to help you decide how to do your own cataloging.

COLOR

Color was not an issue (for the most part) when insulators were made, and as a result, we are blessed with a complete spectrum of colors with seemingly endless variation. The authors have tried to reach a compromise solution in describing colors, and the problem has been approached this way. Descriptive adjectives have been used which identify the depth of color going from lightest to darkest as follows: Tint (Blue Tint), Ice (Ice Blue), Light (Light Green), "No Descriptor" (Green), Dark (Dark Green), Royal (Royal Purple), and Blackglass (Olive Blackglass).

Over the years, various other color descriptors have been associated with insulators that are used to distinguish the subtle differences between shades. For example, amber insulators have been described as "honey amber", "orange amber", "red amber", "golden amber", "olive amber", etc. As specialty collectors seek out every minor color variation, the descriptors tend to proliferate. The authors have attempted to list clearly distinguishable colors, although it must be conceded

that not everyone will agree on colors or what insulators those colors represent. Generally, minor differences in color will not affect price, although there are always exceptions. As a rule, insulators with more vibrant colors will be valued higher than lighter or weaker colors.

The presence of milk swirling has become a popular color feature among insulator collectors. In the 1995 edition, listings and pricing were included for some of the more commonly occurring or well-known milky insulators. Some additional listings have been included in the current edition. Insulators with significant swirling (30% or more) have been designated using the phrase "**w/ Milk Swirls**". Insulators with very heavy swirling (75% or more) have been identified as "**Milky**". The term "**Jade**" has been reserved for insulators with a solid color and an opaque appearance.

Some concern has been raised about the physical or mechanical alteration of colors. Unfortunately, some fakes have been introduced into the hobby. This subject cannot be adequately treated in this guide, but three areas of concern need to be highlighted. First, purple insulators can be altered to become a range of colors which include burgundy, straw, peach and yellow. These are difficult, if not impossible, to distinguish from legitimate specimens that have been found in these colors. Examples include insulators embossed "California" and "W.G.M.", but potential alteration is certainly not limited to these embossing categories.

Second, a surface treatment has been applied to some clear insulators (not as part of the original factory process) to give them a carnival glass appearance. While carnival glass insulators were manufactured by both Hemingray and Pyrex (in large numbers), some altered pieces have been identified and are known to exist.

Third, the color of an insulator can be altered through irradiation. Aqua insulators can be altered to become a range of colors from blue to purple, and clear insulators can be altered to colors that range from smoky to brown. Pricing in this guide assumes that the specimens are original manufacture, not alterations. For further information on altered insulators, contact the National Insulator Association.

PRICING

The price ranges that are included in this book are the authors' best estimate of the fair market value of the insulators described. Prices are based on the assumption that the condition of the insulator is Very Near Mint (VNM) or better and that the embossing is clear. Mint insulators usually command a premium price, but that is a function of the buyer, the seller and the insulator itself. A VNM insulator is considered an excellent collector's specimen, and using VNM as the standard avoids the confusion of establishing a premium for a flawless insulator.

We consulted a number of experts in the area of evaluating insulators and compared their opinions. On most of the insulators, the range of estimates of fair market value was very small. In those cases pricing was easy. However, on a number of insulators, the estimates ranged widely. All of the input was considered, but in the final analysis, the authors made the decision on the prices that are included in the guide. The prices DO NOT represent an average of the opinions expressed.

Price ranges rather than specific prices are used for several reasons. First, since market value is a function of supply and demand, it is hard to put an "exact" value on an insulator. Also, although we have tried to list all of the color variations known within a single color listing, there can be enough difference in the appearance of two insulators that they may have different value in the marketplace. Finally, prices may vary from place to place based on availability. The same ranges of price have been used throughout the book. Very common insulators valued at less than $1.00 are listed with a price of "x1".

The listings in the price guide are identical to our reference book with the following four exceptions: (1) New embossings and colors that have been verified since the reference book was published have been included. (2) Some color adjustments have been made for the purposes of consistency (e.g. "Green Jade" was changed to "Jade Green Milk"). (3) With only a few exceptions, references to "Amber Swirls" have been eliminated from the price guide. (4) Additions have been made to include insulators with milk swirling, as mentioned above.

Other irregularities such as foreign objects embedded in the glass, mold deformities, etc., have become very desirable in recent years and will certainly affect the market value of an insulator, sometimes dramatically. Differences in value for these reasons have been left to the discretion of the buyer and the seller.

Due to the recent popularity of catalog and internet sales of insulators, it seems appropriate to comment on the prices realized from these sales, some of which have been substantially above the prices suggested in this guide. The authors believe that there are two primary reasons for this. The first is that the buyer is not well-informed about the insulator market. This usually occurs with low-to medium-priced insulators that have a well-established market price, and the buyer is simply not aware of the availability of these pieces at regular market prices. This price guide is designed to help alleviate this problem.

The second reason is that many collectors, particularly specialty collectors, are primarily interested in adding an unusual piece to their personal collection, and they are willing to pay a premium, sometimes a substantial one, to acquire the new piece. While prices realized from these sales have been observed and recorded, and they may indicate a trend in the demand for certain pieces, these prices have **not** been used as the primary input for the determination of the prices in this guide.

CATALOG NUMBER

This edition of the price guide includes catalog numbers for use when referencing specific listings in the price guide. This catalog numbering system was designed so that existing catalog numbers will not change over time when new price guides are produced and new insulators cataloged. At the beginning of each listing in the price guide you will find a three-digit number, referred to as the "embossing index", enclosed in square brackets, e.g. [010]. Initially, these numbers were originally created in numerical intervals of 10 (i.e. 010, 020, 030, etc.) to provide room later as additional listings were identified and cataloged.

Since this is the second edition of the price guide to incorporate the embossing index concept, a few changes from the prior edition should be noted. First, many (but not all) of the new listings will appear with an index number that ends in something other than zero (e.g. [035]). These are new listings that appear between listings from the 1995 guide, based on the sorting sequence that is used for embossings. Second, you may find a few cases where the index number appears to be out of sequence (e.g. CD 145 Brookfield [050], [120], [060]). This is caused by a clarification in the embossing that resulted in a resequencing of the individual embossing. We felt that it was important to retain the embossing index number, and that the new placement in the price guide would be a minor inconvenience.

A full catalog number is formed by taking the individual insulator's CD number, appending the abbreviated primary embossing (found in Appendix IV), and finally appending the embossing index. This creates a unique catalog number for each listing. An example using the first listing in the price guide would be "100SURGE010". In the age of computers, this kind of systematic approach should be helpful for identification in many facets of the insulator-collecting hobby. In addition, with more emphasis being placed on embossing variations in the hobby, the catalog number (or even just the embossing index when the CD and primary embossing are known) is a precise way to communicate an exact embossing to someone over the phone, on sales lists, etc.

ACKNOWLEDGMENTS

We want to acknowledge the assistance of the following people in the preparation of this price guide: Peter Abbott, Marilyn Albers, Charlie Allmon, Dwayne Anthony, Ross Baird, Jim Bates, Ora Beary, Bob Berry, Mike Bliss, Dick Bowman, Ray Curiel, David Dahle, Duane Davenport, Morgan Davis, James Doty, Dudley Ellis, Jim Fielding, Dave French, Jim Frustieri, Mike Gay, Clarice Gordon, Paul Greaves, Mike Guthrie, Denny Hackthorne, Greg Hafer, Dave Hall, Butch and Eloise Haltman, Chris Hedges, Bill Heitkotter, Bob Hendricks, Peter Hoffman, Ken House, Don Hutchinson, Ted Ingram, Mike Issler, Bob Jones, Jeff Katchko, Tom Katonak, Stanley Klein, Gary Kline, Terry Kornberg, Shaun Kotlarsky, Mark Lauckner, Kevin Lawless, Milt Livesey, Dean Lommen, Roger Lucas, Steve Marks, Chip McElwee, Jim Meyer, Mike Miller, Tom Moulton, George Nowacki, Fred Padgett, Pat Patocka, Robin Plewes, Paul Plunkett, William Plunkett, Matt Poage, Bill Reid, John Rajpolt, Mike Roediger, Keith Roloson, Bill Rosato, Larry Shumaker, Bob Stahr, Dave Sztramski, Win Trueblood, Walter Ward, Dennis Weber, Rich Wentzel, Doug Williams, N.R. Woodward, Ron Yuhas and countless others who have continued to provide us with new information for inclusion in this edition. Their input was absolutely essential in making this price guide a reality. Many thanks to all of you.

The beautiful cover artwork was prepared by Gus Stafford. Gus shared his thoughts behind its design: "I remember picking up Woodward's original *Glass Insulator in America Report*. I was fascinated with the sectional drawing of the CD154 on the cover. I was in high school at the time and deeply involved in mechanical drawing. Years later I would re-enter the hobby with the McDougald's books and the wonderful line drawings of insulators on the covers. For thirty years I have stared out the windows of motor transportation, marveling at the glass and wire as I raced to my future. This cover is intended to span these memories, and to serve as a bridge to the future for the collectors of tomorrow."

Special thanks again go to Bill and Jill Meier who have provided an incredible amount of assistance in updating this third edition of the price guide. They provided us with dozens of new Hemingray listings and colors, and verified and clarified many more listings. Many of the features in this edition are a direct result of Bill and Jill's work. They developed a unique new font that allows most of the special symbols, embossing errors and logos found on insulators to be reproduced exactly in the listings. This removed the need for many comments in brackets "['W/T' in a triangle]", and enabled them to be replaced with "characters" such as ▽, ◇, ⌗, "HƎMINϽAY-ᗄƧ", etc. In addition, they created a more logical breakout of primary embossing categories and the appendices. To our knowledge, this price guide is the first time an exhaustive chart cross referencing patent dates embossed on insulators with Primary Embossings and CD's has ever been published. They also designed and developed the catalog numbering system. Bill and Jill spent countless hours developing a more efficient database for managing the nearly 11,000 listings we now have and also provided an interface between the new database and our desktop publishing program. Their efforts have saved us a mountain of work, and we are thankful to be able to count them among our insulator friends.

We hope you will find this revised price guide a helpful reference in your continued pursuit of the insulator-collecting hobby.

Good collecting to everyone,

John and Carol McDougald

CD 101
THREADED GLASS PINTYPES

CD 100

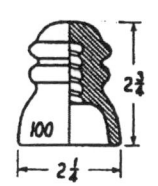

SURGE

[010] (F-Skirt) SURGE/REG. U.S. PAT. OFF.
(R-Skirt) BABSON BROS.CO./CHICAGO U.S.A. RDP
Clear, Light Straw 3-5

[020] (F-Skirt) SURGE/T.M. REQ. PAT. PEND {Note spelling} (R-Skirt) BABSON BROS.CO./ CHCIAGO U.S.A. {Note spelling} RDP
Clear 10-15

[030] (F-Skirt) SURGE/T.M. REQ. PAT. PEND {Note spelling} (R-Skirt) BABSON BROS.CO./ CHICAGO U.S.A. RDP
Clear 3-5

[040] (F-Skirt) SURGE/T.M. REQ. PAT. PEND/[Blotted out embossing] {Note spelling} (R-Skirt) BABSON BROS.CO./CHCIAGO U.S.A. {Note spelling} RDP
Clear 10-15

CD 100.2

SURGE

[010] (F-Skirt) SURGE (R-Skirt) BABSON BROS.CO. RDP
Clear 3,000-3,500

CD 100.5

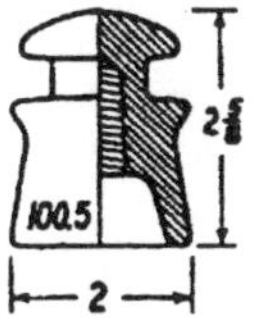

PYREX

[010] (F-Skirt) PYREX/REG. U.S. PAT. OFF.
(R-Skirt) MADE IN U.S.A. SB
Clear 75-100

[020] (F-Skirt) PYREX/REG. U.S. PAT. OFF.
(R-Skirt) ['1', '2', '3' or '4']/MADE IN U.S.A. SB
Clear 75-100

CD 100.6

HEMINGRAY

CD 100.6
$2\,1/_8 \times 3$

[010] (F-Skirt) HEMINGRAY E.1
(R-Skirt) MADE IN U.S.A. SB
Off Clear 3,000-3,500

NO EMBOSSING

[010] [No embossing] {Hemingray product} SB
Blue Tint 3,000-3,500

CD 101

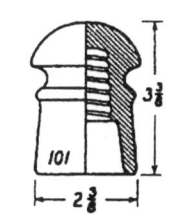

BROOKFIELD

[010] (Dome) [Number] (F-Skirt) BROOKFIELD SB
Aqua, Dark Aqua, Green Aqua 1-2

[020] (Dome) [Number] (F-Skirt) BROOKFIELD SDP
Aqua, Blue Aqua 1-2

[030] (Dome) [Number] (F-Skirt) BROOKFIELD/[Number] SDP
Dark Yellow Amber 75-100

[040] (F-Skirt) BROOKFIELD SB
Aqua, Dark Aqua 1-2
Light Blue 3-5
Light Green 5-10
Emerald Green 10-15
Yellow Green 15-20

[050] (F-Skirt) BROOKFIELD SDP
Aqua 1-2
Yellow Green 20-30
Dark Yellow Green 30-40
Olive Amber 75-100

[060] (F-Skirt) BROOKFIELD (R-Skirt) NO.9 SDP
Aqua 1-2

[070] (F-Skirt) BROOKFIELD (R-Skirt) NO.9 {Slug embossing} SB
Light Blue 3-5

1

CD 101

[080] (F-Skirt) BROOKFIELD/[Number] SB
Aqua 1-2
Olive Green 20-30
Olive Amber, Yellow Olive Green 50-75
Dark Yellow Amber 75-100

[090] (F-Skirt) BROOKFIELD/[Number]
(R-Skirt) NO.9 SB
Aqua, Dark Aqua 1-2
Light Blue 3-5
Green, Lime Green 10-15
Yellow Green 20-30

[100] (F-Skirt) BROOKFIELD/[Number]
(R-Skirt) NO.9 SDP
Aqua, Blue Aqua 1-2

[110] (F-Skirt) BROOKFIELD/[Number]
(R-Skirt) NO.9 {Slug embossing} SB
Aqua 1-2
Green 10-15

[120] (F-Skirt) BROOKFIELD/[Number]
(R-Skirt) NO.9 {Slug embossing} SDP
Aqua 1-2

CD 102

B

[010] (F-Skirt) B SB
Aqua, Light Aqua 3-5
Light Green 5-10
Green 15-20

B.G.M.CO.

[010] (Dome) [Letter or number]
(F-Skirt) B.G.M.CO. SB
Light Purple 40-50
Purple 50-75
Clear 125-150

[020] (F-Skirt) B.G.M.CO. SB
Light Purple 40-50
Purple 50-75

B.T.C.

[010] (F-Skirt) B.T.C./CANADA SB
Purple 20-30

[020] (F-Skirt) B.T.C./CANADA/['MONTREAL'
blotted out] SB
Ice Blue 1-2
Blue Gray 5-10
Light Green 10-15
Lavender, Off Clear, Purple Tint 15-20
Clear, Purple, Smoke 20-30

[030] (F-Skirt) B.T.C./MCANADAL SB
Ice Aqua, Light Aqua 1-2
Steel Blue 5-10
Light Purple 15-20
Purple, Royal Purple 20-30

[040] (F-Skirt) B.T.C./MONTREAL SB
Aqua, Ice Blue, Light Aqua, Light Blue 3-5
Light Gray, Light Green 10-15
Dark Purple, Royal Purple 20-30
Pink 30-40

[050] (F-Skirt) B.T.Co. OF CAN. SB
Aqua, Blue, Blue Aqua, Green Aqua 5-10
Light Green 10-15
Yellow Green 20-30
Emerald Green 30-40

[060] (F-Skirt) B.T.Co. OF CAN./◇ SB
Dark Olive Green 100-125

[070] (F-Skirt) B.T.Co. OF CAN./◇
(R-Skirt) ['B.T.Co. OF CAN.' blotted out]/◇
SB
Ice Aqua, Light Aqua, Light Blue Aqua 3-5
Teal Blue 350-400
Midnight Blue 500-600

[080] (F-Skirt) B.T.Co. OF CAN./◇
(R-Skirt) ◇ SB
Aqua, Ice Aqua, Ice Blue 1-2
Light Steel Blue 3-5
Light Green 10-15
Purple, Royal Purple 20-30

[090] (F-Skirt) ['B.T.C.' blotted out]/
MONTREAL SB
Aqua, Ice Aqua, Light Blue Aqua 3-5
Dark Purple, Purple 20-30

[100] (F-Skirt) ['B.T.Co. OF CAN.' blotted out]/
['◇' blotted out] (R-Skirt) ['B.T.Co. OF CAN.'
blotted out]/['◇' blotted out] SB
Dark Olive Amber, Dark Olive Green 50-75

[110] (F-Skirt) ['B.T.Co. OF CAN.' blotted out]/
◇ (R-Skirt) [Blotted out embossing]/◇ SB
Gray 10-15
Light Lavender 15-20

[120] (F-Skirt) ['B.T.Co. OF CAN.' blotted out]/
◇ (R-Skirt) ['B.T.Co. OF CAN.' blotted out]/
◇ SB
Light Aqua, Light Blue, Light Steel Blue 3-5
Gray, Steel Blue 5-10
Lavender 10-15
Light Green 15-20
Royal Purple 20-30
Brown Amber Blackglass,
Green Blackglass, Olive Blackglass 50-75

CD 102

[130] (F-Skirt) ['B.T.Co. OF CAN.' blotted out]/
◇ (R-Skirt) ◇ SB
Aqua, Ice Blue, Light Aqua,
Light Blue, Light Blue Aqua 3-5
Light Green, Light Lavender 10-15
Lavender 15-20
Dark Olive Green 50-75
Teal Blue 350-400

BROOKFIELD

[010] (Dome) [Letter or number]
(F-Crown) (Arc)W. BROOKFIELD/(Arc)NEW
YORK SB
Aqua, Light Aqua, Light Blue 3-5
Purple 800-1,000

[020] (Dome) [Letter or number]
(F-Skirt) BROOKFIELD SB
Aqua, Dark Aqua, Green Aqua x1
Emerald Green, Green, Light Green 1-2
Olive Green, Yellow Green 5-10
Teal Aqua 30-40
Milky Light Aqua 125-150
Milky Green 200-250

[030] (Dome) [Letter or number]
(F-Skirt) BROOKFIELD SDP
Aqua, Blue, Dark Aqua x1
Dark Green, Green 1-2
Lime Green, Yellow Green 5-10
Dark Olive Green 20-30
Dark Olive Amber 75-100
Dark Yellow Amber 100-125
Milky Aqua 150-175
Dark Red Amber 700-800

[040] (Dome) [Letter or number]
(F-Skirt) BROOKFIELD
(R-Skirt) BROOKFIELD SB
Aqua, Light Aqua 20-30
Green 40-50

[050] (Dome) [Letter or number]
(F-Skirt) BROOKFIELD (R-Skirt) NEW YORK
SDP
Aqua, Green Aqua x1
Green 1-2
Apple Green, Yellow Green 5-10

[060] (Dome) [Letter or number]
(F-Skirt) BROOKFIELD (R-Skirt) NEW YORK/
['B.G.M.CO.' blotted out] SB
Aqua, Light Aqua 5-10

[065] (Dome) [Letter or number]
(F-Skirt) BROOKFIELD (R-Skirt) [Blotted out embossing] SB
Dark Aqua x1

[070] (Dome) [Letter or number]
(F-Skirt) RROOKFIELD {Note spelling}
(R-Skirt) NEW YORK SDP
Aqua 5-10

[080] (Dome) [Letter or number] (F-Skirt) W.
BROOKFIELD (R-Skirt) NEW YORK SB
Aqua, Ice Aqua, Ice Blue, Light Aqua,
Light Blue x1
Light Green 1-2
Light Purple 30-40
Clear, Purple 40-50
Cornflower Blue 300-350

[090] (Dome) [Letter, number or glass dot]
(F-Skirt) BROOKFIELD (R-Skirt) NEW YORK
SB
Aqua, Dark Aqua, Light Aqua, Light Blue x1
Green, Ice Green 1-2
Yellow Green 5-10
Light Purple 20-30
Purple 30-40
Clear 40-50
Cornflower Blue 300-350
Ink Blue 400-500

[100] (F-Crown) (Arc)W. BPOOKFIELD/NEW
YOPK {Note spelling} {MLOD} SB
Aqua 5-10

[110] (F-Crown) (Arc)W. BROOFIELD/45
CLIFF ST./N.Y. {Note spelling}
(R-Crown) [Number] {MLOD} SB
Light Aqua 5-10

[120] (F-Crown) (Arc)W. BROOFIELD/45
CLIFF ST./N.Y. {Note spelling} {MLOD} SB
Light Aqua 5-10

[130] (F-Crown) (Arc)W. BROOKFIELD/
(Arc)45 CLIFF ST. (R-Crown) [Number]
{MLOD} SB
Light Blue 3-5

[140] (F-Crown) (Arc)W. BROOKFIELD/
(Arc)45 CLIFF ST./N.Y. (R-Crown) [Number]
{MLOD} SB
Light Aqua 3-5
Light Green 50-75
Green 200-250
Yellow Green 300-350

[150] (F-Crown) (Arc)W. BROOKFIELD/45
CLIFF ST./N.Y. (R-Crown) [Letter or number]
{MLOD} SB
Green Aqua, Light Aqua, Light Blue 3-5
Apple Green, Lime Green 100-125
Light Olive Green 250-300
Dark Green 700-800
Dark Teal Green 800-1,000

[160] (F-Crown) (Arc)W. BROOKFIELD/45
CLIFF ST./N.Y. {MLOD} SB
Aqua 1-2
Blue Aqua, Ice Blue 3-5

CD 102

[170] (F-Crown) (Arc)W. BROOKFIELD/45 CLIFF ST./И.Y. {Note backwards letter} (R-Crown) [Number] {MLOD} SB
Light Aqua 5-10

[180] (F-Crown) (Arc)W. BROOKFIELD/45 CLIFF ST./И.Y. {Note backwards letter} {MLOD} SB
Light Aqua 5-10
Green 150-175
Yellow Green 300-350

[190] (F-Crown) (Arc)W. BROOKFIELD/NEW YORK (R-Crown) [Letter or number] {MLOD} SB
Light Aqua 1-2
Green 150-175
Light Yellow Green 200-250

[200] (F-Crown) (Arc)W. BROOKFIELD/NEW YORK {MLOD} SB
Aqua, Light Aqua, Light Blue Aqua 1-2
Olive Green 350-400

[210] (F-Crown) (Arc)W<u>M</u> BROOKFIELD/NEW YORK (R-Crown) [Letter or number] {MLOD} SB
Light Aqua 3-5
Green 200-250

[220] (F-Crown) W. BROOKFIELD/NEW YORK {MLOD} SB
Aqua, Light Aqua 1-2

[230] (F-Crown) [Number]/(Arc)W. BROOKFIELD/(Arc)45 CLIFF ST./N.Y. (R-Crown) [Number] {MLOD} SB
Light Aqua 3-5

[240] (F-Crown) [Number]/(Arc)W. BROOKFIELD/(Arc)45 CLIFF ST./N.Y. {MLOD} SB
Light Aqua 3-5

[250] (F-Crown) [Number]/(Arc)W. BROOKFIELD/45 CLIFF ST./N.Y. {MLOD} SB
Light Aqua 3-5

[260] (F-Skirt) BROOFIELD {Note spelling} (R-Skirt) NEW YORK SB
Light Blue 5-10

[270] (F-Skirt) BROOKFIELD SB
Aqua x1
Dark Green, Emerald Green 1-2
Yellow Green 5-10

[280] (F-Skirt) BROOKFIELD SDP
Aqua x1
Emerald Green, Green 1-2
Yellow Green 5-10
Dark Olive Green 20-30
Dark Yellow Amber 100-125

[290] (F-Skirt) BROOKFIELD (R-Skirt) NEW YORK {MLOD} SB
Aqua x1
Green 1-2
Light Purple 30-40
Clear 40-50

[300] (F-Skirt) W. BROOKFIELD (R-Skirt) NEW YORK SB
Aqua, Light Aqua, Light Blue x1
Light Green 1-2
Clear 40-50

C.G.I.CO.

[010] (F-Skirt) C.G.I.CO. SB
Light Green, Light Purple,
Sage Green, Smoke 5-10
Light Plum, Rose, Straw 10-15
Aqua, Clear, Light Yellow Green, Purple 20-30
Green 30-40
Citrine 50-75
Yellow, Yellow/Peach Two Tone 100-125
Purple/Green Two Tone 150-175

CALIFORNIA

[010] (F-Skirt) CALIFORNIA SB
Aqua 15-20
Blue 20-30

DIAMOND ◇

[010] (Dome) 13 (F-Skirt) ◇ SB
Light Aqua 15-20
Light Purple, Off Clear 20-30
Royal Purple 30-40

[020] (F-Skirt) [1/3 Diamond] SB
Dark Green, Dark Yellow Amber,
Ice Aqua, Olive Green 5-10

[023] (F-Skirt) [Vertical bar]/['◇' blotted out] (R-Skirt) [Vertical bar]/['◇' blotted out] SB
Turquoise Blue 3-5
Dark Yellow Amber,
Olive Green Blackglass 10-15
Royal Purple 20-30

[027] (F-Skirt) [Vertical bar]/◇ SB
Royal Purple 20-30

[030] (F-Skirt) [Vertical bar]/◇ (R-Skirt) [Vertical bar]/['◇' blotted out] SB
Light Aqua 1-2
Blue, Gray Blue 3-5
Dark Olive Green, Dark Yellow Green 15-20
Dark Purple, Royal Purple 20-30
Teal Blue 300-350
Ink Blue, Midnight Blue 400-500

CD 102

[040] (F-Skirt) [Vertical bar]/◇
(R-Skirt) [Vertical bar]/◇ SB
Ice Blue, Light Aqua 1-2
Light Steel Blue 3-5
Light Green, Steel Blue 5-10
Light Lavender, Off Clear 10-15
Clear, Green, Lavender 15-20
Light Aqua w/ Milk Swirls,
Olive Blackglass, Royal Purple 20-30
Teal Blue 300-350
Teal Green 350-400
Midnight Blue 400-500

[050] (F-Skirt) [Vertical bar]/◇
(R-Skirt) [Vertical bar]/◇ SDP
Light Aqua 10-15
Light Aqua w/ Milk Swirls 30-40

[060] (F-Skirt) ◇ SB
Aqua 1-2
Blue Aqua, Blue Tint, Dark Olive Green,
Green Tint, Ice Blue, Ice Blue Aqua, Ice Green,
Ice Yellow Green, Light Green, Light Steel Blue,
Olive Blackglass, Sage Green 3-5
Blue Gray, Brown Amber, Root Beer Amber,
Yellow Green, Yellow Tint 5-10
Dark Yellow Green, Emerald Green,
Gray Green, Lime Green, Turquoise Blue ... 10-15
Dark Lime Green, Dark Orange Amber,
Gray, Green, Light Olive Green,
Sky Blue, Yellow Amber Blackglass 15-20
7-up Green, Light Purple, Off Clear, Purple,
Smoky Gray, Violet Gray 20-30
Dark Violet, Violet 30-40
Cornflower Blue,
Honey Yellow Amber, Light Teal Green 200-250
Teal Blue 300-350
Chartreuse, Mustard Yellow,
Olive Mustard Yellow, Teal Green 350-400
Midnight Blue 400-500
Cobalt Blue 1,250-1,500

[070] (F-Skirt) ◇ (R-Below wire groove) [Small number] SB
Aqua, Ice Blue, Light Green 10-15
Gray, Olive Blackglass, Purple, Steel Gray .. 20-30

[080] (F-Skirt) ◇ (R-Below wire groove) [Small number] (R-Skirt) [Backwards number] SB
Light Purple, Purple 20-30

[090] (F-Skirt) ◇ (R-Skirt) [1/3 Diamond] SB
Ice Green 3-5
Blue Aqua, Dark Olive Green, Olive Amber ... 5-10
Olive Green 10-15
Cornflower Blue 200-250
Mustard Yellow, Olive Mustard Yellow 350-400

[100] (F-Skirt) ◇ (R-Skirt) [1/3 Diamond]/[Number] SB
Olive Amber, Root Beer Amber 5-10
Teal Blue 350-400
Mustard Yellow 400-500

[110] (F-Skirt) ◇ (R-Skirt) [1/3 Diamond]/[Number] {Drip points are very small} SDP
Amber Blackglass,
Dark Olive Amber, Olive Blackglass 30-40
Orange Amber 40-50

[120] (F-Skirt) ◇ (R-Skirt) [Backwards number] SB
Olive Green Blackglass 3-5
Dark Yellow Amber, Root Beer Amber 5-10
Light Purple, Orange Amber 15-20
Purple 20-30
Teal Blue, Teal Green 350-400
Midnight Blue 400-500

[130] (F-Skirt) ◇ (R-Skirt) [Backwards number] {Drip points are very small} SDP
Olive Blackglass, Root Beer Amber 30-40
Dark Yellow Amber, Orange Amber 40-50

[135] (F-Skirt) ◇ (R-Skirt) [Blotted out embossing] SB
Ice Green 3-5
Midnight Blue 400-500

[140] (F-Skirt) ¤ (R-Skirt) [Number] SB
Ice Green, Light Green 3-5
Amber Blackglass, Light Aqua,
Light Blue, Light Yellow Green 5-10
Blue, Dark Green, Straw 10-15
Light Purple 15-20
Gray, Purple 20-30
Teal Blue 300-350
Teal Green 350-400
Midnight Blue 400-500

[150] (F-Skirt) ◇ (R-Skirt) ['◇' blotted out] SB
Light Aqua, Light Green,
Root Beer Amber, Straw 15-20

[160] (F-Skirt) ◇ (R-Skirt) ◇ SB
Aqua, Ice Blue, Light Aqua,
Light Blue, Light Green 3-5
Brown Amber, Dark Olive Amber, Steel Blue . 5-10
Emerald Green 10-15
Amber Blackglass, Clear, Gray,
Gray Blue, Light Purple 15-20
Royal Purple 20-30
Dark Purple 30-40

[165] (F-Skirt) ◇ {3 1/2" tall; modern style} SB
Light Peach, Light Straw 5-10

[170] (F-Skirt) ◇ {Drip points are very small} SDP
Brown Amber, Olive Amber 30-40

[180] (F-Skirt) ◇ {Solid raised diamond} SB
Light Aqua, Light Green 50-75

H.G.W.

[010] (F-Skirt) ['HGW' logo] SB
Light Green 200-250

5

CD 102

HAWLEY ⌀

[010] (F-Skirt) ⌀/HAWLEY, PA. U.S.A. SB
Aqua, Light Aqua, Light Blue 3-5
Ice Green 40-50
Light Jade Aqua Milk, Milky Green Aqua .. 125-150
Milky Blue Aqua 150-175

[020] (F-Skirt) ⌀/HAWLEY, PA. U.S.A.
(R-Skirt) ['STERLING' blotted out] SB
Aqua 3-5

[030] (F-Skirt) ⌀/HAWLEY, PA. U.S.A.
(R-Skirt) ['STERLING' blotted out]/['£' blotted
out] SB
Aqua 3-5
Light Green 50-75

[040] (F-Skirt) ⌀/HAWLEY, PA. U.S.A./
['STERLING' blotted out] SB
Aqua 3-5

[050] (F-Skirt) ⌀/HAWLEY, PA. U.S.A./['£'
blotted out] (R-Skirt) ['£' blotted out] SB
Aqua, Blue Aqua, Light Aqua 3-5
Ice Green 40-50
Milky Blue Aqua 150-175

[060] (F-Skirt) ⌀/HAWLEY, PA./U.S.A.
(R-Skirt) [Number] SB
Aqua 30-40

[070] (F-Skirt) ⌀/HAWLEY, PA./U.S.A.
(R-Skirt) ['STERLING' blotted out]/[Number]
SB
Aqua 30-40

HEMINGRAY

[010] (F-Skirt) HEMINGRAY/№ 14 SB
Aqua, Blue Aqua 5-10
Light Blue Aqua 10-15

MONTREAL

[010] (F-Skirt) MONTREAL SB
Aqua, Ice Blue 5-10

[020] (F-Skirt) MONTREAL {Drip points are
tipped in} SDP
Light Blue 75-100

[030] (F-Skirt) [Blotted out embossing]/
MONTREAL {Drip points are tipped in} SDP
Light Blue 75-100

N.E.G.M.CO.

[010] (F-Skirt) N.E.G.M.CO. SB
Dark Aqua, Green Aqua 5-10
Blue Aqua, Green 10-15
Yellow Green 20-30
Sapphire Blue 75-100

N.W.& B.I.T.CO.

[010] (F-Skirt) N.W.& B.I.T.CO SB
Light Purple, Purple 200-250
Clear, Smoke 300-350
Light Aqua, Light Green 700-800

[020] (F-Skirt) N.W.& B.I.T.CO
(R-Skirt) [Comma]/[Blotted out embossing] SB
Light Purple 200-250
Clear 300-350

NO EMBOSSING

[010] [No embossing] {Large pony; Pennycuick
style} SB
Aqua, Light Aqua, Light Blue Aqua,
Light Green Aqua 5-10
Milky Aqua 15-20
Green 50-75

NO EMBOSSING - CANADA

[010] [No embossing] RB
Aqua, Green Aqua, Light Blue 100-125
Ice Green 150-175

[020] [No embossing] RDP
Light Blue Aqua 30-40

[030] [No embossing] SB
Aqua, Blue, Blue Aqua, Light Aqua,
Steel Blue 3-5
Sage Green 5-10
Gray Lavender, Lavender 10-15
Green, Off Clear 15-20
Dark Olive Green, Purple 20-30

[040] [No embossing] SDP
Aqua 20-30

[050] [No embossing] {Drip points tipped in,
threaded to base} SDP
Light Blue 20-30

[060] [No embossing] {MLOD} SB
Aqua, Light Green 5-10
Clear, Straw 10-15
Gray Blue, Steel Gray 20-30
Blue 30-40
Light Lavender 100-125
Light Purple 125-150

[065] [No embossing] {Threaded to base} SB
Blue 15-20

[070] [No embossing] {Threaded to base}
{MLOD} SB
Aqua, Ice Aqua, Light Aqua 15-20
Lime Green 30-40
Dark Lime Green 50-75

CD 102

NO NAME

[010] (F-Crown) † (R-Crown) [Blotted out embossing] {MLOD} SB
Aqua 5-10

NO NAME - CANADA

[010] (Dome) [Backwards number] SB
Royal Purple 350-400
Dark Purple, Light Purple 500-600

[020] (F-Skirt) [Comma] {Erased N.W.& B.I.T.} SB
Aqua 125-150
Light Green 150-175
Light Purple 175-200

[030] (F-Skirt) [Number] SB
Light Purple 20-30

[040] (F-Skirt) [Vertical bar] SB
Gray, Lavender 10-15
Olive Blackglass 15-20
Royal Purple 20-30
Dark Yellow Amber 100-125

[050] (F-Skirt) [Vertical bar] (R-Skirt) [Vertical bar] SB
Ice Aqua, Light Aqua 1-2
Blue, Blue Aqua, Light Green Aqua 3-5
Ice Green, Light Gray, Sage Green,
Smoky Blue 5-10
Gray Lavender, Lavender,
Light Purple, Olive Blackglass 15-20
Purple, Royal Purple 20-30
Comflower/Purple Two Tone 150-175
Teal Blue 300-350
Midnight Blue 400-500

[055] (F-Skirt) [Vertical bar] (R-Skirt) [Vertical bar] {Both bars are short and heavy} SB
Light Blue Aqua 15-20

[060] (F-Skirt) [Vertical bar] (R-Skirt) [Vertical bar] {Threaded to base} SB
Light Gray, Light Green,
Light Green Aqua 20-30

[070] (F-Skirt) [Vertical bar] {Bar is short and heavy} SB
Ice Aqua, Steel Blue 15-20
Light Purple 50-75

PAT'D

[010] (Base) PATD {MLOD} BE
Lime Green 300-350

[020] (Dome) [Number in a circle] (Base) PATD BE
Aqua, Light Aqua 150-175

PATENT - OTHER

[010] (F-Crown) PAT.JAN. 14 1879 {MLOD} SB
Aqua 3-5

[020] (F-Crown) PAT/JAN 25TH 1870/JAN 14TH 1879 (R-Crown) [Letter or number] {MLOD} SB
Aqua, Light Aqua 3-5
Purple 800-1,000

[030] (F-Crown) PAT/JAN 25TH 1870 (R-Crown) FEB 22ND 1870/MCH 20TH 1877 {MLOD} SB
Light Aqua 3-5
Light Green 100-125
Lime Green 125-150
Yellow Green 300-350
Purple 800-1,000

[040] (F-Crown) PAT/JAN 25TH 1870 (R-Crown) [Number]/FEB 22ND 1870/MCH 20TH 1877 {MLOD} SB
Aqua 3-5
Off Clear 150-175
Yellow Green 300-350

[050] (F-Crown) PATD/JAN 25TH 1870 (R-Crown) FEB 22ND 1870/MCH 20TH 1877 {MLOD} SB
Light Aqua 3-5

[060] (F-Crown) [Number]/FEB 22ND 1870/MARCH 20TH 1877 (R-Crown) PAT/JAN 25TH 1870 {MLOD} SB
Light Aqua 3-5

[070] (F-Crown) [Number]/PAT/JAN. 25TH 1870 (R-Crown) FEB. 22ND 1870/MCH 20TH 1877 {MLOD} SB
Light Aqua 3-5

[080] (F-Crown) [Number]/PAT/JAN. 25TH 1870/JAN. 14TH 1879 {MLOD} SB
Light Green Aqua 5-10
Light Blue 10-15
Light Yellow Green 150-175

[090] (F-Crown) [Number]/PAT/MAR. 25TH 1870/JAN 25TH 1870 {Note 'MAR. 25TH 1870' date} {MLOD} SB
Light Green 100-125

[100] (F-Crown) †/PAT/JAN 25TH 1870/JAN 14TH 1879 SB
Aqua, Light Aqua 5-10
Light Purple 800-1,000

CD 102

S.F.

[010] (Dome) [Letter or number] (F-Skirt) S.F. SB
Aqua, Green Aqua 1-2
Green 3-5
Yellow Green 5-10

[020] (F-Skirt) S.F. SB
Aqua, Blue, Green Aqua 1-2
Green 3-5
Lime Green, Yellow Green 5-10

SO.MASS.TEL.CO.

[010] (F-Skirt) So. Mass. Tel. Co. SB
Light Aqua 75-100
Clear 100-125

STAR ✱

[010] (Dome) [Number] (F-Skirt) ✱ SB
Aqua, Green Aqua, Light Blue x1
Light Green 1-2
Dark Blue 5-10
Dark Yellow Green 10-15
Gray 40-50

[020] (Dome) [Number] (F-Skirt) ✱ SDP
Aqua 3-5
Dark Blue 5-10

[030] (Dome) [Number] (F-Skirt) ✱/['S.F.' blotted out] SB
Aqua, Light Aqua x1
Green 3-5

[040] (Dome) [Number] (F-Skirt) ✱/['S.F.' blotted out] SDP
Aqua 3-5
Green 5-10
Emerald Green 50-75

[050] (F-Skirt) ✱ SB
Aqua, Blue Aqua, Light Aqua, Light Blue x1
Light Green 1-2
Green 3-5
Dark Blue, Dark Green, Yellow Green 5-10
Olive Green 15-20
Emerald Green 30-40
Milky Lime Green 50-75
Olive Amber 200-250

[060] (F-Skirt) ✱ SDP
Aqua, Light Blue 3-5
Light Green 5-10

[070] (F-Skirt) ✱ (R-Skirt) [Vertical bar] SB
Aqua, Blue 3-5

[080] (F-Skirt) ✱ (R-Skirt) ✱ SB
Aqua, Light Aqua, Light Blue x1
Ice Green, Light Green 3-5
Green, Lime Green 5-10
Dark Green 10-15
Yellow Green 15-20
Yellow Olive Green 175-200
Mustard Yellow 200-250

[090] (F-Skirt) ✱ (R-Skirt) ✱ WDP
Light Aqua 75-100

[100] (F-Skirt) ✱ (R-Skirt) ✱ {Rear '✱' is small} SB
Aqua, Blue Aqua, Light Blue, Light Green 1-2
Lime Green 5-10
Yellow Green 10-15

[110] (F-Skirt) ✱/['S.F.' blotted out] SB
Aqua x1

[120] (F-Skirt) ✱/['S.F.' blotted out] SDP
Green Aqua 3-5

STERLING £

[010] (F-Skirt) STERLING (R-Skirt) £ SB
Aqua, Blue, Light Aqua 5-10

[020] (F-Skirt) £ (R-Skirt) £ SB
Aqua, Light Aqua 5-10

[030] (F-Skirt) £/[Blotted out embossing in a 1/2" circle] (R-Skirt) £ SB
Blue Aqua 5-10

WESTINGHOUSE

[010] (F-Skirt) WESTINGHOUSE/NO.3 SB
Aqua, Light Green 500-600
Green 600-700
Off Clear 800-1,000
Light Sapphire Blue 1,000-1,250
Electric Blue, Peacock Blue 2,000-2,500

CD 102.3

K.C.G.W.

[010] (F-Skirt) [Glass dot] ['KCGW' blotted out] (R-Skirt) [Glass dot] SB
Dark Aqua 30-40

[020] (F-Skirt) [Glass dot]/['KCGW' blotted out] (R-Skirt) [Glass dot]/[Number] SB
Dark Aqua 30-40
Light Green 40-50

[025] (F-Skirt) [Vertical bar] (R-Skirt) [Vertical bar] {K.C.G.W. product} SB
Ice Blue 50-75

CD 104

[030] (F-Skirt) [Vertical bar]/['KCGW' blotted out] (R-Skirt) [Vertical bar] SB
Aqua 30-40
Light Green 40-50
Ice Blue 50-75

NO NAME

[010] (F-Skirt) [Glass dot] (R-Skirt) [Glass dot] {K.C.G.W. product} SB
Dark Aqua 30-40

CD 102.4

NEW ENG.TEL.& TEL.CO.

[010] (Base) N.E.TEL.& TEL.Co. BE
Aqua 800-1,000
Green 1,500-1,750

CD 103

B

[010] (F-Skirt) B SB
Aqua, Dark Aqua 10-15
Dark Green, Green 20-30
Dark Yellow Green 50-75

GAYNER

[010] (F-Skirt) GAYNER (R-Skirt) NO.6/[Number] SB
Aqua 400-500

NO EMBOSSING

[010] [No embossing] {MLOD} SB
Teal Aqua 400-500

CD 103.4

NO EMBOSSING - MEXICO

[010] [No embossing] SB
Dark Green, Emerald Green 40-50

CD 104

BROOKFIELD

[010] (Dome) [Number] (F-Skirt) (Arc)BROOKFIELD/NEW YORK SB
Aqua, Dark Aqua, Light Aqua 5-10
Emerald Green 250-300

[015] (F-Skirt) (Arc)BROOKFIELD/NEW YORK SB
Light Aqua 5-10

[020] (F-Skirt) (Arc)W. BROOKFIELD/45 CLIFF ST/N.Y. SB
Aqua 15-20

[030] (F-Skirt) (Arc)W. BROOKFIELD/83 FULTON ST./N.Y. SB
Aqua, Light Aqua 3-5
Light Green 20-30

[040] (F-Skirt) (Arc)WM BROOFIELD/NEW YORK {Note spelling}{MLOD} SB
Lime Green 125-150

[050] (F-Skirt) (Arc)WM BROOKFIELD/NEW YORK {MLOD} SB
Aqua 20-30
Lime Green 100-125
Yellow Green 150-175

[060] (F-Skirt) BROOKFIELD (R-Skirt) NEW YORK SB
Aqua 30-40
Light Blue 40-50

NATIONAL INSULATOR CO.

[010] (Base) NATIONAL INSULATOR CO./PAT.JAN.1. & OCT.7. 1884 BE
Aqua, Light Aqua 200-250
Light Green 700-800
Lime Green 1,000-1,250

NEW ENG.TEL.& TEL.CO.

[010] (Dome) [Number] (F-Skirt) (Arc)NEW ENG.TEL & TEL./CO. SB
Aqua, Ice Blue 3-5
Blue 5-10
Green, Lime Green 10-15
Yellow Green 20-30
Olive Green 50-75
Purple 2,000-2,500

CD 104

[020] (Dome) [Number] (F-Skirt) (Arc)NEW ENG.TEL & TEL/C.O {Note placement of periods in the center of the embossing} SB
Aqua 10-15
Lime Green 40-50

[030] (F-Skirt) (Arc)NEW ENG.TEL & TEL./CO. SB
Aqua 3-5
Light Green 5-10

[040] (F-Skirt) (Arc)NEW ENG.TEL & TEL./CO. (R-Crown) [Number] {MLOD} SB
Aqua 3-5
Green 20-30

[050] (F-Skirt) (Arc)NEW ENG.TEL & TEL./CO. (R-Skirt) [Blotted out embossing in an oval slug] SB
Blue 5-10
Milky Aqua 75-100

[060] (F-Skirt) (Arc)NEW ENG.TEL & TEL./CO. (R-Skirt) [Oval slug] (Base) NATIONAL INSULATOR CO./PAT.JAN 1.& OCT.7.1884 BE
Light Blue Aqua 400-500
Light Green 800-1,000

[070] (F-Skirt) (Arc)NEW ENG.TEL & TEL./CO. (R-Skirt) ['(Arc)NEW ENG.TEL & TEL./CO.' blotted out] SB
Light Aqua 5-10

[080] (F-Skirt) (Arc)NEW ENG.TEL & TEL./CO. {MLOD} SB
Light Aqua 3-5

[090] (F-Skirt) (Arc)NƎW ƎNG.TEL & TEL./CO. {Note backwards letters} SB
Aqua 10-15

[100] (F-Skirt) (Arc)ИEW EИG.TEL & TEL./CO. {Note backwards letters} SB
Aqua, Ice Blue 5-10

[110] (F-Skirt) (Arc)ИEW ENG.TEL & TEL./CO. {Note backwards letter} SB
Light Aqua 5-10

NO EMBOSSING

[010] [No embossing] SB
Aqua, Dark Aqua, Light Aqua, Light Blue 5-10
Teal Blue 75-100

[020] [No embossing] {Pennycuick style} CD
Aqua, Light Aqua 5-10
Light Green 20-30
Dark Green 125-150

[030] [No embossing] {Pennycuick style} SB
Aqua, Blue Aqua, Dark Aqua, Light Aqua 5-10
Milky Green Aqua 75-100
Dark Green, Dark Teal Aqua 125-150
Yellow Green 200-250
Dark Yellow Green 250-300
Emerald Green 350-400

[040] [No embossing] {Unembossed National} SB
Blue Aqua 50-75

NO NAME

[010] (F-Skirt) [Blotted out embossing - 6 blots] {Pennycuick style} SB
Aqua 5-10

PATENT - DEC. 19, 1871

[010] (F-Crown) PATENT/DEC. 19. 1871 (R-Crown) 4 SB
Aqua, Green Aqua, Ice Green,
Light Aqua, Off Clear 800-1,000
Purple Tint 1,000-1,250
Light Purple 1,500-1,750

STANDARD GLASS INSULATOR CO.

[010] (Base) THE STANDARD GLASS INSULATOR CO./BOSTON MASS. BE
Light Aqua 350-400

STAR ✶

[010] (F-Skirt) ✶ SB
Blue, Light Aqua, Light Blue 10-15
Green Aqua 15-20
Green 100-125

CD 105

AM.INSULATOR CO.

[010] (Base) AM INS CO./PAT SEP 13 1881 BE
Aqua, Blue Aqua, Light Aqua,
Light Blue Aqua 75-100

[020] (Base) AM.INS.CO. PAT.SEPT. 13. 81 BE
Green Aqua, Light Green Aqua 75-100
Light Green 125-150

[030] (Base) AM.INS.CO./PAT.SEP.13 1881 {Lines and dots over dome} BE
Aqua, Light Aqua 100-125
Green 200-250

CD 106

[040] (Base) AM.INS.CO./PAT.SEP.13 1881 {Lines over dome} BE
Aqua, Blue Aqua, Light Aqua 100-125

NO EMBOSSING

[010] [No embossing] SB
Aqua, Blue Aqua, Ice Blue 50-75

CD 106

AM.TEL.& TEL.CO.

[010] (F-Skirt) AM.TEL.& TEL.CO. SDP
Aqua 2,000-2,500

AYALA

[010] (F-Skirt) R.AYALA.L {Embossed in a concave rectangle} RDP
Light Green 30-40

[020] (F-Skirt) R.AYALA.L {Embossed in a concave rectangle} SDP
Light Green 30-40
Dark Honey Amber 500-600

BIRMINGHAM

[010] (F-Skirt) BIRMINGHAM (R-Skirt) NO.10 RDP
Straw 30-40

[020] (F-Skirt) BIRMINGHAM (R-Skirt) NO.10 SDP
Straw 30-40

[030] (F-Skirt) BIRMINGHAM (R-Skirt) [Number]/NO.10 SDP
Straw 30-40

CIA COMERCIAL

[010] (F-Skirt) *Cia Comercial/Ericsson, S.A.* RDP
Light Green 10-15
Ice Green, Off Clear 20-30

[020] (F-Skirt) *Cia Comercial/Ericsson, S.A.* SDP
Aqua, Celery Green, Light Aqua,
Light Green 10-15
Green 20-30
Dark Cornflower Blue 400-500

[030] (F-Skirt) *Cia Comercial/Ericsson, S.A.* (R-Skirt) [Blotted out embossing] SDP
Light Green 10-15
Dark Cornflower Blue 400-500

DIAMOND ◇

[010] (F-Skirt) [Vertical diamond] (R-Skirt) [Vertical diamond] SDP
Pink Tint, Straw Tint 10-15

[020] (F-Skirt) ◇ SDP
Peach Tint, Pink Tint 20-30

[030] (F-Skirt) ◇ (R-Skirt) ◇ SDP
Clear, Pink Tint, Straw Tint 15-20

ERICSSON

[010] (F-Skirt) ERICSSON RDP
Clear 10-15

[020] (F-Skirt) ERICSSON (R-Skirt) TELEFONOS RDP
Blue 10-15
Cornflower Blue 300-350
Yellow Olive Green 400-500

[030] (F-Skirt) ERICSSON (R-Skirt) TELEFONOS SDP
Aqua 10-15
Root Beer Amber 500-600

[040] (F-Skirt) ERICSSON (R-Skirt) Ⓜ RDP
Pink Tint 10-15

[050] (F-Skirt) ERICSSON (R-Skirt) Ⓜ SDP
Clear, Light Pink 10-15

GAYNER

[010] (F-Skirt) GAYNER (R-Skirt) NO.90 SDP
Aqua 3-5
Green 100-125

[020] (F-Skirt) GAYNER (R-Skirt) NO.90/[Number] SB
Aqua 5-10

[030] (F-Skirt) GAYNER (R-Skirt) NO.90/[Number] SDP
Light Aqua 3-5

GOOD

[010] (F-Skirt) GOOD SB
Aqua 10-15

HEMINGRAY

[010] (Dome) 9 (F-Skirt) HEMINGRAY (R-Skirt) № 9 SDP
Aqua, Blue Aqua, Green Aqua, Light Aqua 3-5

[020] (Dome) 9 (F-Skirt) HEMINGRAY-9 (R-Skirt) MADE IN U.S.A. SDP
Light Hemingray Blue 5-10

[030] (Dome) 9 (F-Skirt) HEMINGRAY/MADE IN U.S.A. (R-Skirt) № 9 SDP
Aqua 3-5

11

CD 106

[040] (F-Skirt) HEMIGRAY/MADE IN U.S.A. {Note spelling} (R-Skirt) № 9 SDP
Aqua 10-15

[050] (F-Skirt) HEMIGRAY/№ 9 {Note spelling} (R-Skirt) PATENT/MAY 2 1893 SDP
Aqua 10-15

[060] (F-Skirt) HEMIGAY/№ 9 {Note spelling} (R-Skirt) PATENT/MAY 2 1893 {'P' is engraved over an 'H'} SDP
Aqua 10-15

[070] (F-Skirt) HEMINGRAY SB
Aqua 15-20

[080] (F-Skirt) HEMINGRAY SDP
Aqua, Blue Aqua, Dark Aqua 5-10
Hemingray Blue 10-15

[090] (F-Skirt) HEMINGRAY (R-Skirt) № 9 SB
Aqua 15-20

[100] (F-Skirt) HEMINGRAY (R-Skirt) № 9 SDP
Aqua, Blue Aqua x1
Blue 1-2
Green 15-20
Dark Green 30-40
Dark Yellow Green, Yellow Green 40-50
Emerald Green, Olive Green 75-100

[110] (F-Skirt) HEMINGRAY-9 SDP
Aqua 15-20

[120] (F-Skirt) HEMINGRAY-9 (R-Skirt) MADE IN U.S.A. RDP
Aqua, Clear, Hemingray Blue x1
Blue, Ice Blue, Ice Green, Light Aqua 1-2
Green 10-15
Light 7-up Green 15-20
7-up Green, Light Apple Green 20-30
Apple Green, Yellow Green 30-40
Aqua/Green Two Tone,
Clear/Aqua Two Tone,
Clear/Blue Two Tone 50-75

[130] (F-Skirt) HEMINGRAY-9 (R-Skirt) MADE IN U.S.A. SDP
Blue Aqua x1
Hemingray Blue 1-2
Light Hemingray Blue 3-5

[140] (F-Skirt) HEMINGRAY-9 (R-Skirt) MADE IN U.S.A./[Numbers and dots] RDP
Clear, Green Tint, Light Straw, Pink Tint x1
Ice Blue 1-2

[150] (F-Skirt) HEMINGRAY-9 (R-Skirt) MADE IN U.U.A./[Numbers and dots] {Note spelling} RDP
Clear 5-10

[160] (F-Skirt) HEMИNGRAY-9 {Note backwards letter} (R-Skirt) MADE IИ U.S.A. {Note backwards letter} RDP
Aqua 5-10
Light Hemingray Blue 10-15

[170] (F-Skirt) HEMИNGRAY-9 {Note backwards letter} (R-Skirt) MADE IИ U.S.A. {Note backwards letter} SDP
Aqua 5-10
Light Hemingray Blue 10-15

[180] (F-Skirt) HEMINGRAY-9/[Numbers and dots] (R-Skirt) MADE IN U.S.A./[Number] RDP
Clear x1
Ice Blue 1-2

[190] (F-Skirt) HEMИNGRAY-9/[Numbers and dots] (R-Skirt) [Number]/MADE IN U.S.A. RDP
Clear x1

[200] (F-Skirt) HEMINGRAY-9/[Number] (R-Skirt) MADE IN U.S.A. RDP
Clear x1
Ice Blue 1-2

[210] (F-Skirt) HEMINGRAY/M № 9 {Note extra 'M'} (R-Skirt) PATENT/MAY 2 1893 SDP
Aqua 5-10

[220] (F-Skirt) HEMИNGRAY/MADE IN U.S.A. (R-Skirt) № 9 SDP
Aqua 1-2
Light Blue 3-5

[230] (F-Skirt) HEMINGRAY/№ 9 {Note no line under 'O'} SB
Aqua, Blue Aqua 30-40

[240] (F-Skirt) HEMINGRAY/№ 9 (R-Skirt) PATENT/MAY 2 1893 SB
Aqua, Blue 50-75

[250] (F-Skirt) HEMINGRAY/№ 9 (R-Skirt) PATENT/MAY 2 1893 SDP
Aqua x1
Light Aqua 1-2
Blue Aqua 3-5
Light Blue 5-10
Hemingray Blue 10-15
Blue, Light Jade Green Milk 15-20
Jade Green Milk, Milky Aqua 20-30
Aqua Tint, Ice Aqua, Ice Green Aqua 100-125
Bubbly Hemingray Blue, Gray, Light Purple,
Lilac, Milky Hemingray Blue 125-150
Clear, Gray Green, Ice Blue,
Light Cornflower Blue, Light Gray Purple,
Light Green, Purple, Sage Gray 150-175
Dark Purple, Gray Purple 175-200
Dark Lavender, Jade Blue Milk 200-250
Gray Blue, Green, Lime Green 250-300
Dark Green 300-350
Apple Green, Light Depression Glass Green,
Yellow Green 350-400
Dark Yellow Green 400-500

CD 106

[260] (F-Skirt) HEMINGRAY/№ 9
(R-Skirt) PATENT/MAY 2 1893 {'PA' is engraved over 'PAN'} SDP
Aqua 5-10

[270] (F-Skirt) HEMINGRAY/№ 9
(R-Skirt) PATENT/MAY 2 1893/[Blotted out embossing] SDP
Jade Green Milk 20-30

[280] (F-Skirt) HEMINGRAY/№ 9
(R-Skirt) PATENT/MAY 2 18Ɛ3 {Note backwards number} SDP
Aqua 10-15
Light Aqua 20-30
Light Green 50-75
Ice Blue 150-175

[290] (F-Skirt) HEMINGRAY/№ 9
(R-Skirt) PATENTE/MAY 2 1893 {Note spelling} SDP
Aqua 5-10

[300] (F-Skirt) HEMINGRAY/№ 9
(R-Skirt) PATENTED RDP
Aqua 1-2

[310] (F-Skirt) HEMINGRAY/№ 9
(R-Skirt) PATENTED SDP
Aqua x1

[320] (F-Skirt) HEMINGRAY/№ 9
(R-Skirt) PATENTED/MAY 2 1893 RDP
Aqua 1-2

[330] (F-Skirt) HEMINGRAY/№ 9
(R-Skirt) PATENTED/MAY 2 1893 SDP
Aqua x1
Hemingray Blue 5-10
Milky Aqua, Milky Blue Aqua 20-30

[340] (F-Skirt) HEMIИGRAY/№ 9 {Note backwards letter} (R-Skirt) PATENT/MAY 1893 {Note no day in date} SDP
Aqua, Blue Aqua 50-75
Off Clear 175-200

[350] (F-Skirt) HEMIИGRAY/№ 9 {Note backwards letter} (R-Skirt) PATENT/MAY 2 1893 SDP
Aqua 5-10

[360] (F-Skirt) HEMINGRAY/['PATENT' blotted out]/№ 9/['MAY 2 1893' blotted out] (R-Skirt) PATENT/MAY 2 1893 SDP
Aqua x1
Jade Green Milk 20-30

[370] (F-Skirt) HEMNGRAY-9 {Note spelling} (R-Skirt) MADE IN U.S.A. SDP
Hemingray Blue 5-10

LYNCHBURG Ⓛ

[003] (F-Skirt) LYNCHBURG (R-Skirt) NO.10/MADE IN {Note no 'U.S.A.'} RDP
Straw 10-15

[007] (F-Skirt) LYNCHBURG
(R-Skirt) [Number]/NO.10/MADE IN/U.S.A. RDP
Light Aqua 3-5

[010] (F-Skirt) LYNCHBURG
(R-Skirt) [Number]/NO.10/MADE IN/U.S.A. SDP
Aqua 1-2
Green, Yellow Green 10-15

[020] (F-Skirt) LYNCHBURG/[Blotted out embossing]/Ⓛ (R-Skirt) NO.10/[Blotted out embossing]/MADE IN/U.S.A. SDP
Light Aqua 1-2
Light Green 5-10

[030] (F-Skirt) LYNCHBURG/Ⓛ
(R-Skirt) NO.10/MADE IN [Number]/U.S.A. {'[Number]' may appear in a variety of positions relative to the other embossing} SDP
Aqua, Light Aqua 1-2
Light Green 5-10
Green 10-15

[040] (F-Skirt) LYNCHBURG/Ⓛ
(R-Skirt) NO.10/MADE IN/U.S.A. SDP
Aqua, Light Aqua 1-2
Light Green, Light Sage Green 5-10
Dark Green 15-20

[050] (F-Skirt) LYNCHBURG/Ⓛ [Number]
(R-Skirt) NO.10/MADE IN/U.S.A. SDP
Light Aqua 1-2
Green, Yellow Green 10-15

[060] (F-Skirt) LYNCHBURG/Ⓛ [Number]
(R-Skirt) NO.10/MADE IN/[Number] U.S.A. SDP
Light Aqua 1-2

[070] (F-Skirt) LYNCHBURᎮ/Ⓛ {Note backwards letter} (R-Skirt) NO.10/MADE IN/U.Ƨ.A. {Note backwards letter} SDP
Aqua 3-5
Green 10-15

[080] (F-Skirt) Ⓛ (R-Skirt) NO.10/MADE IN/U.S.A. RDP
Aqua 3-5

[090] (F-Skirt) Ⓛ (R-Skirt) NO.10/MADE IN/['U.S.A.' blotted out]/['LYNCHBURG' blotted out] RDP
Straw 10-15

13

CD 106

[100] (F-Skirt) Ⓛ (R-Skirt) [Number]/NO.10/ MADE IN/ A. {Note no 'U.S.'} RDP
Dark Straw 10-15

[105] (F-Skirt) Ⓛ/LYNCHBURG (R-Skirt) NO.10 [Number]/MADE IN/U.S.A. SDP
Light Green 5-10
Dark Blue Aqua 10-15

[110] (F-Skirt) Ⓛ/LYNCHBURG (R-Skirt) NO.10/MADE IN {Note no 'U.S.A.'} RDP
Aqua 3-5

[120] (F-Skirt) Ⓛ/LYNCHBURG (R-Skirt) NO.10/MADE IN/U.S.A. RDP
Aqua 3-5
Gingerale, Straw 15-20

[130] (F-Skirt) Ⓛ/LYNCHBURG (R-Skirt) NO.10/MADE IN/U.S.A. SDP
Aqua, Light Blue Aqua 1-2
Light Green 5-10

[140] (F-Skirt) Ⓛ/LYNCHBURG (R-Skirt) No.10/[Number]/MADE IN/U.Ƨ.A. {Note backwards letter} SDP
Aqua 3-5
Green 10-15
Yellow Green 15-20

[150] (F-Skirt) Ⓛ/LYNCHBURG (R-Skirt) [Number]/NO.10/MADE IN {Note no 'U.S.A.'} RDP
Pink Tint 5-10
Straw 10-15
Clear, Gingerale 15-20

[160] (F-Skirt) Ⓛ/LYNCHBURG (R-Skirt) [Number]/NO.10/MADE IN {Note no 'U.S.A.'} SDP
Straw 15-20

[165] (F-Skirt) Ⓛ/LYNCHBURG (R-Skirt) [Number]/NO.10/MADE IN/U.S.A. RDP
Aqua 5-10
Straw 15-20

[170] (F-Skirt) Ⓛ/LYNCHBURG (R-Skirt) [Number]/NO.10/MADE IN/U.S.A. SDP
Aqua, Green Aqua 1-2
Green, Yellow Green 10-15
Straw 15-20

MAYDWELL

[010] (F-Skirt) MAYDWELL-9 (R-Skirt) U.S.A. SDP
Clear, Straw 1-2
Light Green, Pink 3-5
Gingerale, Gray, Steel Green 10-15
Purple Tint 20-30
Sky Blue 250-300

[020] (F-Skirt) MAYDWELL-9 (R-Skirt) U.S.A./ ['MᶜLAUGHLIN' blotted out] SDP
Ice Green, Light Smoke, Light Straw 1-2
Light Green 3-5

[030] (F-Skirt) MAYDWELL-9/['U.S.A.' blotted out] (R-Skirt) U.S.A./['MᶜLAUGHLIN-9' blotted out] SDP
Light Straw 1-2
Light Green, Light Yellow Green 3-5

MᶜLAUGHLIN

[010] (F-Skirt) MᶜLAUGHLIN № 9 FDP
Light Green 1-2
Aqua, Light Blue 3-5
Green 5-10
7-up Green, Dark Green, Dark Olive Green, Dark Yellow Green,
Green Blackglass, Yellow Green 15-20
Light Cornflower Blue, Light Olive Green 20-30
Olive Amber Blackglass 30-40

[020] (F-Skirt) MᶜLAUGHLIN № 9 RDP
Light Green 1-2
Aqua, Light Blue, Steel Blue 3-5
Green 5-10
Emerald Green 10-15
Apple Green 15-20
Gingerale 20-30

[030] (F-Skirt) MᶜLAUGHLIN № 9 SB
Ice Blue, Light Aqua, Light Gray, Light Green, Sage Green, Smoke, Steel Blue 5-10

[040] (F-Skirt) MᶜLAUGHLIN № 9 SDP
Light Blue, Light Green, Light Sage Green 1-2
Blue Aqua 3-5
Sky Blue 5-10

[050] (F-Skirt) MᶜLAUGHLIN № 9 (R-Skirt) U.S.A. FDP
Light Green 1-2
Aqua, Light Blue 3-5
Dark Green, Dark Yellow Green, Emerald Green, Green Blackglass,
Light 7-up Green, Lime Green 10-15
Gingerale, Light Cornflower Blue, Light Olive Green 20-30
Amber Blackglass 125-150

CD 106

[060] (F-Skirt) MᶜLAUGHLIN Nº 9
(R-Skirt) U.S.A. RDP
Ice Green, Light Green 1-2
Aqua, Light Blue, Light Green Aqua 3-5
Green, Light Yellow Green 5-10
7-up Green, Emerald Green, Lime Green ... 10-15
Delft Blue, Gingerale 20-30
Light Straw 30-40

[070] (F-Skirt) MᶜLAUGHLIN Nº 9
(R-Skirt) U.S.A. SDP
Light Green 1-2
Light Blue, Light Green Aqua 3-5
Green, Light Apple Green 5-10
Light Straw 20-30

NO EMBOSSING

[010] [No embossing] {Unembossed Birmingham} RDP
Straw 15-20

[020] [No embossing] {Unembossed Birmingham} SDP
Straw 15-20

NO NAME

[010] (F-Skirt) 7 (R-Skirt) 7 SB
Aqua, Green 20-30

[020] (F-Skirt) NO.9 SB
Blue, Light Aqua 5-10

[030] (F-Skirt) NO.9/[Number] SB
Blue 5-10

[040] (F-Skirt) NO.90/[Number] SB
Aqua 10-15

[050] (F-Skirt) NO.90/[Number] SDP
Aqua 20-30

[060] (F-Skirt) [Vertical bar] (R-Skirt) [Vertical bar] SB
Green, Light Green 15-20

NO NAME - MEXICO

[010] (Dome) 🔺 (F-Skirt) 🔺 RDP
Light Green 15-20

[015] (F-Skirt) SM-1 RDP
Green Blackglass 1,000-1,250

[020] (F-Skirt) [Blotted out embossing] RDP
Green 10-15

[030] (F-Skirt) [Blotted out embossing] SDP
Sage Green, Straw 10-15
Blue 15-20

[040] (F-Skirt) 🔺 RDP
Light Green 15-20

O.V.G.CO.

[010] (F-Skirt) O.V.G.CO. SB
Aqua, Light Aqua, Light Blue 1-2
Light Green 3-5
Blue 5-10
Green 30-40
Dark Green 100-125

P.S.S.A. Ⓥ

[010] (F-Skirt) Ⓥ/PSSA #9 RDP
Green, Ice Blue, Light Green,
Light Sage Green, Straw 15-20
Yellow Green 20-30

[020] (F-Skirt) Ⓥ/PSSA #9 SB
Light Blue Gray, Straw 15-20

PATENT - OTHER

[010] (F-Skirt) PATENT/MAY 2 1893
{Hemingray product} SDP
Aqua 50-75

STAR ✶

[010] (F-Skirt) ✶ SB
Aqua, Blue Aqua, Dark Aqua,
Light Aqua, Light Blue 1-2
Green, Light Green 5-10
7-up Green, Lime Green 10-15
Olive Green, Yellow Green 20-30

V.B.

[010] (F-Skirt) Ⓥ SB
Aqua 15-20

W.F.G.CO.

[010] (F-Skirt) W.F.G.CO./DENVER, COLO. SB
Blue Aqua, Light Aqua 10-15
Ice Aqua 15-20
Clear, Light Steel Blue 20-30
Ice Green 40-50

W.G.M.CO.

[010] (F-Skirt) W.G.M.Co. SB
Light Purple, Purple 30-40
Royal Purple 40-50
Off Clear 100-125
Purple/Peach Two Tone 150-175
Light Peach 200-250
Burgundy, Clear, Light Citrine,
Light Green, Peach, Pink, Straw 300-350
Light Lime Green 400-500

15

CD 106.1

CD 106.1

DUQUESNE

[010] (Dome) [Number] (F-Skirt) DUQUESNE (R-Skirt) GLASS CO. SB
Blue, Blue Aqua, Dark Aqua, Light Aqua 40-50
Cornflower Blue 100-125

[020] (Dome) [Number] (F-Skirt) DUQUESNE/ G.CO. (R-Skirt) DUQUESNE/G.CO. SB
Aqua 40-50
Cornflower Blue 100-125

[030] (F-Skirt) DUQUESNE (R-Skirt) GLASS CO. {No ribs under wire groove} SB
Cornflower Blue 125-150

[040] (F-Skirt) DUQUESNE./GLASS.CO. (R-Skirt) DUQUESNE./G.CO. SB
Cornflower Blue 100-125

CD 106.2

P.S.S.A. ⓢ

[010] (F-Skirt) P S S A #9 SB
Gray, Ice Blue, Light Aqua, Off Clear 50-75
Celery Green, Yellow, Yellow Green 75-100

[020] (F-Skirt) ⓢ /P S S A #9 SB
Ice Green 50-75

CD 106.3

DUQUESNE

[010] (Dome) [Number] (F-Skirt) DUQUESNE. (R-Skirt) GLASS.CO. SB
Aqua, Dark Aqua, Light Blue 10-15
Cornflower Blue 75-100

[020] (Dome) [Number] (F-Skirt) DUQUESNE./ GLASS.CO. (R-Skirt) DUQUESNE./G.CO. SB
Aqua 10-15
Light Green 40-50

[030] (F-Skirt) DUQUESNE. (R-Skirt) GLASS.CO. SB
Aqua 10-15

[040] (F-Skirt) DUQUESNE./G.CO. (R-Skirt) DUQUESNE./G.CO. SB
Aqua, Light Blue 10-15
Light Green 40-50

[050] (F-Skirt) DUQUESNE./GLASS.CO. (R-Skirt) DUQUESNE./G.CO. SB
Aqua, Sky Blue 10-15
Light Green 40-50

NO EMBOSSING

[010] [No embossing] {Duquesne product} SB
Aqua, Light Aqua 5-10
Cornflower Blue 75-100

CD 106.4

NO EMBOSSING - MEXICO

[010] [No embossing] SDP
Aqua, Light Green 10-15
Straw 20-30

CD 106.5

CRISOL TEXCOCO ▽

[010] (F-Skirt) CRISOL/▽/TEXCOCO (R-Skirt) No 9 {Tall variant} SDP
Green, Green Aqua 100-125
Yellow Green 125-150

[020] (F-Skirt) CRISOL/▽/TEXCOCO (R-Skirt) № 9 {Short style} SDP
Dark Green, Green 125-150
Yellow Green 150-175

CD 107

ARMSTRONG Ⓐ

[010] (F-Skirt) Ⓐrmstrong No.9 (R-Skirt) MADE IN U.S.A. [Numbers and dots] SB
Blue Tint, Clear 1-2

[020] (F-Skirt) *Armstrong's* No.9
(R-Skirt) MADE IN U.S.A. Ⓐ [Numbers and dots] SB
Clear, Off Clear 1-2

[030] (F-Skirt) *Armstron'g* No.9
(R-Skirt) MADE IN U.S.A. Ⓐ [Numbers and dots] {Note location of apostrophe} SB
Clear 10-15

CRISA

[010] (F-Skirt) CRISA (R-Skirt) [Crisa logo] 9 SB
Clear 30-40

GAYNER

[010] (F-Skirt) GAYNER (R-Skirt) NO.91/[Number] {Previously shown as CD 106 in the 1991 edition} SB
Aqua 500-600

HEMINGRAY

[010] (F-Skirt) HEMINGRAY-9 (R-Skirt) MADE IN U.S.A./[Numbers and dots] CB
Clear x1

[020] (F-Skirt) HEMINGRAY-9 (R-Skirt) MADE IN U.S.A./[Numbers and dots] SB
Clear x1

P.S.S.A.

[010] (F-Skirt) P S S A № 9 SB
Aqua, Light Green, Light Olive Green,
Light Yellow Green, Sage Green 20-30
Gingerale, Yellow Olive Green 30-40

WHITALL TATUM 🝩

[010] (F-Skirt) (Arc)WHITALL TATUM/№ 9 (R-Skirt) (Arc)MADE IN U.S.A./[Numbers]/Ⓐ/[Number] {Dots may appear around the 'Ⓐ'} SB
Clear 1-2
Dark Straw 5-10
Light Olive Green 20-30

[020] (F-Skirt) WHITALL TATUM CO. № 9/[Number] (R-Skirt) MADE IN U.S.A. 🝩 SB
Clear, Straw 1-2
Ice Aqua, Light Aqua 3-5

[025] (F-Skirt) WHITALL TATUM № 9/[Number] (R-Skirt) MADE IN U.S.A. 🝩 SB
Clear 1-2

[030] (F-Skirt) WHITALL TATUM № 9/[Number] (R-Skirt) MADE IN U.S.A. 🝩/[Number] {Dots may appear around the '🝩'} SB
Clear 1-2

CD 108

CD 107.2

NO EMBOSSING - MEXICO

[010] [No embossing] SB
Light Straw 15-20
Clear 20-30

CD 108

DIAMOND ◇

[010] (F-Skirt) ◇ RDP
Straw 50-75

DOMINION ◉

[005] (F-Skirt) DOMINION - {Note no '9'} (R-Skirt) [Number] ◉ /[Number] SDP
Clear, Straw x1

[010] (F-Skirt) DOMINION {Note no '-9'} (R-Skirt) [Dots]/◇ SDP
Clear 10-15

[020] (F-Skirt) DOMINION-6 {Note '6' style number} (R-Skirt) [Number] ◉ /[Number] RDP
Straw 40-50

[030] (F-Skirt) DOMINION-6 {Note '6' style number} (R-Skirt) [Number] ◉ /[Number] SDP
Straw 40-50

[040] (F-Skirt) DOMINION-6 {Note '6' style number} (R-Skirt) ◇/[Number] RDP
Clear, Straw 40-50

[050] (F-Skirt) DOMINION-9 (R-Skirt) ◉/[Number] RDP
Straw x1
Delft Blue 125-150

[055] (F-Skirt) DOMINION-9 (R-Skirt) ◉/[Number] (Top of pinhole) [Number] SDP
Ice Green 3-5

[060] (F-Skirt) DOMINION-9 (R-Skirt) [Dots]/◇ SDP
Straw x1
Light Green 1-2

[065] (F-Skirt) DOMINION-9 (R-Skirt) [Dots]/◇/[Number] SDP
Clear 1-2

CD 108

[070] (F-Skirt) DOMINION-9 (R-Skirt) [Number] ◇/[Number] SB
Clear, Straw 1-2

[080] (F-Skirt) DOMINION-9 (R-Skirt) [Number] ◇/[Number] SDP
Clear, Straw x1
Light Green, Peach 1-2

[090] (F-Skirt) DOMINION-9 (R-Skirt) ◇ RDP
Clear, Straw x1
Light Green 1-2
Light Peach 3-5
Delft Blue 125-150

[100] (F-Skirt) DOMINION-9 (R-Skirt) ◇ SDP
Light Peach 3-5

[110] (F-Skirt) DOMINION-['6' blotted out] {Note '6' style number} (R-Skirt) [Number] ◇/[Number] SDP
Light Peach 15-20

WHITALL TATUM ▽

[010] (F-Skirt) № 9/WHITALL TATUM CO. (R-Skirt) MADE IN U.S.A./[Number] SB
Aqua, Light Aqua, Light Straw 1-2
Light Purple 20-30

[015] (F-Skirt) № 9/WHITALL TATUM CO. (R-Skirt) [Number] MADE IN U.S.A. SB
Light Blue 1-2

[020] (F-Skirt) № 9/WHITALL TATUM CO. (R-Skirt) ▽/MADE IN U.S.A./[Number] SB
Clear, Pink 1-2
Peach 3-5

[030] (F-Skirt) WHITALL TATUM CO. № 9 (R-Skirt) MADE IN U.S.A. ▽ SB
Aqua 1-2

[040] (F-Skirt) WHITALL TATUM CO. № 9/[Number] (R-Skirt) MADE IN U.S.A. ▽ SB
Light Aqua, Pink, Straw 1-2
Light Steel Blue, Peach 3-5

CD 109

CHICAGO INSULATING CO.

[010] (Base) CHICAGO INSULATING CO./PAT.OCT.16, 1883. BE
Aqua 2,500-3,000
Milky Aqua 3,000-3,500

CD 109.5

HARLOE ⌘

[010] (F-Skirt) HARLOE'S PATENT/MAR 21. '99/MAR 12. '01/DEC 9. '02/12PI (R-Skirt) ⌘/HAWLEY PA./U.S.A. SB
Light Aqua 2,500-3,000

[020] (F-Skirt) HARLOE'S PATENT/MAR 21. '99/MAR 12. '01/DEC 9. '02/P.I. SB
Aqua 2,500-3,000

CD 109.7

CD 109.7 $2\ ^3/_8$ x 4

JOHNSON & WATSON

[010] (F-Skirt) JOHNSON & WATSON/PAT'D SB
Aqua .... 20,000+

CD 109.9

CD 109.9 $2\ ^3/_8$ x $3\ ^3/_4$

PAT APP FOR

[010] (F-Skirt) PAT/APL'D FOR {Written vertically on skirt} SB
Light Green .... 20,000+

CD 110

BROOKFIELD

[010] (F-Skirt) BROOKFIELD/PAT.OCT. 8-1907 SB
Aqua 150-175
Green 300-350
Yellow Green 400-500

CD 112

CD 110.5

NATIONAL INSULATOR CO.

[010] (Base) +NAT.INSUL.CO.+/PAT.MAY 1.DEC.25.1883. JAN.1. OCT.7.1884 {Segmented threads} BE
Light Aqua 150-175

[020] (Base) +NAT.INSUL.CO.+/PAT.MAY.1.DEC.25.1883 JAN.1 OCT.7.1884 BE
Aqua 150-175
Light Green 600-700

[030] (Base) +NAT.INSULATOR CO.+/PAT.MAY.1.DEC.25.1883 JAN.1.OCT.7.1884. BE
Light Aqua 150-175

[040] (Dome) 1 (Base) +NAT.INSUL.CO.+/PAT.MAY 1.DEC.25.1883. JAN.1. OCT.7.1884 {Segmented threads} BE
Light Aqua 150-175

[050] (Dome) 1 (Base) +NAT.INSULATOR CO.+/PAT.MAY.1.DEC.25.1883 JAN.1.OCT.7.1884. BE
Light Aqua 150-175

NEW ENG.TEL.& TEL.CO.

[010] (F-Skirt) (Arc)NEW ENG.TEL & TEL/CO (Base) +NAT.INSULATOR CO.+/PAT.MAY.1.DEC.25.1883 JAN.1.OCT.7.1884. BE
Light Aqua 300-350

[020] (F-Skirt) NEW ENG TEL./&/TEL.CO. (Base) +NAT.INSULATOR CO.+/PAT.MAY 1.DEC.25.1883 JAN.1.OCT.7.1884. BE
Light Aqua 400-500

[030] (F-Skirt) NEW ENG TEL/&/TEL.CO. (Base) +NAT.INSUL.CO.+/PAT.MAY 1.DEC.25.1883. JAN.1.OCT.7.1884 {Segmented threads} BE
Light Aqua 400-500
Light Green 800-1,000

CD 110.6

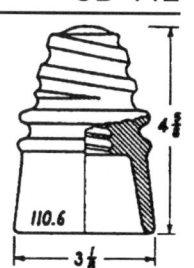

NATIONAL INSULATOR CO.

[010] (Base) NATIONAL INSULATOR CO./PAT.MAY 1. DEC.25.1883 JAN.1. OCT.7.1884 BE
Aqua, Light Aqua, Light Blue Aqua 1,750-2,000
Light Green 2,000-2,500
Green 5,000-7,500

CD 111

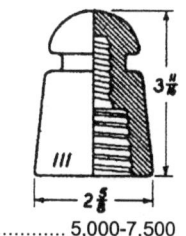

PYREX

[010] (F-Skirt) PYREX SB
Clear 5,000-7,500

CD 112

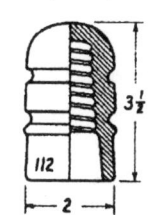

B

[010] (Dome) [Number] (F-Center between wire grooves) B SB
Light Aqua 3-5

[020] (Dome) [Number] (F-Skirt) B SB
Aqua x1

[030] (Dome) [Number] (F-Skirt) B SDP
Aqua, Dark Aqua 1-2
Green 5-10

[040] (F-Center between wire grooves) B SB
Light Aqua 3-5
Light Green 10-15
Apple Green 20-30

[050] (F-Skirt) B SB
Light Aqua x1

[060] (F-Skirt) B/['BROOKFIELD' blotted out] {The 'B' from 'BROOKFIELD' has been reused} SB
Dark Aqua 1-2

19

CD 112

BROOKFIELD

[010] (Dome) [Letter or number]
(F-Skirt) BROOKFIELD SB
Aqua, Dark Aqua, Dark Green Aqua,
Light Aqua x1
Green 3-5
Olive Green 10-15

[020] (Dome) [Letter or number]
(F-Skirt) BROOKFIELD SDP
Aqua, Green Aqua x1
Green 3-5
Emerald Green, Yellow Green 5-10
Olive Green 20-30
Olive Amber 125-150

[030] (Dome) [Letter or number]
(F-Skirt) BROOKFIELD (R-Skirt) NEW YORK SB
Light Aqua, Light Blue x1
Blue, Light Green 1-2
Green 3-5
Yellow Green 5-10

[040] (Dome) [Letter or number]
(F-Skirt) BROOKFIELD (R-Skirt) NEW YORK SDP
Olive Amber 125-150

[050] (F-Center between wire grooves) B
(F-Skirt) BROOKFIELD (R-Skirt) NEW YORK SB
Aqua, Ice Blue, Ice Green 5-10
Yellow Green 15-20

[060] (F-Skirt) BROOKFIELD SB
Green 3-5

[070] (F-Skirt) BROOKFIELD (R-Skirt) № 31 SB
Aqua, Light Aqua 30-40

CALIFORNIA

[010] (F-Skirt) CALIFORNIA SB
Smoke 15-20
Light Purple 20-30
Clear, Light Green, Light Peach,
Sage Green Tint 50-75
Aqua {Style variant},
Blue {Style variant}, Yellow 150-175

HAWLEY ⌗℗

[010] (F-Skirt) ⌗℗ (R-Skirt) HAWLEY. PA. U.S.A. SB
Aqua 5-10

[020] (F-Skirt) ⌗℗ /['STERLING' blotted out]
(R-Skirt) HAWLEY, PA. U.S.A. SB
Aqua, Light Green Aqua 5-10

[030] (F-Skirt) ⌗℗ /['STERLING' blotted out]
(R-Skirt) HAWLEY, PA. U.S.A. /['£' blotted out] SB
Aqua 5-10
Light Green 15-20

LYNCHBURG ℒ

[010] (Dome) [Letter or number]
(F-Skirt) LYNCHBURG No 31 (R-Skirt) ℒ
[Number] MADE IN/U.S.A. RDP
Green 5-10
Apple Green, Clear, Pink Tint,
Straw, Yellow Green 15-20
Emerald Green 30-40

[020] (Dome) [Letter or number]
(F-Skirt) LYNCHBURG No 31 (R-Skirt) ℒ
[Number] MADE IN/U.S.A. SDP
Aqua, Light Green Aqua 1-2
Green, Sage Green 5-10

[030] (F-Skirt) LYNCHBURG No 31
(R-Skirt) ℒ [Number] MADE IN/U.S.A. RDP
Pink 15-20

[040] (F-Skirt) LYNCHBURG No 31
(R-Skirt) ℒ [Number] MADE IN/U.S.A. SDP
Aqua 1-2
Yellow Green 15-20

NEW ENG.TEL.& TEL.CO.

[010] (F-Skirt) NEW ENG (R-Skirt) TEL & TEL CO. {MLOD} SB
Light Aqua 50-75

[020] (F-Skirt) NEW ENG (R-Skirt) TEL & TEL.Co. SB
Aqua 50-75
Green 125-150

NO EMBOSSING

[010] [No embossing] {Unembossed S.B.T.& T.; Pennycuick style} CD
Aqua 15-20
Dark Aqua 30-40
Light Gray 75-100
Green 100-125
Olive Green 200-250

O.V.G.CO.

[010] (F-Skirt) #11 O.V.G.CO. SB
Green Aqua, Light Aqua 1-2
Blue 3-5
Light Green 5-10
Green 30-40
Light Purple 2,000-2,500

CD 113

S.B.T.& T.CO.

[010] (F-Center between wire grooves) S.B.T.& T.Co. SB
Dark Aqua, Ice Blue, Ice Green,
Light Green Aqua 40-50
Off Clear 50-75
Gray Green, Green, Light Yellow Green ... 150-175

[020] (F-Skirt) S.B.T.& T.Co. SB
Aqua, Ice Blue, Light Aqua,
Light Green, Off Clear 30-40
Green 50-75
Forest Green 100-125

[030] (F-Skirt) S.B.T.& T.Co. {Pennycuick style} CD
Aqua, Light Aqua 30-40

S.F.

[010] (F-Skirt) S.F. SB
Aqua 15-20
Light Green 50-75

STAR ✷

[010] (F-Skirt) ✷ SB
Aqua, Blue Aqua, Light Aqua, Light Blue 1-2
Light Green 3-5
Apple Green, Green,
Hemingray Blue, Lime Green 5-10
Olive Green, Yellow Green 15-20
Chartreuse 75-100
Light Gray Lavender 100-125
Yellow Olive Green 125-150
Olive Amber 250-300

[020] (F-Skirt) ✷ (R-Skirt) ✷ SB
Aqua, Dark Aqua, Light Blue 3-5
Green, Light Yellow Green, Lime Green 20-30
Olive Green 50-75

STERLING £

[010] (F-Skirt) STERLING (R-Skirt) £ SB
Light Aqua 10-15

CD 112.4

B.T.C.

[010] (F-Skirt) B.T.CO. OF CAN./◇ SB
Light Aqua 125-150
Olive Green Blackglass 150-175

[020] (F-Skirt) ['B.T.CO. OF CAN.' blotted out]/ ◇ (R-Skirt) ['B.T.CO. OF CAN.' blotted out]/ ◇ SB
Light Aqua 100-125

DIAMOND ◇

[010] (F-Skirt) [Vertical bar]/◇ (R-Skirt) [Vertical bar]/◇ SB
Aqua, Light Blue 20-30
Olive Blackglass 50-75

HEMINGRAY

[010] (F-Skirt) HEMINGRAY/N° 8 SB
Aqua, Light Aqua 30-40
Light Blue Aqua 40-50
Ice Green, Light Blue, Light Green 50-75
Blue, Ice Blue 75-100
Off Clear 200-250
Purple Tint 300-350
Light Purple 3,000-3,500
Golden Amber 4,000-4,500

[020] (F-Skirt) HEMINGRAY/N° 8/['S.B.T.& T.Co.' blotted out] SB
Aqua 30-40
Ice Aqua, Light Blue Aqua 40-50
Ice Green 50-75

CD 112.5

DIAMOND ◇

[010] (F-Skirt) [Vertical bar]/◇ (R-Skirt) [Vertical bar] SB
Light Aqua 20-30

CD 112.5
$2\ {}^{1}/_{2} \times 3\ {}^{3}/_{8}$

NO EMBOSSING - CANADA

[010] [No embossing] SB
Dark Olive Green 30-40

NO NAME - CANADA

[010] (F-Skirt) [Vertical bar] (R-Skirt) [Vertical bar] SB
Blue Aqua 20-30
Brown Amber, Dark Green,
Dark Olive Green, Dark Yellow Amber 30-40

CD 113

ARMSTRONG Ⓐ

[010] (F-Skirt) *Armstrong's* No.13 (R-Skirt) MADE IN U.S.A. Ⓐ [Numbers and dots] SB
Clear, Straw 5-10

21

CD 113

HEMINGRAY

[010] (F-Skirt) HEMIGRAY/N⁰ 12 {Note spelling} (R-Skirt) PAT∃NT/MAY 2 1893 {Note backwards letter} SDP
Aqua 10-15

[020] (F-Skirt) HEMIGRAY/N⁰ 12 {Note spelling} (R-Skirt) PATENTED SDP
Aqua, Blue Aqua 10-15

[025] (F-Skirt) HEMINGAY-12 {Note spelling} (R-Skirt) MADE IN U.S.A. RDP
Hemingray Blue 10-15

[030] (F-Skirt) HEMINGAY-12 {Note spelling} (R-Skirt) MADE IN U.S.A. SDP
Aqua, Hemingray Blue,
Light Hemingray Blue 10-15

[040] (F-Skirt) HEMINGRAY (R-Skirt) N⁰ 12 SDP
Aqua x1
Blue Aqua 1-2
Hemingray Blue 5-10
Dark Green, Green 75-100
Dark Yellow Green 100-125
Dark Yellow Olive Green,
Emerald Green 150-175

[045] (F-Skirt) HEMINGRAY (R-Skirt) N⁰ 12/ [Vertical 'N⁰' blotted out] SDP
Aqua 5-10

[050] (F-Skirt) HEMINGRAY (R-Skirt) PATENTED SDP
Aqua 1-2

[060] (F-Skirt) HEMINGRAY-12 (R-Skirt) MADE IN U.S.A. RDP
Aqua, Clear x1
Hemingray Blue, Ice Blue 1-2
Light Hemingray Blue 5-10

[070] (F-Skirt) HEMINGRAY-12 (R-Skirt) MADE IN U.S.A. SDP
Aqua x1
Hemingray Blue 1-2
Light Blue 3-5
Green 75-100
Emerald Green 100-125

[080] (F-Skirt) HEMINGRAY-12 (R-Skirt) MADE IN U.S.A./[Numbers and dots] RDP
Clear x1

[090] (F-Skirt) HEMINGRAY-12 (R-Skirt) MADE IN U.S.A./[Numbers and dots] {Mold variation has a larger lower wire ridge and tapers more from there to the lower wire groove} RDP
Clear x1
Blue Tint, Clear/Yellow Two Tone,
Light Yellow 10-15

[100] (F-Skirt) HEMINGRAY-12/[Numbers and dots] (R-Skirt) MADE IN U.S.A./[Number] RDP
Clear x1
Ice Blue, Pink Tint 1-2

[110] (F-Skirt) HEMINGRAY/HE N⁰ 12/ [Upside down 'N⁰' blotted out] {Note extra 'HE'} (R-Skirt) PATENT/MAY 2 1893 SDP
Aqua, Light Aqua 10-15
Steel Blue 100-125
Lavender 175-200

[115] (F-Skirt) HEMINGRAY/MA N⁰ 12 {Note extra 'MA'} (R-Skirt) PATENT/MAY 2 1893 SDP
Aqua 10-15

[120] (F-Skirt) HEMINGRAY/MADE IN U.S.A. (R-Skirt) N⁰ 12 SDP
Aqua 1-2
Hemingray Blue, Light Blue 3-5

[130] (F-Skirt) HEMINGRAY/N⁰ 12 (R-Skirt) PATENT/MAY 2 1893 SDP
Aqua, Light Aqua x1
Hemingray Blue 5-10
Milky Aqua 20-30
Jade Green Milk 30-40
Light Blue, Milky Blue Aqua 50-75
Ice Aqua, Light Green, Light Sage Green ... 75-100
Light Gray Purple, Sage Gray,
Sage Green 150-175
Lavender, Pink, Purple/Sage Two Tone 175-200
Jade Blue Milk, Purple 200-250
Dark Purple, Emerald Green 250-300

[140] (F-Skirt) HEMINGRAY/N⁰ 12 (R-Skirt) PAT∃NT/MAY 2 1893 {Note backwards letter} SDP
Aqua 10-15

[150] (F-Skirt) HEMINGRAY/N⁰ 12 (R-Skirt) PATENTED SDP
Aqua 1-2
Milky Aqua 30-40

[160] (F-Skirt) HEMINGRAY/N⁰ 12 {'A' is engraved over a 'Y'} (R-Skirt) PATENT/MAY 2 1893 SDP
Aqua 3-5

[170] (F-Skirt) HEMINGRAY/N⁰ 12/['9' blotted out] (R-Skirt) PATENT/MAY 2 1893 SDP
Aqua 1-2
Blue Aqua 3-5
Milky Aqua 20-30
Jade Green Milk 30-40

[175] (F-Skirt) HEMINGRAY/N⁰ 12/['MAY 2 1893' blotted out] (R-Skirt) PATENT/MAY 2 1893 RDP
Aqua 10-15

CD 114

[180] (F-Skirt) HEMINGRAY/N⁰ 12/['MAY 2 1893' blotted out] (R-Skirt) PATENT/MAY 2 1893 SDP
Aqua x1
Milky Aqua 20-30
Jade Green Milk 30-40

[190] (F-Skirt) HEMINGRAY/N⁰ 9 {Note 'N⁰ 9' style number} (R-Skirt) PATENT/MAY 2 1893 SDP
Aqua 125-150

[200] (F-Skirt) HEMINGRAY/[Blotted out embossing]/N⁰ 12 (R-Skirt) PATENT/MAY 2 1893 SDP
Aqua x1

NO NAME

[010] (Dome) H SB
Light Aqua 50-75

NO NAME - CANADA

[010] (F-Skirt) [Dots] (R-Skirt) [Dots] {Several variations that appear like braille} SDP
Royal Purple 175-200
Light Purple, Milky Royal Purple 300-350
Milky Lavender, Milky Off Clear 700-800

PATENT - OTHER

[010] (F-Skirt) PATENTED {Hemingray product} SDP
Aqua 50-75

STAR ✯

[010] (F-Skirt) ✯ SB
Aqua, Blue Aqua, Light Blue 3-5
Green, Lime Green 10-15
Emerald Green, Yellow Green 20-30
Olive Green 30-40

WESTINGHOUSE

[010] (F-Skirt) WESTINGHOUSE/NO.4 SB
Light Green, Off Clear 500-600
Blue Aqua 800-1,000
Brooke's Blue 1,750-2,000

WHITALL TATUM ⱳ

[010] (F-Skirt) WHITALL TATUM CO. No.13/[Number] (R-Skirt) MADE IN U.S.A. ⱳ SB
Clear, Light Straw 3-5

[020] (F-Skirt) WHITALL TATUM CO. No.13/[Number] (R-Skirt) MADE IN U.S.A. ⱳ / [Number] SB
Clear 3-5

[030] (F-Skirt) WHITALL TATUM No.13/[Number] (R-Skirt) MADE IN U.S.A. Ⓐ/[Numbers and dots] SB
Clear 3-5

[040] (F-Skirt) WHITALL TATUM No.13/[Number] (R-Skirt) MADE IN U.S.A. ⱳ SB
Clear 3-5

[050] (F-Skirt) WHITALL TATUM No.13/[Number] (R-Skirt) MADE IN U.S.A. ⱳ / [Number] SB
Clear 3-5

CD 113.2

DUQUESNE

[010] (F-Skirt) DUQUESNE (R-Skirt) GLASS.CO. SB
Aqua 150-175
Cornflower Blue 250-300

[020] (F-Skirt) DUQUSNE {Note spelling} (R-Skirt) GLASS.CO. SB
Blue Aqua 175-200
Cornflower Blue 250-300

NO NAME

[010] (F-Skirt) GLASS.CO. {Duquesne product} SB
Blue 200-250

CD 114

HEMINGRAY

[010] (F-Skirt) HEMINGRAY/N⁰ 11 (R-Skirt) PATENT MAY 2 1893 SDP
Aqua, Green Aqua, Light Aqua 5-10
Light Blue Aqua 10-15
Hemingray Blue 40-50
Light Green 50-75
Ice Aqua, Ice Blue 75-100
Off Clear 100-125
Jade Aqua Milk, Lime Green 200-250
Jade Blue Milk 300-350
Red Amber 5,000-7,500

23

CD 114.2

CD 114.2

STANDARD GLASS INSULATOR CO.

[010] (Base) THE STANDARD GLASS INSULATOR CO./BOSTON MASS. BE
Blue Aqua, Light Aqua 2,000-2,500

CD 115

ARMSTRONG Ⓐ

[010] (F-Skirt) Armstrong No 3 (R-Skirt) MADE IN U.S.A. [Number and dots] SB
Clear 3-5

[020] (F-Skirt) Ⓐrmstrong No 3 (R-Skirt) MADE IN U.S.A. [Number and dots] SB
Blue Tint, Clear 3-5

[030] (F-Skirt) Ⓐrmstrong No 3 Ⓐ (R-Skirt) MADE IN U.S.A. [Number and dots] SB
Clear, Off Clear 3-5

[040] (F-Skirt) *Armstrong's* No 3 (R-Skirt) MADE IN U.S.A. Ⓐ [Numbers and dots] SB
Clear 3-5

BROOKFIELD

[010] (Dome) [Number] (F-Skirt) BROOKFIELD SB
Aqua 1-2
Green 15-20

[020] (F-Skirt) BROOKFIELD SB
Aqua, Green Aqua 1-2
Green 15-20
Emerald Green 30-40
Yellow Green 50-75
Olive Green 100-125

[030] (F-Skirt) BROOKFIELD SDP
Aqua 30-40
Green 100-125

DIAMOND ◇

[010] (F-Skirt) ◇ (R-Skirt) ◇ RDP
Straw 20-30

[020] (F-Skirt) ◇ (R-Skirt) ◇ SB
Aqua, Blue, Ice Blue, Light Aqua,
Light Green, Light Steel Blue, Straw 10-15
Peach 20-30

DOMINION ◇

[010] (F-Skirt) DOMINION-10 (R-Skirt) ◇ RDP
Clear, Straw 3-5
Light Green 5-10
Dark Straw, Yellow Straw 10-15

HEMINGRAY

[010] (F-Skirt) HEMIGRAY {Note spelling} (R-Skirt) PATENTED/MAY 2 1893 SDP
Aqua 10-15

[020] (F-Skirt) HEMINGRAY (R-Skirt) № 10 SDP
Aqua, Blue Aqua, Light Aqua 1-2
Hemingray Blue 3-5
Light Green 5-10
Green 10-15

[030] (F-Skirt) HEMINGRAY (R-Skirt) PATENTED SDP
Hemingray Blue 3-5

[040] (F-Skirt) HEMINGRAY (R-Skirt) PATENTED/MAY 2 1893 SDP
Aqua, Blue Aqua 1-2
Green 20-30

[050] (F-Skirt) HEMINGRAY-10 (R-Skirt) MADE IN U.S.A. RDP
Clear x1
Aqua, Green Aqua, Ice Blue 1-2
Hemingray Blue, Ice Green 3-5

[060] (F-Skirt) HEMINGRAY-10 (R-Skirt) MADE IN U.S.A. SDP
Aqua 1-2
Hemingray Blue 3-5

[070] (F-Skirt) HEMINGRAY-10 (R-Skirt) MADE IN U.S.A./[Numbers and dots] CB
Clear x1
Pink Tint 1-2

[080] (F-Skirt) HEMINGRAY-10 (R-Skirt) MADE IN U.S.A./[Numbers and dots] SB
Clear x1

CD 115

[090] (F-Skirt) HEMINGRAY-10
(R-Skirt) [Numbers and dots]/MADE IN U.S.A.
SB
Clear 10-15

[100] (F-Skirt) HEMINGRAY-10
(R-Skirt) [Numbers] SB
Clear 10-15

[110] (F-Skirt) HEMINGRAY-10/[Numbers and dots] (R-Skirt) MADE IN U.S.A./[Number] RDP
Clear x1
Ice Aqua, Ice Blue 1-2

[120] (F-Skirt) HEMINGRAY-10/[Number]
(R-Skirt) MADE IN U.S.A. RDP
Ice Blue 1-2

[130] (F-Skirt) HEMINGRAY-10/['[Number]'
blotted out] (R-Skirt) MADE IN U.S.A./
[Numbers and dots] SB
Clear x1

[140] (F-Skirt) HEMINGRAY-16 {Note '16' style number} (R-Skirt) MADE IN U.S.A./[Number] SB
Clear 15-20

[150] (F-Skirt) HEMINGRAY/MADE IN U.S.A.
(R-Skirt) № 10 SDP
Aqua 1-2
Hemingray Blue 3-5

[160] (F-Skirt) HENNGRAY-10 {Note spelling}
(R-Skirt) MADE IN U.S.A. SDP
Hemingray Blue 10-15

[170] (F-Skirt) HEWINGRAY {Note spelling}
(R-Skirt) № 10 SDP
Aqua 10-15

[180] (F-Skirt) HEWINGRAY/MADE IN U.S.A.
{Note spelling} (R-Skirt) № 10 SDP
Aqua 10-15

MAYDWELL

[010] (F-Skirt) MAYDWELL-10 (R-Skirt) U.S.A. RDP
Clear, Gingerale, Light Green,
Off Clear, Straw 3-5
Peach 15-20

[020] (F-Skirt) MAYDWELL-10 (R-Skirt) U.S.A. SDP
Straw 3-5
Smoke 10-15
Peach, Purple Tint 15-20

[030] (F-Skirt) MAYDWELL-10 (R-Skirt) U.S.A./
['McLAUGHLIN' blotted out] RDP
Clear, Ice Green, Straw 3-5
Smoke 10-15

McLAUGHLIN

[010] (F-Skirt) McLAUGHLIN {Note no line under 'c'} (R-Skirt) № 10 RDP
Light Green, Sage Aqua 10-15
Apple Green, Sky Blue 20-30
Light Cornflower Blue 50-75
Clear 175-200

[020] (F-Skirt) McLAUGHLIN (R-Skirt) № 10 FDP
Aqua, Ice Blue, Light Green, Sage Green 3-5
Apple Green 20-30
Gingerale 40-50
Light Cornflower Blue 50-75
Light Straw 75-100
Clear 175-200

[030] (F-Skirt) McLAUGHLIN (R-Skirt) № 10 RDP
Light Aqua, Light Blue, Light Green,
Sage Green 5-10
Green 15-20
Ice Mint Green 20-30
Light Cornflower Blue 50-75
Clear 175-200

[040] (F-Skirt) McLAUGHLIN (R-Skirt) № 10 SB
Light Green, Sage Green 5-10

[050] (F-Skirt) McLAUGHLIN (R-Skirt) № 10 SDP
Light Green 5-10
Green 15-20

WHITALL TATUM ▽

[010] (F-Skirt) (Arc)WHITALL TATUM/N⁰ 3
(R-Skirt) (Arc)MADE IN U.S.A./[Numbers and dots] Ⓐ SB
Clear, Light Straw 1-2
Light Peach 3-5

[013] (F-Skirt) (Arc)WHITALL TATUM/N⁰ 3
(R-Skirt) (Arc)MADE IN U.S.A./[Numbers and dots]/Ⓐ/[Number] SB
Clear 1-2
Light Peach 3-5

[017] (F-Skirt) WHITALL TATUM CO. N⁰ 3
(R-Skirt) MADE IN U.S.A./[Number] SB
Light Pink 3-5

[020] (F-Skirt) WHITALL TATUM CO. N⁰ 3
(R-Skirt) [Number] MADE IN U.S.A. SB
Light Pink 3-5

[030] (F-Skirt) WHITALL TATUM CO. N⁰ 3/
[Number] (R-Skirt) MADE IN U.S.A. SB
Clear, Light Aqua 1-2
Peach 3-5

CD 115

[040] (F-Skirt) WHITALL TATUM CO. N⁰ 3/ [Number] (R-Skirt) MADE IN U.S.A. ▽ SB
Clear, Ice Blue, Light Aqua,
Off Clear, Straw 1-2
Ice Green Aqua, Light Peach 3-5
Light Olive Green 15-20

[050] (F-Skirt) WHITALL TATUM CO. N⁰ 3/ [Number] (R-Skirt) MADE IN U.S.A. ▽ / [Number] SB
Straw 1-2

[060] (F-Skirt) WHITALL TATUM CO. N⁰ 3/ [Number] (R-Skirt) [Number] MADE IN U.S.A. SB
Aqua, Light Blue 1-2

[070] (F-Skirt) WHITALL TATUM N⁰ 3/ [Number] (R-Skirt) MADE IN U.S.A. ▽ SB
Clear, Light Aqua, Straw Tint 1-2
Pink Tint 3-5

[080] (F-Skirt) WHITALL TATUM N⁰ 3/ [Number] (R-Skirt) MADE IN U.S.A. ▽ / [Number] SB
Clear 1-2

CD 115.1

NO EMBOSSING - CANADA

[010] [No embossing] {Previously shown as CD 115 in the 1991 edition} SB
Light Aqua 700-800

CD 116

A.U.

[010] (F-Crown) PATENT/DEC. 19. 1871 (F-Skirt) ['A.U.' blotted out] (R-Skirt) JULY 1ST 1882/PAT APPLIED FOR SB
Aqua 100-125

BROOKFIELD

[010] (F-Skirt) W. BROOKFIELD SB
Aqua, Light Aqua, Light Blue 20-30
Green 125-150
Dark Green 175-200

[020] (F-Skirt) W. BROOKFIELD {MLOD} SB
Aqua, Dark Aqua, Light Aqua, Light Blue 20-30
Green 125-150

PATENT - DEC. 19, 1871

[010] (F-Crown) PATENT/DEC. 19. 1871 (F-Skirt) PATENT/MAY 2 1893 (R-Skirt) JULY 1ST 1882/PAT APPLIED FOR SDP
Aqua 75-100
Ice Aqua 100-125
Light Green 200-250

[020] (F-Crown) PATENT/DEC. 19. 1871 (F-Skirt) PATENT/MAY 2 1893 (R-Skirt) PATENT APPLIED FOR SB
Aqua 150-175

[030] (F-Crown) PATENT/DEC. 19. 1871 (F-Skirt) [Letter] (R-Skirt) JULY 1ST 1882/PAT APPLIED FOR SB
Aqua 75-100
Cornflower Blue 1,750-2,000

PATENT - OTHER

[010] (F-Crown) ['PAT JAN ...' blotted out] SB
Light Aqua 75-100

[020] (F-Skirt) JULY 1ST 1882/PAT APPLIED FOR (R-Skirt) PATENT/MAY 2 1893 {Hemingray product} SDP
Aqua 50-75
Ice Aqua 100-125
Light Green 200-250

[030] (F-Skirt) JULY 1ST 1882/PAT APPLIED FOR (R-Skirt) [Letter] SB
Aqua 50-75

CD 116.5

B

[010] (F-Skirt) B (R-Skirt) B SB
Light Aqua 75-100
Blue 125-150
Light Green 400-500
Apple Green 800-1,000

CD 120

CD 117

NO NAME

[010] (F-Skirt) & SB
Aqua, Dark Aqua 30-40
Green 125-150
Emerald Green 500-600
Olive Green 1,000-1,250

CD 118

NO EMBOSSING

[010] [No embossing] {The outside of the skirt is threaded and fitted with a copper sheath} {Hemingray product} SB
Carnival (without copper sheath) 300-350
Carnival (with copper sheath) 500-600

CD 119

BROOKFIELD

[010] (F-Skirt) W. BROOKFIELD. N.Y.
(R-Skirt) PAT.APRIL 28 1885 SB
Aqua, Blue Aqua, Light Aqua 3,000-3,500

CD 120

BROOKE'S

[010] (Dome) [Letter] (F-Skirt) H. BROOKE'S PAT. (R-Skirt) AUG 14 1883. SB
Brooke's Blue 50-75

[020] (F-Skirt) [Letter]/H. BROOKE'S PAT. (R-Skirt) AUG 14 1883. SB
Light Blue, Light Blue Aqua 40-50
Brooke's Blue 50-75
Emerald Green 1,250-1,500

[030] (F-Skirt) [Letter]/H. BROOKE'S PAT. (R-Skirt) AUG 14 188E. {Note backwards number} SB
Blue Aqua 50-75

C.E.W.

[010] (F-Skirt) C.E.W. SB
Aqua, Blue 125-150
Ice Aqua, Light Green Aqua 400-500
Cornflower Blue 1,250-1,500
Olive Green, Purple 5,000-7,500
Cobalt Blue, Mustard Yellow 7,500-10,000

[020] (F-Skirt) C.E.W. {MLOD} SB
Light Aqua, Light Blue 100-125
Blue 125-150

K.C.G.W.

[010] (F-Skirt) [Glass dot]/K.C.G.W.
(R-Skirt) [Glass dot]/5 SB
Aqua, Light Green 75-100

NO EMBOSSING

[010] [No embossing] {Hemingray style} SB
Light Aqua 50-75
Light Green 250-300
Gray Tint, Off Clear 400-500
Lavender 500-600
Purple 800-1,000
Peacock Blue 4,500-5,000

[020] [No embossing] {MLOD} SB
Teal Blue 1,750-2,000

[030] [No embossing] {Unembossed C.E.W.} SB
Purple 400-500
Clear, Dark Purple, Light Purple,
Purple/Burgundy Two Tone, Straw 500-600
Aqua 600-700

NO EMBOSSING - CANADA

[010] [No embossing] {Double threaded} {MLOD} RB
Aqua, Light Blue 500-600

PATENT - DEC. 19, 1871

[005] (Dome) [Letter] (F-Skirt) PATENT/DEC. 19. 1871 (R-Skirt) 5 SB
Lime Green 50-75

[010] (Dome) ['1' in a circle] (F-Skirt) PATENT/DEC. 19. 1871 (R-Skirt) 5/PATENT MAY 2 1893 SDP
Aqua, Blue Aqua 5-10
Ice Yellow Green 50-75

27

CD 120

[020] (F-Crown) PAT.DEC. 19. 1871
(F-Skirt) [Letter or number] (R-Crown) 5 SB
Aqua, Blue Aqua, Green Aqua,
Light Aqua, Light Green Aqua 5-10
Dark Aqua, Ice Green, Light Blue,
Light Green 10-15
Off Clear 40-50
Light Yellow Green 75-100
Milky Green 400-500
Purple 1,250-1,500

[030] (F-Crown) PAT.DEC. 19. 1871
(R-Crown) 5 SB
Aqua, Green Aqua, Light Blue Aqua 5-10

[040] (F-Crown) PAT.DEC. 19. 1871
(R-Crown) 5 (R-Skirt) [Letter] SB
Aqua, Blue Aqua, Green Aqua, Light Aqua 5-10
Light Green 10-15
Off Clear 40-50

[050] (F-Crown) PATENT/DEC. 19. 1871
(R-Crown) 5 (R-Skirt) [Letter] SB
Aqua, Green Aqua 15-20

[060] (F-Skirt) PATENT/DEC. 19. 1871
(R-Skirt) 5 SB
Aqua, Blue Aqua, Light Aqua 5-10
Ice Aqua, Light Blue, Light Green 10-15
Ice Blue 20-30
Off Clear 40-50
Light Yellow 100-125
Clear 125-150
Light Gray 200-250
Teal Blue, Yellow Green 600-700
Cornflower Blue, Light Purple 800-1,000
Olive Green 1,750-2,000

[070] (F-Skirt) PATENT/DEC. 19. 1871
(R-Skirt) 5/PATENT MAY 2 1893 SDP
Aqua, Blue Aqua, Green Aqua, Light Aqua 5-10
Light Green 30-40
Off Clear, Yellow Green Tint 50-75

[080] (F-Skirt) PATENT/DEC. 19. 1871
(R-Skirt) 5/[Letter] SB
Aqua, Green Aqua 5-10
Ice Aqua, Light Green 10-15

CD 120.2

PAT APP FOR

[010] (F-Skirt) PAT APP FOR (R-Skirt) [Letter] SB
Aqua, Blue, Light Aqua 50-75
Teal Green 500-600
Emerald Green 1,250-1,500

PONY

[010] (F-Skirt) PONY (R-Skirt) PAT APP FOR SB
Blue 50-75
Teal Green 500-600
Jade Green Milk 1,500-1,750

CD 121

A.T.& T.CO.

[010] (Dome) [Letter or number] (F-Skirt) A.T.& T.Co. SB
Aqua, Dark Aqua, Green Aqua,
Light Blue, Light Blue Aqua 3-5
Light Green 5-10
Green 10-15
Yellow Green 20-30
Olive Green 40-50

[020] (F-Skirt) A.T.& T.Co. SB
Light Green Aqua 10-15

AM.TEL.& TEL.CO.

[010] (Dome) [Letter or number]
(F-Skirt) AM.TEL.& TEL.Co. SB
Aqua, Blue Aqua, Light Green Aqua x1
Light Blue 1-2
Ice Blue 3-5
Light Green 5-10
Green, Hemingray Blue 10-15
Lime Green 30-40
Purple Tint 50-75
Cornflower Blue 150-175
Clear 175-200

[030] (Dome) [Number] (F-Crown) [Number]
(F-Skirt) AM.TEL.& TEL.Co. SB
Aqua, Light Aqua, Light Blue 3-5
Light Green 5-10
Light Yellow Green 30-40

[035] (Dome) [Number] (F-Skirt) AM TEL TEL Co {Note no '&'} SB
Aqua 20-30

[040] (Dome) [Number] (F-Skirt) AM.TEL. TEL.Co. {Note no '&'} SB
Aqua 20-30
Green 50-75

[050] (Dome) [Number] (F-Skirt) AM.TEL.& TE.Co. {Note spelling} SB
Green 30-40

CD 121

[060] (Dome) [Number] (F-Skirt) ['AM.TEL.&
TEL.Co.' blotted out] SB
Light Aqua 5-10
Peach 50-75

[065] (Dome) ['[Number]' blotted out]
(F-Skirt) ['AM.TEL.& TEL.Co.' blotted out] SB
Light Aqua 5-10

[020] (F-Crown) AM.TEL & TEL.Co.
(R-Crown) [Number] {MLOD} SB
Light Aqua 10-15
Green Aqua 20-30
Green 100-125

[070] (F-Crown) AM.TEL & TEL.Co. {MLOD}
SB
Aqua, Light Aqua 10-15
Dark Green, Milky Aqua 200-250
Jade Green Milk 400-500

[080] (F-Crown) [Number] (F-Skirt) AM.TEL.&
TEL.Co. SB
Aqua, Light Blue 3-5
Light Green 5-10

[090] (F-Skirt) AM TEL TEL Co {Note no '&'}
SB
Blue Aqua 20-30
Blue 30-40
Green 50-75
Yellow Green 100-125

[100] (F-Skirt) AM.TEL & TEL.Co.
(R-Skirt) [Blotted out embossing] SB
Aqua 5-10
Green 20-30

[105] (F-Skirt) AM.TEL & TEL.Co. {The 'M' is
shaped like a 'Y' with two extra legs; large
embossing} SB
Aqua, Green Aqua 10-15

[110] (F-Skirt) AM.TEL.& TEL.Co. SB
Aqua, Light Blue Aqua, Light Green Aqua x1
Light Blue 1-2
Aqua {Denver mold}, Hemingray Blue 5-10
Green, Jade Green Milk 20-30
Lime Green, Purple, Yellow Green 30-40
Light Straw, Sage Green 50-75
Light Purple {Denver mold} 75-100
Aqua w/ Light Amber Swirls, Off Clear,
Off Clear {Denver mold},
Yellow Olive Green 100-125
Sage Gray 175-200
Depression Glass Green 250-300
Aqua w/ Amber Swirls 300-350
Aqua w/ Dark Amber Swirls 500-600

[120] (F-Skirt) AM.TEL.& TEL.Co. SDP
Aqua 150-175

[125] (F-Skirt) AM.TEL.& TEL.Co. (R-Skirt) A
SB
Hemingray Blue 5-10

[130] (F-Skirt) AM.TEL.& TEL.Co.
(R-Skirt) ['AM.TEL.& TEL.Co.' blotted out] SB
Aqua 10-15

[140] (F-Skirt) AM.TEL.& TEL.Co. {MLOD} SB
Light Aqua 3-5
Light Blue 5-10

[150] (F-Skirt) ['AM.TEL.& TEL.Co.' blotted out]
SB
Light Aqua 5-10

B.T.C.

[010] (F-Skirt) B.T.C./CANADA SB
Ice Aqua, Light Aqua, Light Blue, Off Clear 1-2
Clear, Ice Green 5-10
Light Purple 15-20
Royal Purple 20-30
Light Aqua w/ Milk Swirls 30-40

[020] (F-Skirt) B.T.C./CANADA/['MONTREAL'
blotted out] SB
Aqua, Ice Blue, Light Aqua, Off Clear 1-2
Steel Blue 3-5
Light Purple 15-20

[030] (F-Skirt) B.T.C./MONTREAL SB
Ice Aqua, Ice Blue, Light Aqua 3-5
Dark Blue 15-20

[040] (F-Skirt) B.T.Co. OF CAN. SB
Aqua, Blue Aqua 3-5
Dark Blue Aqua 10-15
Green 30-40
Yellow Green 75-100
Olive Green 250-300

[050] (F-Skirt) B.T.Co./CANADA SB
Blue Aqua, Green Aqua, Light Aqua,
Light Blue Aqua 3-5
Light Aqua w/ Milk Swirls, Light Purple 30-40
Royal Purple 40-50
Lavender w/ Milk Swirls 50-75
Milky Green 175-200
Yellow Green 250-300

[060] (F-Skirt) B.T.Co./CANADA/
['MONTREAL' blotted out] SB
Aqua, Ice Blue 5-10

[070] (F-Skirt) ['B.T.C./CANADA' blotted out]
(R-Skirt) ◇ SB
Green Tint 15-20
Royal Purple 30-40
Light Yellow Green 50-75

[080] (F-Skirt) ['B.T.C.' blotted out] SB
Light Aqua 15-20

29

CD 121

[090] (F-Skirt) ['B.T.C.' blotted out]/CANADA SB
Ice Aqua, Light Aqua, Off Clear 1-2
Gray, Light Lavender, Steel Blue 10-15
Lavender 20-30
Dark Purple,
Light Aqua w/ Milk Swirls, Royal Purple 30-40
Milky Green 150-175
Milky Lavender 400-500

[100] (F-Skirt) ['B.T.C.' blotted out]/CANADA (R-Skirt) <> SB
Royal Purple 30-40

[110] (F-Skirt) ['B.T.C.' blotted out]/CANADA/ ['MONTREAL' blotted out] SB
Ice Blue, Light Aqua 1-2
Light Gray 5-10
Gray, Light Lavender 10-15
Lavender 20-30
Light Aqua w/ Milk Swirls, Royal Purple 30-40

BROOKFIELD

[010] (Dome) [Number] (F-Skirt) BROOKFIELD SB
Aqua, Green Aqua x1
Green 1-2
Emerald Green 5-10
Yellow Green 10-15
Dark Olive Green 15-20
Light Jade Aqua Milk,
Light Jade Green Milk 40-50
Jade Green Milk 250-300

[020] (Dome) [Number] (F-Skirt) BROOKFIELD SDP
Dark Aqua x1
Green 1-2
Emerald Green 3-5
Yellow Green 15-20
Olive Green 30-40
Olive Amber 200-250
Dark Orange Amber 500-600

[030] (Dome) [Number] (F-Skirt) BROOKFIELD (R-Skirt) [Five blots] SB
Aqua, Dark Aqua x1
Emerald Green, Green, Light Green 1-2
Light Aqua 3-5
Dark Yellow Green 10-15
Dark Olive Green 15-20
Yellow Olive Green 100-125
Dark Olive Amber 200-250

[040] (Dome) [Number] (F-Skirt) BROOKFIELD (R-Skirt) [Five blots] SDP
Aqua x1
Green 1-2
Yellow Green 15-20
Olive Green 30-40
Olive Amber 200-250
Amber Blackglass 400-500

[050] (Dome) [Number] (F-Skirt) W. BROOKFIELD SB
Blue Aqua, Light Aqua 3-5
Light Green 5-10
Light Lime Green 10-15

[055] (F-Skirt) BROOKFIELD SB
Blue 3-5

[060] (F-Skirt) BROOKFIELD (R-Skirt) [Five blots] SB
Blue Aqua, Light Aqua x1

[070] (F-Skirt) W. BROOKFIELD SB
Aqua, Light Green Aqua 3-5

C.& P.TEL.CO.

[010] (Dome) [Number] (F-Skirt) C.& P.TEL Co. SB
Aqua, Light Aqua 5-10
Light Green 10-15
Green 40-50
Yellow Green 250-300

[020] (F-Skirt) C.& P.TEL Co. SB
Aqua, Light Aqua 5-10
Blue 10-15
Green 40-50

C.D.& P.TEL.CO.

[010] (Dome) [Letter or number] (F-Skirt) C.D.& P.TEL Co SB
Aqua, Light Aqua, Light Green Aqua 3-5
Green 20-30
Milky Aqua 175-200
Yellow Green 300-350

[020] (F-Skirt) C.D.& P.T.Co SB
Aqua, Blue Aqua, Ice Blue 5-10
Light Green 10-15
Lime Green 50-75
Yellow Green 200-250
Light Yellow Olive Green 400-500

CALIFORNIA

[010] (F-Skirt) A 007/CALIFORNIA SB
Sage Green 5-10
Smoke, Yellow Sage Green 10-15
Light Purple, Purple 15-20
Burgundy, Steel Blue 20-30
Peach 150-175
Purple/Sage Two Tone 175-200

CANADA

[010] (F-Skirt) CANADA/['MONTREAL' blotted out] SB
Ice Aqua, Light Aqua 1-2
Light Steel, Off Clear 3-5
Clear 5-10
Royal Purple 20-30

CD 121

DIAMOND ◇

[005] (F-Skirt) ['B.T.C.' blotted out]/◇ (R-Skirt) SB
Light Yellow Green 75-100

[010] (F-Skirt) ['◇' blotted out] SB
Light Aqua 15-20

[015] (F-Skirt) ['◇' blotted out] (R-Skirt) ['◇' blotted out] SB
Light Blue Aqua 15-20

[020] (F-Skirt) ◇ SB
Light Aqua 1-2
Aqua, Green Aqua 3-5
Light Green, Off Clear 5-10
Steel Blue 10-15
Light Lavender 15-20
Blue 20-30
Purple, Royal Purple 30-40
Lemon, Straw 50-75
Peach 125-150
Yellow Green 150-175

[030] (F-Skirt) ◇ (R-Skirt) ['B.T.C.' blotted out] SB
Aqua 5-10
Steel Blue 10-15
Light Purple 20-30
Royal Purple 30-40
Straw 50-75
Light Yellow Green, Lime Green 75-100
Yellow Green 175-200

[035] (F-Skirt) ◇ (R-Skirt) ['◇' blotted out] SB
Aqua 15-20

[040] (F-Skirt) ◇ (R-Skirt) ◇ SB
Aqua, Blue Aqua, Light Aqua 3-5
Steel Blue 10-15
Light Green 30-40
Lemon, Light Purple 50-75
Royal Purple 75-100

GAYNER

[010] (F-Skirt) GAYNER (R-Skirt) NO.160/[Number] SDP
Aqua, Ice Blue, Light Aqua 10-15
Green Aqua 20-30

GOOD

[010] (F-Skirt) R. GOOD JR./DENVER, COLO./[0, 1, 2 or 3 dots] (R-Skirt) 16 {2 dot molds have a period between 'DENVER' and 'COLO.'} SB
Aqua, Ice Blue, Light Aqua 5-10
Blue Aqua, Steel Blue 10-15
Light Sage Aqua 20-30
Steel Gray 30-40
Light Green 125-150
Purple 150-175
Light Purple, Purple Tint 175-200
Light Yellow Green, Royal Purple 200-250
Aqua/Purple Two Tone 300-350
Blue/Purple Two Tone 350-400
Yellow Green 500-600
Green w/ Amber Swirls 700-800
Dark Yellow Green 800-1,000

HAWLEY ⌗

[010] (F-Skirt) ⌗/HAWLEY, PA./U.S.A. SB
Aqua, Blue Aqua 50-75

[020] (F-Skirt) ⌗/HAWLEY, PA./U.S.A. (R-Skirt) [Number] SB
Blue Aqua 50-75

HEMINGRAY

[010] (Dome) 1 (F-Skirt) HEMINGRAY (R-Skirt) № 16 SDP
Aqua, Blue Aqua, Light Aqua x1
Blue, Hemingray Blue 1-2
Emerald Green 5-10
Yellow Green 20-30

[020] (Dome) 1 (F-Skirt) HEMINGRAY/№ 16 (R-Skirt) HEMINGRAY SDP
Aqua, Blue Aqua, Hemingray Blue,
Light Blue Aqua 1-2
Blue 3-5
Dark Green 5-10
Yellow Green 30-40
Olive Green 50-75

[025] (Dome) [Blotted out embossing] (F-Skirt) HEMINGRAY (R-Skirt) № 16 SDP
Hemingray Blue 1-2

[030] (F-Skirt) HEMIGRAY {Note spelling} (R-Skirt) PATENT/MAY 2 1893 SDP
Aqua 10-15

[040] (F-Skirt) HEMIGRAY {Note spelling} (R-Skirt) PATENTED/MAY 2 1893 SDP
Aqua 10-15

[050] (F-Skirt) HEMINGAY/[Blotted out embossing] {Note spelling} (R-Skirt) PATENT/MAY 2 1893/[Blotted out embossing] SDP
Aqua 10-15

CD 121

[060] (F-Skirt) HEMINGAY/[Blotted out embossing] {Note spelling}
(R-Skirt) PATENTED/[Blotted out embossing] SDP
Aqua 10-15

[070] (F-Skirt) HEMINGRAY (R-Skirt) № 16 SDP
Aqua, Blue Aqua x1
Hemingray Blue 1-2
Emerald Green 5-10
Green 15-20
Yellow Green 20-30
Olive Green 30-40
Olive Amber 75-100

[080] (F-Skirt) HEMINGRAY
(R-Skirt) PATENT/MAY 2 1893 SB
Aqua 20-30
Hemingray Blue 40-50
Milky Aqua 300-350
Jade Aqua Milk 400-500

[090] (F-Skirt) HEMINGRAY
(R-Skirt) PATENT/MAY 2 1893 SDP
Aqua, Blue Aqua, Light Aqua x1
Hemingray Blue 1-2
Milky Aqua 150-175
Jade Aqua Milk 250-300

[100] (F-Skirt) HEMINGRAY
(R-Skirt) PATENTED SDP
Aqua, Light Aqua 1-2

[110] (F-Skirt) HEMINGRAY
(R-Skirt) PATENTED/MAY 2 1893 RDP
Aqua 50-75

[120] (F-Skirt) HEMINGRAY
(R-Skirt) PATENTED/MAY 2 1893 SDP
Aqua, Blue Aqua x1
Yellow Green 20-30

[130] (F-Skirt) HEMINGRAY
(R-Skirt) PATENTED/['AM.TEL.& TEL.Co.' blotted out] SDP
Aqua 3-5
Light Green 10-15

[140] (F-Skirt) HEMINGRAY
(R-Skirt) PATETED/MAY 2 1893 {Note spelling} SDP
Aqua, Blue, Blue Aqua 10-15

[150] (F-Skirt) HEMINGRAY/[Blotted out embossing] (R-Skirt) PATENTED/[Blotted out embossing] SDP
Aqua 5-10

[160] (F-Skirt) HEMINGRAY/['AM.TEL.& TEL.Co.' blotted out] (R-Skirt) PATENTED SDP
Aqua 3-5

LYNCHBURG Ⓛ

[010] (Dome) Ⓛ (F-Skirt) LYNCHBURG
(R-Skirt) No.30/MADE IN U.S.A. SDP
Blue Aqua 10-15

[020] (Dome) Ⓛ (F-Skirt) LYNCHBURG
(R-Skirt) No.30/[Number]/MADE IN U.S.A. SDP
Aqua, Ice Blue, Light Aqua 10-15
Light Green 40-50
Aqua w/ Milk Swirls 50-75

MAYDWELL

[010] (F-Skirt) MAYDWELL-16W
(R-Skirt) U.S.A. SDP
Straw 30-40

[020] (F-Skirt) MAYDWELL-16W/
['MᶜLAUGHLIN' blotted out] (R-Skirt) U.S.A. RDP
Off Clear 30-40

[030] (F-Skirt) MAYDWELL-16W/
['MᶜLAUGHLIN' blotted out] (R-Skirt) U.S.A. SDP
Off Clear 30-40

MᶜLAUGHLIN

[010] (F-Skirt) MᶜLAUGHLIN (R-Skirt) № 16 RDP
Light Green 3-5
Ice Aqua, Light Aqua, Light Blue 5-10
Green 10-15
Dark Green, Emerald Green Blackglass,
Green Blackglass, Lime Green, Mint Green,
Olive Blackglass, Teal Green Blackglass 15-20
Amber Blackglass,
Olive Amber Blackglass, Teal Green 20-30
Light Straw, Off Clear, Straw 30-40
Apple Green, Yellow Green 40-50
7-up Green, Delft Blue 50-75
Light Olive Green 350-400
Cornflower Blue 800-1,000
Citrine 1,000-1,250
Mustard Yellow, Yellow Olive Green ... 1,250-1,500

[020] (F-Skirt) MᶜLAUGHLIN (R-Skirt) № 16 SB
Gray Blue, Light Blue, Light Sage Green 5-10
Light Steel Blue 10-15
Sage Green 40-50
Yellow Green 700-800

[030] (F-Skirt) MᶜLAUGHLIN (R-Skirt) № 16 SDP
Aqua, Light Gray Blue, Steel Green 5-10
Steel Blue 10-15
Emerald Green 15-20
Amber Blackglass,
Dark Olive Amber, Dark Olive Green 20-30
Delft Blue, Delft Cornflower Blue 50-75

CD 121

NEW ENG. TEL. & TEL. CO.

[010] (F-Skirt) (Arc)NEW EEG.TEL.& TEL.CO. {Note spelling} SB
Aqua, Green Aqua 10-15
Green 20-30

[020] (F-Skirt) (Arc)NEW ENG.TEL.& TEL.CO. SB
Aqua, Light Green 3-5
Green 20-30
Yellow Green 50-75

NO EMBOSSING

[010] [No embossing] {Four fins under the wire groove; unembossed Duquesne; comes in round and square dome variations} SB
Light Aqua, Light Blue 5-10

[020] [No embossing] {Glass dot in dome distinguishes it from the unembossed AM.TEL.& TEL.CO.; possible Pennycuick} SB
Light Aqua 30-40

[030] [No embossing] {Unembossed AM.TEL.& TEL.CO.} SB
Aqua, Light Aqua 10-15
Purple 30-40
Clear 100-125

NO EMBOSSING - CANADA

[010] [No embossing] {Diamond style} SB
Light Aqua, Steel Blue 40-50

NO NAME

[010] (F-Skirt) [Glass dot] (R-Skirt) [Glass dot] SB
Gray, Ice Blue, Light Aqua 50-75
Light Green 75-100

NO NAME - CANADA

[010] (F-Skirt) [Blotted out embossing] SB
Light Blue Aqua 10-15

O.V.G.CO.

[010] (F-Skirt) O.V.G.CO. SB
Aqua, Blue Aqua, Light Aqua 3-5
Blue 10-15
Green 30-40
Light Jade Green Milk 175-200

PATENT - OTHER

[010] (Dome) (Circle)PAT'D SEPT.19TH 1899 {Pleated skirt extends around base giving the appearance of SDP} {Canadian product} SDP
Aqua, Blue, Blue Aqua,
Green Aqua, Light Aqua 20-30
Green 75-100
Yellow Green 200-250
Olive Green 300-350

[020] (Dome) (Circle)PAT'D SEP'T.19TH 1899 {Pleated skirt extends around base giving the appearance of SDP} {Canadian product} SDP
Aqua, Blue Aqua, Light Aqua 20-30
Green 75-100

STANDARD GLASS INSULATOR CO.

[010] (Base) THE STANDARD GLASS INSULATOR CO./BOSTON MASS. BE
Light Aqua, Light Blue 1,500-1,750

U.S.TEL.CO.

[010] (Dome) [Number] (F-Skirt) U.S.TEL.CO. SB
Aqua, Blue Aqua 5-10
Dark Aqua, Light Green 10-15
Green 20-30
Yellow Green 30-40
Olive Green 40-50

W.F.G.CO.

[010] (F-Skirt) W.F.G.CO./DENVER, COLO./ [1, 2 or 3 dots] (R-Skirt) 16 {2 dot molds have a period between 'DENVER' and 'COLO.'} SB
Aqua, Ice Blue, Light Aqua, Steel Aqua 5-10
Ice Aqua 10-15
Blue Aqua, Gray Blue 15-20
Steel Blue, Steel Gray 20-30
Off Clear 40-50
Light Lavender, Purple Tint 50-75
Clear, Delft Blue 75-100
Lavender, Milky Light Aqua 175-200
Blue/Purple Two Tone 250-300
Cornflower Blue 300-350
Straw 400-500
Dark Lavender 500-600
Green, Jade Aqua Milk, Milky Purple 800-1,000
Dark Green, Emerald Green 1,000-1,250

W.G.M.CO.

[010] (F-Skirt) W.G.M.CO. SB
Light Purple, Purple 30-40
Royal Purple 40-50
Off Clear, Straw/Peach Two Tone 200-250
Peach, Straw 300-350
Straw/Clear Two Tone 350-400
Burgundy, Burgundy/Straw Two Tone 400-500
Citrine, Light Lime Green,
Milky Purple 800-1,000

33

CD 121.4

CD 121.4

A.U.

[010] (F-Crown) PATENT/DEC. 19. 1871 (F-Skirt) A.U. SB
Light Aqua 30-40
Light Green 50-75

[020] (F-Crown) PATENT/DEC. 19. 1871 (F-Skirt) A.U. (R-Skirt) [Letter] SB
Aqua, Green Aqua, Light Aqua 30-40
Light Green 50-75

[030] (F-Crown) PATENT/DEC. 19. 1871 (F-Skirt) A.U./[Letter] (R-Skirt) [Letter] SB
Light Green 50-75

[040] (F-Crown) PATENT/DEC. 19. 1871 (F-Skirt) ['A.U.' blotted out] SB
Aqua 30-40

[050] (F-Crown) PATENT/DEC. 19. 1871 (F-Skirt) ['A.U.' blotted out] (R-Skirt) [Blotted out embossing] SB
Green Aqua 30-40

[060] (F-Crown) PATENT/DEC. 19. 1871 (F-Skirt) ['A.U.' blotted out] (R-Skirt) [Letter or number] SB
Aqua, Green Aqua 30-40
Yellow Green 400-500

[070] (F-Crown) PATENT/DEC. 19. 1871 (F-Skirt) ['A.U.' blotted out]/[Letter] SB
Aqua 30-40

NO EMBOSSING

[010] [No embossing] SB
Light Green 100-125

CD 122

ARMSTRONG Ⓐ

[010] (F-Skirt) Ⓐrmstrong No.2 (R-Skirt) MADE IN U.S.A. SB
Clear 1-2

[020] (F-Skirt) Ⓐrmstrong No.2 (R-Skirt) MADE IN U.S.A. [Numbers and dots] SB
Blue Tint, Clear 1-2

[030] (F-Skirt) Ⓐrmstrong No.2 (R-Skirt) MADE IN U.S.A./[Numbers and dots] SB
Clear, Smoke 1-2

[040] (F-Skirt) *Armstrong's* No.2 (R-Skirt) MADE IN U.S.A. Ⓐ SB
Off Clear, Smoke, Straw 1-2
Pink 3-5

[050] (F-Skirt) *Armstrong's* No.2 (R-Skirt) MADE IN U.S.A. Ⓐ [Numbers and dots] CB
Off Clear 1-2

[060] (F-Skirt) *Armstrong's* No.2 (R-Skirt) MADE IN U.S.A. Ⓐ [Numbers and dots] SB
Green Tint, Off Clear, Olive Green Tint 1-2

[070] (F-Skirt) *Armstrong's* № 2 (R-Skirt) MADE IN U.S.A. Ⓐ [Numbers and dots] SB
Off Clear, Straw 1-2

DIAMOND ◇

[010] (F-Skirt) ◇ RDP
Straw 20-30

[020] (F-Skirt) ◇ SB
Light Green, Straw 10-15

DOMINION ⊙

[010] (F-Skirt) DOMINIOIN-16 {Note spelling} (R-Skirt) ◇ RDP
Light Peach 10-15
Light Green 15-20

[020] (F-Skirt) DOMINION-16 (R-Skirt) ⊙ SDP
Light Green 1-2

[030] (F-Skirt) DOMINION-16 (R-Skirt) ⊙ [Number] SB
Clear x1

[035] (F-Skirt) DOMINION-16 (R-Skirt) ⊙ [Number]/[Number] {Number may appear before or after '⊙'} SB
Clear, Light Straw x1

[040] (F-Skirt) DOMINION-16 (R-Skirt) ⊙/[Number] (Top of pinhole) [Number] SB
Green Tint 1-2

[050] (F-Skirt) DOMINION-16 (R-Skirt) ◇ RDP
Clear, Straw x1
Light Green, Light Peach 1-2

[060] (F-Skirt) DOMINION-16 (R-Skirt) ◇ SB
Clear, Straw x1

CD 122

[070] (F-Skirt) DOMINION-16 (R-Skirt) ◇ SDP
Light Green 1-2

[075] (F-Skirt) DOMINION-16 (R-Skirt) ◇/[Number] SB
Clear x1
Light Green, Light Peach 1-2

[080] (F-Skirt) DOMINOIN-16 {Note spelling} (R-Skirt) ◇ RDP
Peach 15-20

HEMINGRAY

[010] (F-Skirt) HEMIGRAY-16 {Note spelling} (R-Skirt) MADE IN U.S.A. RDP
Aqua 10-15
Green 20-30

[020] (F-Skirt) HEMINGRAY-16 (R-Skirt) MADE IN U S A {Note no periods} SDP
Hemingray Blue 5-10

[030] (F-Skirt) HEMINGRAY-16 (R-Skirt) MADE IN U.S.A. RDP
Aqua, Blue Tint, Clear x1
Hemingray Blue, Ice Blue 1-2
7-up Green 30-40
Clear/Blue Two Tone 50-75

[040] (F-Skirt) HEMINGRAY-16 (R-Skirt) MADE IN U.S.A. SDP
Aqua, Hemingray Blue 1-2

[050] (F-Skirt) HEMINGRAY-16 (R-Skirt) MADE IN U.S.A./[Numbers and dots] CB
Clear, Straw x1
Ice Blue 1-2

[060] (F-Skirt) HEMINGRAY-16 (R-Skirt) MADE IN U.S.A./[Numbers and dots] SB
Clear, Green Tint x1
Dark Straw 1-2

[070] (F-Skirt) HEMINGRAY-16 (R-Skirt) [Numbers and dots]/MADE IN U.S.A. SB
Clear x1

[080] (F-Skirt) HEMINGRAY-16 (R-Skirt) [Numbers] SB
Clear 10-15

[090] (F-Skirt) HEMINGRAY-16/[Number and dots] (R-Skirt) MADE IN U.S.A. RDP
Clear x1
Ice Blue 1-2

[100] (F-Skirt) HEMINGRAY-16/[Number and dots] (R-Skirt) MADE IN U.S.A./[Number] RDP
Clear x1
Ice Blue 1-2
Ice Green 3-5
Light Opalescent, White Milk 500-600
Opalescent 800-1,000

[110] (F-Skirt) HEMINGRAY-16/[Number and dots] (R-Skirt) MADE IN U.S.A./[Number] SB
Clear, Light Straw x1
Light Pink 1-2

[120] (F-Skirt) HEMINGRAY-17 (R-Skirt) MADE IN U.S.A./[Numbers and dots] CB
Clear, Light Straw x1
Green Tint, Pink Tint 1-2

[130] (F-Skirt) HEMINGRAY-17 (R-Skirt) MADE IN U.S.A./[Numbers and dots] SB
Clear x1

[135] (F-Skirt) HEMINGRAY=16 (R-Skirt) MADE IN = U.S.A. {Note '='} SDP
Hemingray Blue 5-10

[140] (F-Skirt) HEMINRAY-16 {Note spelling} (R-Skirt) MADE IN U.S.A./[Numbers and dots] SB
Clear 10-15

KERR

[010] (F-Skirt) KERR No.2 (R-Skirt) MADE IN U.S.A. [Number and dots] SB
Blue Tint, Clear 3-5

[020] (F-Skirt) KERR No.22 {Note '22' style number} (R-Skirt) MADE IN U.S.A. [Number and dots] SB
Clear 5-10

LYNCHBURG Ⓛ

[010] (Dome) Ⓛ (F-Skirt) LYNCHBURG (R-Skirt) NO.30/[Number]/MADE IN U.S.A. SDP
Apple Green 40-50

[020] (F-Skirt) Ⓛ/LYNCHBURG (R-Skirt) NO.30 [Number]/MADE IN U.S.A. RDP
Aqua 5-10
Ice Blue 10-15
Green 15-20

[030] (F-Skirt) Ⓛ/LYNCHBURG (R-Skirt) NO.30 [Number]/MADE IN U.S.A. SDP
Aqua, Light Aqua 3-5
Smoky Sage Green 10-15
Green 20-30

35

CD 122

[040] (F-Skirt) Ⓛ/LYNCHBURG (R-Skirt) [Blotted out embossing]/NO.30 [Number]/MADE IN U.S.A. SDP
Aqua 3-5
Blue Aqua, Light Green 5-10
Ice Blue, Smoke, Straw 10-15
Green, Off Clear 15-20
Yellow Green 30-40
Aqua w/ Milk Swirls 50-75

MAYDWELL

[010] (F-Skirt) MAYDWELL-16 (R-Skirt) U.S.A. SDP
Clear, Ice Green, Light Green, Straw 3-5
Gray, Light Yellow Green 5-10
Gingerale, Purple Tint 10-15
Light Olive Green 15-20
Smoky Purple 20-30

[020] (F-Skirt) MAYDWELL-16 (R-Skirt) U.S.A./ ['MᶜLAUGHLIN' blotted out] {Round wire groove variant} SDP
Clear, Light Green, Off Clear 3-5
Dark Straw, Gingerale 10-15

[030] (F-Skirt) ['NO.16' blotted out]/ MAYDWELL-16 (R-Skirt) ['MᶜLAUGHLIN' blotted out]/U.S.A. SDP
Clear 3-5
Dark Straw 5-10

MᶜLAUGHLIN

[010] (F-Skirt) MᶜLAUGHLIN (R-In wire groove) [Number] (R-Skirt) NO.16 RDP
Aqua, Ice Green, Light Green 3-5
Light Apple Green, Lime Green,
Yellow Green 20-30
7-up Green 40-50
Light Citrine 100-125
Citrine 400-500

[020] (F-Skirt) MᶜLAUGHLIN (R-Skirt) NO.16 RDP
Aqua, Blue Tint, Green Tint, Light Green 3-5
Light Yellow Green 10-15
Mint Green 15-20
Dark Green 20-30
Apple Green, Lime Green 30-40
7-up Green 40-50
Olive Green 250-300

[030] (F-Skirt) MᶜLAUGHLIN (R-Skirt) NO.16 SDP
Aqua, Ice Green 3-5

[040] (F-Skirt) MᶜLAUGHLIN (R-Skirt) NO.16 {Round wire groove variant} SDP
Green Aqua, Light Green, Mint Green 20-30

NO EMBOSSING

[010] [No embossing] {Hemingray product} CB
Clear 50-75

WHITALL TATUM ▽

[020] (F-Skirt) (Arc)WHITALL TATUM/Nº 2 (R-Skirt) (Arc)MADE IИ U.S.A./[Numbers and dots]/Ⓐ {Note backwards letter} SB
Clear 5-10
Light Blue, Light Sage Green 10-15
Light Olive Green 20-30

[010] (F-Skirt) (Arc)WHITALL TATUM/Nº 2 (R-Skirt) (Arc)MADE IN U.S.A./[Numbers and dots]/Ⓐ/[Number] SB
Clear, Off Clear x1
Light Sage Green 1-2
Light Pink, Light Steel Blue, Peach 5-10
Light Olive Green, Light Yellow Green 20-30

[030] (F-Skirt) WHITALL TATUM CO. Nº 2 (R-Skirt) MADE IN U.S.A./[Number] SB
Light Aqua 1-2

[040] (F-Skirt) WHITALL TATUM CO. Nº 2 (R-Skirt) MADE IN U.S.A./▽ SB
Light Straw 1-2

[050] (F-Skirt) WHITALL TATUM CO. Nº 2 (R-Skirt) [Number] MADE IN U.S.A. SB
Light Blue Aqua 3-5

[060] (F-Skirt) WHITALL TATUM CO. Nº 2 (R-Skirt) ▽/MADE IN U.S.A./[Number] SB
Clear, Smoky Straw 1-2

[070] (F-Skirt) WHITALL TATUM CO. Nº 2/ [Number] (R-Skirt) MADE IN U.S.A. ▽ SB
Off Clear x1
Light Aqua, Light Green Aqua, Straw 1-2
Peach 3-5

[080] (F-Skirt) WHITALL TATUM CO. Nº 2/ [Number] (R-Skirt) MADE IN U.S.A. ▽ / [Number] SB
Light Straw 1-2

[090] (F-Skirt) WHITALL TATUM Nº 2/ [Number] (R-Skirt) MADE IN U.S.A. ▽ / [Number] SB
Clear, Off Clear x1
Ice Blue, Light Straw, Straw 1-2
Pink 3-5

CD 122.4

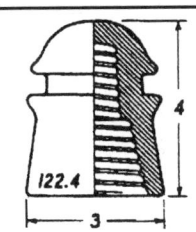

HEMINGRAY

[010] (F-Skirt) HEMINGRAY-E2 (R-Skirt) MADE IN U.S.A. {Inside of skirt is threaded} SB
Light Lemon, Off Clear 75-100
Lemon 100-125
Ice Blue w/ Carnival inside skirt 400-500

PYREX

[010] (F-Skirt) PYREX/REG. U.S. PAT. OFF. (R-Skirt) [Letter or number]/MADE IN U.S.A./ PAT.APPD.FOR. {Inside of skirt is threaded} SB
Clear, Light Lemon 3-5

[020] (F-Skirt) PYREX/REG. U.S. PAT. OFF. MADE IN U.S.A. (R-Skirt) PAT.APPD.FOR {Inside of skirt is threaded} SB
Clear 3-5

[030] (F-Skirt) PYREX/REG. U.S. PAT. OFF. MADE IN U.S.A. (R-Skirt) [Number] {Inside of skirt is threaded} SB
Clear 3-5
Peach 5-10

[050] (F-Skirt) PYREX/REG. U.S. PAT. OFF. MADE IN U.S.A. (R-Skirt) [Number]/ PAT.APPD.FOR SB
Clear, Light Lemon 3-5

[040] (F-Skirt) PYREX/REG. U.S. PAT. OFF. MADE IN U.S.A. {Inside of skirt is threaded} SB
Clear, Light Lemon, Off Clear 3-5

CD 123

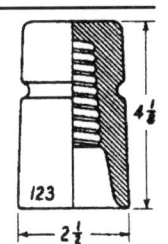

E.C.& M.CO.

[010] (F-Crown) E.C & M C$\underline{o}$ S.F. {There may be a glass button on the front or rear skirt} {MLOD} SB
Aqua, Green Aqua 100-125
Light Blue 250-300
Blue 500-600
Light Green 700-800
Lime Green 800-1,000
Green 1,000-1,250
Sage Green 1,500-1,750
Apple Green, Dark Green,
Olive Green Amber 1,750-2,000
Cobalt Blue, Dark Yellow Amber,
Ink Blue, Light Cobalt Blue,
Olive Amber, Yellow Green 2,000-2,500
Dark Olive Green, Emerald Green,
Forest Green, Teal Green 2,500-3,000
Chartreuse, Smoke 3,500-4,000
Midnight Blue,
Olive Amber Blackglass, Teal Blue 5,000-7,500
Purple 7,500-10,000

[020] (F-Crown) E.C & M.C$\underline{o}$ S.F. {Embossing is upside down} (F-Skirt) [Glass dot] {MLOD} SB
Aqua 5,000-7,500
Emerald Green, Green 7,500-10,000

[030] (F-Skirt) E.C & M C$\underline{o}$ S.F. {MLOD} SB
Gray Blue, Smoky Gray,
Yellow Green 7,500-10,000

CD 123.2

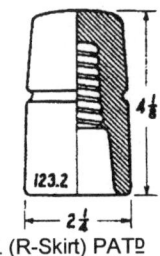

CHESTER

[005] (F-Skirt) CHESTER/N.Y. (R-Skirt) PAT$\underline{D}$ JAN 25 1870 SB
Aqua 5,000-7,500

[010] (F-Skirt) CHESTER/N.Y. (R-Skirt) PAT$\underline{D}$ JAN 25$\underline{TH}$ 1870 SB
Aqua 5,000-7,500

CD 124

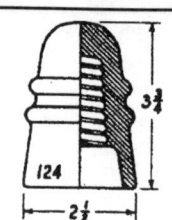

CD 124

HEMINGRAY

[010] (Dome) [Number] (F-Skirt) A/ HEMINGRAY/№ 4 (R-Crown) PAT.DEC. 19. 1871 (R-Skirt) 4/PATENT/MAY 2 1893 SDP
Aqua, Light Aqua 20-30

[020] (Dome) [Number] (F-Skirt) [Vertical bar]/ HEMINGRAY/№ 4 (R-Skirt) PATENT/MAY 2 1893 SDP
Blue Aqua 10-15

[030] (F-Crown) 4 (F-Skirt) HEMINGRAY/№ 4 (R-Skirt) PATENT/MAY 2 1893 SDP
Aqua 5-10

[040] (F-Crown) ↳ {Note backwards number} (F-Skirt) [Vertical bar]/HEMINGRAY/№ 4 (R-Skirt) PATENT/MAY 2 1893 SDP
Aqua 10-15

[050] (F-Skirt) A/HEMINGRAY/№ 4 (R-Crown) PAT.DEC. 19. 1871 (R-Skirt) 4/ PATENT/MAY 2 1893 SDP
Aqua, Light Aqua 10-15

[060] (F-Skirt) HEMINGRAY/4/№ 4 (R-Crown) PAT.DEC. 19. 1871 (R-Skirt) PATENT/A/MAY 2 1893 SDP
Aqua 20-30

[070] (F-Skirt) HEMINGRAY/№ 13/['4' blotted out] (R-Skirt) PATENT/MAY 2 1893 SDP
Aqua, Blue Aqua, Light Aqua 40-50
Hemingray Blue 50-75

[080] (F-Skirt) HEMINGRAY/№ 4 (R-Skirt) PATENT/MAY 2 1893 SDP
Aqua, Light Aqua 5-10
Light Blue 20-30
Green 100-125
Aqua w/ Amber Swirls,
Dark Yellow Green, Milky Aqua 200-250
Emerald Green, Olive Green 400-500

[090] (F-Skirt) HEMINGRAY/№ ↳ {Note backwards number} (R-Skirt) PATENT/MAY 2 1893 SDP
Aqua 10-15
Green 100-125
Olive Green 400-500

NO NAME

[010] (Dome) 4 (F-Skirt) [Vertical bar] (R-Skirt) [Crescent] SB
Aqua 15-20
Light Blue 20-30
Off Clear 30-40

[015] (Dome) ↳ {Note backwards number} (F-Skirt) [Vertical bar] (R-Skirt) [Crescent] SB
Aqua, Light Aqua 15-20
Light Green 30-40

[020] (F-Skirt) [Vertical bar] (R-Skirt) [Crescent] SB
Off Clear 30-40

[030] (Dome) Γ [Vertical bar] {Note backwards number} (R-Skirt) Γ [Vertical bar] {Note backwards number} SB
Ice Aqua, Ice Green 30-40

[040] (F-Crown) Γ {Note backwards number} (F-Skirt) Γ {Note backwards number} SB
Aqua 20-30

PATENT - DEC. 19, 1871

[010] (Dome) [Letter] (F-Crown) PAT.DEC. 19. 1871 (F-Skirt) 4 (R-Skirt) [Letter] SB
Aqua 10-15
Light Green 20-30

[020] (F-Crown) PAT.DEC. 19. 1871 (F-Skirt) 4 (R-Skirt) [Letter] SB
Aqua, Light Aqua 10-15
Green Tint, Ice Green, Light Blue,
Light Green 20-30
Clear 75-100

[030] (F-Crown) PAT.DEC. 19. 1871 (F-Skirt) [Letter] (R-Skirt) 4 SB
Light Aqua 10-15

CD 124.2

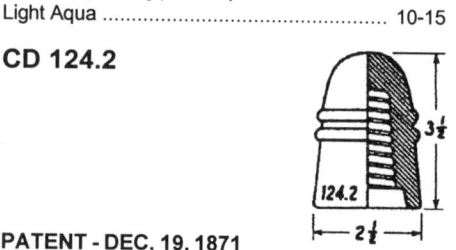

PATENT - DEC. 19, 1871

[010] (F-Crown) PATENT/DEC. 19. 1871 (R-Crown) 4 SB
Aqua, Dark Aqua, Light Aqua 50-75
Ice Aqua, Light Green 75-100
Dark Blue Aqua 100-125
Milky Aqua, Off Clear 200-250
Teal Blue 600-700

CD 124.3

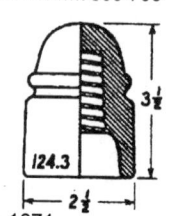

PATENT - DEC. 19, 1871

[010] (F-Crown) PAT.DEC. 19. 1871 (F-Skirt) [Letter] (R-Crown) 4 SB
Green Aqua 40-50

CD 125

[020] (F-Crown) PATENT/DEC. 19. 1871
(F-Skirt) [Letter] (R-Crown) 4 SB
Aqua, Light Aqua, Light Green Aqua 20-30
Blue, Light Green 30-40
Olive Green 1,750-2,000
Purple 2,000-2,500

[030] (F-Crown) PATENT/DEC. 19. 1871
(R-Crown) 4 (R-Skirt) [Letter] SB
Aqua, Blue Aqua, Green Aqua, Light Aqua .. 20-30
Blue, Dark Aqua, Light Green 30-40
Cornflower Blue 400-500

CD 124.5

CHAMBERS

[010] (F-Crown) PATENT/DEC. 19. 1871
(F-Skirt) CHAMBERS/PAT. AUG. 14. 1877
(R-Crown) 4 SB
Ice Green, Light Aqua 2,000-2,500

CD 125

HEMINGRAY

[010] (F-Skirt) HEMINGRAY (R-Skirt) №15 SDP
Aqua, Blue Aqua 10-15
Green 50-75

[020] (F-Skirt) HEMINGRAY-15
(R-Skirt) MADE IN U.S.A. RDP
Aqua 5-10
Hemingray Blue 10-15
Ice Blue 15-20

[030] (F-Skirt) HEMINGRAY-15
(R-Skirt) MADE IN U.S.A. SDP
Aqua 10-15

[040] (F-Skirt) HEMINGRAY/MADE IN U.S.A.
(R-Skirt) №15 SDP
Aqua, Light Aqua 10-15
Hemingray Blue 15-20

[050] (F-Skirt) HEMINGRAY/MADE IN U.S.A.
(R-Skirt) PATENT/DEC. 19. 1871/[Letter] SB
Light Green 50-75

[055] (F-Skirt) HEMINGRAY/W U № 5 {Note no periods} (R-Skirt) PATENT/MAY 2 1893 SDP
Aqua 5-10
Light Aqua 15-20
Purple 800-1,000

[060] (F-Skirt) HEMINGRAY/W.U. № 5
(R-Skirt) PATENT/MAY 2 1893 SDP
Aqua, Dark Aqua 5-10
Light Green 10-15
Hemingray Blue, Light Aqua 15-20
Ice Aqua, Ice Blue, Ice Green 20-30
Green 100-125
Sage Green 175-200
Light Apple Green, Sage Gray 300-350
Light Purple, Purple/Gray Two Tone,
Purple/Sage Two Tone, Teal Green 500-600
Smoky Purple 600-700
Purple 800-1,000

NO EMBOSSING

[010] [No embossing] SB
Aqua 5-10

PAT APP FOR

[010] (F-Skirt) PAT.APP.FOR (R-Skirt) 5/
[Letter] SB
Aqua 15-20
Brooke's Blue 50-75
Dark Sapphire Blue 200-250
Milky Green 350-400

PATENT - OTHER

[010] (F-Skirt) ['PAT JAN 25 1870' blotted out]
{MLOD} SB
Light Aqua 5-10

W.U.

[010] (F-Skirt) W.U./5 (R-Skirt) PAT.APP.FOR SB
Blue 30-40
Blue w/ Milk Swirls 50-75

[020] (F-Skirt) W.U./5 (R-Skirt) PATENT/DEC. 19. 1871 SB
Aqua, Blue Aqua, Light Aqua, Light Green 5-10
Ice Blue 15-20
Lemon 20-30
Apple Green 50-75
Green 100-125
Yellow Green 400-500
Christmas Tree Green 800-1,000
Yellow Olive Green 1,250-1,500

CD 125

[030] (F-Skirt) W.U./5 (R-Skirt) PATENT/DEC. 19. 1871/[Letter] SB
Aqua, Light Aqua 5-10
Light Blue Aqua 10-15
Blue 15-20
Off Clear 20-30
Lime Green 50-75
Clear 150-175
Yellow Olive Green 1,250-1,500

[040] (F-Skirt) W.U./5 (R-Skirt) PATENT/DEC. 19. 1871/[Letter] {Wide dome style} SB
Aqua, Light Aqua, Light Green Aqua 15-20
Aqua Tint 30-40
Brooke's Blue 50-75
Gray Tint 125-150

[050] (F-Skirt) W.U./5 (R-Skirt) PATENT/DEC. 19. 1871/[Letter] {Wide dome style} SDP
Light Blue Aqua 75-100

[060] (F-Skirt) W.U./5/PATENT/MAY 2 1893 (R-Skirt) PATENT/DEC. 19. 1871/[Letter] {Wide dome style} SDP
Aqua 75-100

[070] (F-Skirt) W.U./5/[Letter] (R-Skirt) PATENT/DEC. 19. 1871 SB
Green Aqua 3-5
Aqua, Light Green Aqua 5-10
Ice Green 10-15
Lemon 20-30
Lime Green 50-75
Green 100-125

[080] (F-Skirt) W.U./PATENT/MAY 2 1893 (R-Skirt) PATENT/DEC. 19. 1871/[Letter] SDP
Aqua 15-20
Ice Blue 40-50
Lime Green 75-100
Yellow Green 400-500

CD 126

BROOKFIELD

[010] (F-Crown) (Arc)W. BROOKFIELD/45 CLIFF ST./N.Y. (R-Crown) (Arc)CAUVET/ JULY 25, 1865/PAT.JAN. ", 1870/" FEB. 22," SB
Light Aqua 5-10

[020] (F-Crown) (Arc)W. BROOKFIELD/45 CLIFF ST./N.Y. (R-Crown) [Number] SB
Aqua 5-10

[030] (F-Crown) (Arc)W. BROOKFIELD/NEW YORK {MLOD} SB
Aqua, Blue, Green Aqua, Light Aqua 10-15

[040] (F-Crown) PAT.MARCH 20, 1870/W. BROOKFIELD {Note 'MARCH 20, 1870' date} (R-Crown) PAT.JAN. 25, 1870/PAT.FEB. 20, 1870/55 FULTON ST. N.Y. {Note 'FEB. 20, 1870' date} {MLOD} SB
Blue Aqua 5-10

[042] (F-Crown) PAT.MARCH 20, 1870/W. BROOKFIELD {Note 'MARCH 20, 1870' date} (R-Crown) PAT.JAN. 25, 1870/PAT.FEB. 22, 1870/55 FULTON ST. N.Y. {MLOD} SB
Aqua 5-10

[045] (F-Crown) PAT.MARCH 20, 1877/W. BROOKFIELD (R-Crown) PAT.FEB. 22, 1870/ JAN. 25, 1870/55 FULTON ST. N.Y. {MLOD} SB
Aqua, Light Aqua 5-10

[048] (F-Crown) PAT.MARCH 20, 1877/W. BROOKFIELD (R-Crown) PAT.FEB. 22, 1870/ PAT JAN. 25, 1870/55 FULTON ST. N.Y. {MLOD} SB
Light Aqua 5-10

[050] (F-Crown) PAT.MARCH 20, 1877/W. BROOKFIELD (R-Crown) PAT.JAN. 25, 1870/ PAT.FEB. 20, 1870/55 FULTON ST. N.Y. {Note 'FEB. 20, 1870' date} {MLOD} SB
Light Aqua 5-10

[060] (F-Crown) PAT.MARCH 20, 1877/W. BROOKFIELD (R-Crown) PAT.JAN. 25, 1870/ PAT.FEB. 20, 1870/NO.55 FULTON ST. N.Y. {Note 'FEB. 20, 1870' date} {MLOD} SB
Green 100-125

[070] (F-Crown) PAT.MARCH 20, 1877/W. BROOKFIELD (R-Crown) PAT.JAN. 25, 1870/ PAT.FEB. 22, 1870/55 FULTON ST. N.Y. {MLOD} SB
Light Aqua 5-10
Steel Blue 175-200
Yellow Green 250-300

[080] (F-Crown) PAT.MARCH 20, 1877/W. BROOKFIELD (R-Crown) PAT.JAN. 25, 1870/ PAT.FEB. 22, 1870/NO.55 FULTON ST. N.Y. {MLOD} SB
Aqua 5-10
Blue 15-20
Light Green 50-75
Green 100-125
Dark Green 250-300
Olive Green 400-500

CD 126

[090] (F-Crown) PAT.MARCH 20, 1877/W. BROOKFIELD (R-Crown) [Number]/PAT.JAN 25, 1870/PAT.FEB 22, 1870/55 FULTON ST. N.Y. {MLOD} SB
Light Aqua 5-10

[100] (F-Crown) PAT.MARCH 22, 1877/W. BROOKFIELD {Note 'MARCH 22, 1877' date} (R-Crown) PAT.JAN. 25, 1870/PAT.FEB. 22, 1870/55 FULTON ST. N.Y. {MLOD} SB
Aqua 5-10

[110] (F-Crown) W U T CO./CAUVETS PAT./ W. BROOKFIELD (F-Below wire groove) PAT.MARCH 20, 1877 (R-Crown) FEB. 22, 70/NO 55 FULTON ST. N.Y. (R-Below wire groove) PAT.JAN. 25, 1870 {MLOD} SB
Aqua 20-30

[120] (F-Crown) W U T CO./CAUVETS PAT./ W. BROOKFIELD (F-Below wire groove) PAT.MARCH 20, 1877 (R-Crown) [Number]/FEB. 22, 70/NO 55 FULTON ST. N.Y. (R-Below wire groove) PAT.JAN. 25, 1870 {MLOD} SB
Blue Aqua, Light Aqua 20-30
Olive Green 800-1,000

[130] (F-Crown) W U T CO./CAUVETS PAT./ W. BROOKFIELD (R-Crown) [Number]/FEB. 22, 70/NO.55 FULTON ST. N.Y. {MLOD} SB
Aqua, Light Aqua, Light Blue 10-15

[140] (F-Crown) W. BROOKFIELD/45 CLIFF ST. N.Y. SB
Purple 7,500-10,000

[150] (F-Crown) W. BROOKFIELD/45 CLIFF ST. N.Y. (R-Crown) PAT./JAN. 25TH 1870/JAN. 14TH 1879 {MLOD} SB
Blue Aqua 5-10
Purple 7,500-10,000

[160] (F-Crown) W. BROOKFIELD/45 CLIFF ST. N.Y. (R-Crown) [Number]/PAT./JAN. 25TH 1870/JAN. 14TH 1879 {MLOD} SB
Blue Aqua 5-10

[170] (F-Crown) W. BROOKFIELD/45 CLIFF ST./N.Y. SB
Aqua 5-10

[180] (F-Crown) W. BROOKFIELD/45 CLIFF ST./N.Y. (R-Crown) PAT.JAN. 25, 1870/ PAT.JAN. 14, 1879 {MLOD} SB
Aqua 5-10

[190] (F-Crown) W. BROOKFIELD/55 FULTON ST. N.Y. {Note backwards letter} (R-Crown) PAT.FEB 22 1870/ " JAN 25 " / " " 14 1879 {MLOD} SB
Aqua, Light Aqua 5-10
Yellow Green 250-300

[200] (F-Crown) W. BROOKFIELD/55 FULTON ST. N.Y. {Note backwards letter} (R-Crown) PAT.FEB. 22ND, 1870/ " JAN. 25TH, 1870/ " JAN. 14TH, 1879 {MLOD} SB
Blue Aqua, Light Aqua 5-10
Light Green 50-75
Yellow Green 250-300

[205] (F-Crown) W. BROOKFIELD/55 FULTON ST. N.Y. {Note backwards letter} (R-Crown) PAT.FEB. 22ND, 1870/ " " JAN. 25TH, 1870/ " " JAN. 14, 1879 {MLOD} SB
Light Aqua 5-10

[210] (F-Crown) W. BROOKFIELD/55 FULTON ST. N.Y. {Note backwards letter} (R-Crown) [Number]/PAT.FEB. 22ND, 1870/ " JAN. 25TH, 1870/ " JAN. 14TH, 1879 {MLOD} SB
Light Aqua 5-10

[213] (F-Crown) W. BROOKFIELD/55 FULTON ST. N.Y. (R-Crown) PAT FEB 22ND 1870/JAN 25TH 1870/JAN 14TH 1879 SB
Light Aqua 5-10

[217] (F-Crown) W. BROOKFIELD/55 FULTON ST. N.Y. (R-Crown) PAT.FEB. 22, 1870/ JAN. 25, 1870/ JAN. 14, 1879 SB
Light Blue Aqua 5-10

[220] (F-Crown) W. BROOKFIELD/55 FULTON ST. N.Y. (R-Crown) PAT.FEB. 22, 1870/ " JAN. 25, 1870/ " JAN. 14, 1879 {MLOD} SB
Aqua 5-10
Yellow Green 250-300
Olive Green 400-500

[230] (F-Crown) W. BROOKFIELD/55 FULTON ST. N.Y. (R-Crown) PAT.FEB. 22ND, 1870/ " JAN. 25TH, 1870/ " JAN. 14TH, 1879 {MLOD} SB
Aqua, Blue Aqua, Green Aqua 5-10
Light Green 50-75
Green 100-125
Yellow Green 250-300
Yellow Olive Green w/ Amber Swirls .. 1,500-1,750

[235] (F-Crown) W. BROOKFIELD/55 FULTON ST. N.Y. (R-Crown) [Number]/PAT.FEB. 22D 1870/ " JANTH 25, 1870/ " JAN. 14TH 1879 {MLOD} SB
Light Blue 5-10

[240] (F-Crown) W. EROOKFIELD/55 FULTON ST. N.Y. {Note spelling} {Note backwards letter} (R-Crown) PAT.FEB 22 1870/ " JAN 25 " / " " 14 1879 {MLOD} SB
Light Blue Aqua 10-15

41

CD 126

[250] (F-Crown) WM BROOKFIELD/55 FULTON ST. N.Y. (R-Crown) PAT FEB 22ND 1870/JAN 25TH 1870/JAN 14TH 1879 SB
Blue Aqua 5-10

[260] (F-Crown) WM BROOKFIELD/55 FULTON ST. N.Y. (R-Crown) PAT FEB 22ND 1870/JAN 25TH 1879/JAN 14TH 1879 {Note 'JAN 25TH 1879' date} SB
Light Aqua 5-10

[270] (F-Crown) WM BROOKFIELD/55 FULTON ST. N.Y. (R-Crown) PAT.FEB. 22ND 1879/JAN. 25TH 1870/JAN. 14TH 1879 {Note 'FEB. 22ND 1879' date} SB
Aqua 5-10

[280] (F-Crown) WM BROOKFIELD/55 FULTON ST. N.Y. (R-Crown) PAT.FEB. 22ND 1879/JAN. 25TH 1879/JAN. 14TH 1879 {Note 'JAN. 25TH 1879' date} SB
Aqua, Light Aqua 5-10

[290] (F-Crown) [Number]/(Arc)W. BROOFIELD/45 CLIFF ST./N.Y. {Note spelling} (R-Crown) (Arc)CAUVET/JULY 25 1865/PAT JAN. 25 1870/" FEB. 22 " {MLOD} SB
Aqua 5-10

[300] (F-Crown) [Number]/(Arc)W. BROOKFIELD/45 CLIFF ST. N.Y. (R-Crown) PAT.JAN. 25, 1870/PAT.JAN. 14, 1879 {MLOD} SB
Aqua 5-10
Purple 7,500-10,000

[310] (F-Crown) [Number]/(Arc)W. BROOKFIELD/45 CLIFF ST./N.Y. (R-Crown) (Arc)CAUVET/JULY 25 1865/PAT JAN. " 1870/ " FEB. 22 " {MLOD} SB
Aqua 5-10
Yellow Green 250-300
Teal Green 500-600
Yellow Olive Green 600-700

[320] (F-Crown) [Number]/(Arc)W. BROOKFIELD/45 CLIFF ST./N.Y. (R-Crown) (Arc)DAUVET/PAT./JULY 25, 1865/JAN. 25, 1870/FEB. 22, 1870 {Note spelling} {MLOD} SB
Aqua 5-10

[330] (F-Crown) [Number]/(Arc)W. BROOKFIELD/45 CLIFF ST./N.Y. {MLOD} SB
Aqua, Blue Aqua 5-10
Lime Green 50-75
Yellow Green 250-300
Teal Green 500-600

[340] (F-Crown) [Number]/PAT.MARCH 20, 1877/W. BROOKFIELD (R-Crown) PAT.FEB. 22, 1879/JAN. 25, 1870 JAN. 14, 1879 {Note 'FEB. 22, 1879' date} {MLOD} SB
Aqua 5-10
Light Green 50-75

[350] (F-Crown) [Number]/PAT.MARCH 20, 1877/W. BROOKFIELD (R-Crown) PAT.JAN. 25, 1870/PAT.FEB. 22, 1870/55 FULTON ST. N.Y. {MLOD} SB
Light Aqua, Light Blue 5-10
Light Green 50-75
Green, Lime Green 100-125
Yellow Green 250-300
Dark Teal Aqua 600-700

[355] (F-Crown) [Number]/PAT.MARCH 20, 1877/Wm BROOKFIELD (R-Crown) PAT.JAN. 25, 1870/PAT.FEB. 22, 1870/55 FULTON ST. N.Y. {MLOD} SB
Aqua 5-10

[360] (F-Crown) [Number]/PATT.MARCH 20, 1877/W. BROOKFIELD {Note spelling} (R-Crown) PAT.JAN. 25, 1870/PAT.FEB. 22, 1870/55 FULTON ST. N.Y. {MLOD} SB
Blue Aqua 5-10

[370] (F-Crown) [Number]/W. BROOKFIELD/ 45 CLIFF ST. (R-Crown) PAT JAN 25. 1870/ PAT JAN 14. 1870 {Note 'JAN 14. 1870' date} {MLOD} SB
Blue Aqua 5-10

[380] (F-Crown) [Number]/W. BROOKFIELD/ 45 CLIFF ST. (R-Crown) PAT JAN 25. 1870/ PAT JAN 14. 1879 {MLOD} SB
Light Blue 5-10
Olive Green 400-500

[390] (F-Crown) [Number]/W. BROOKFIELD/ 45 CLIFF ST. N.Y. (R-Crown) (Arc)CAUVET/ JULY 25, 1865/PAT.JAN. " , 1870/ " FEB. 22, " {MLOD} SB
Aqua 5-10
Olive Green 400-500

[400] (F-Crown) [Number]/W. BROOKFIELD/ 45 CLIFF ST. N.Y. (R-Crown) PAT JAN 25. 1870/PAT JAN 14. 1879 {MLOD} SB
Aqua, Light Aqua 5-10
Light Green 50-75
Purple 7,500-10,000

[410] (F-Crown) [Number]/W. BROOKFIELD/ 45 CLIFF ST./N.Y. (R-Crown) PAT JAN 25, 1870/PAT JAN 14, 1879 {MLOD} SB
Blue 10-15

CD 126.3

[420] (F-Crown) [Number]/W. BROOKFIELD/ 55 FULTON ST. N.Y. {Note backwards letter} (R-Crown) PAT FEB 22ND 1870/ " " JAN 25TH 1870/ " " JAN 14TH 1879 {MLOD} SB
Blue Aqua, Light Blue Aqua 5-10
Yellow Green 250-300
Olive Green 400-500

[430] (F-Crown) [Number]/W. BROOKFIELD/ 55 FULTON ST. N.Y. (R-Crown) FEB 22 1870/ JAN 25 1870/JAN 14 1879 {MLOD} SB
Light Aqua, Light Green Aqua 5-10
Blue 10-15
Green, Lime Green 100-125
Yellow Green 250-300
Olive Green 400-500
Olive Amber 800-1,000
Dark Yellow Amber 1,750-2,000

[440] (F-Crown) [Number]/W. BROOKFIELD/ 55 FULTON ST. N.Y. (R-Crown) PAT FEB 22 1870/JAN 25 1870/JAN 14 1879 {MLOD} SB
Aqua 5-10

[450] (F-Crown) [Number]/W. BROOKFIELD/ 55 FULTON ST. N.Y. (R-Crown) PAT FEB 22ND 1870/JAN 25TH 1870/JAN 14TH 1879 {MLOD} SB
Aqua, Blue Aqua 5-10
Light Green 50-75
Yellow Green 250-300
Olive Amber 800-1,000
Yellow Olive Amber 1,000-1,250
Dark Red Amber 10,000-15,000

[460] (F-Crown) [Number]/W. BROOKFIELD/ 55 FULTON ST. N.Y. (R-Crown) PAT.FEB. 22, 1870/JAN. 14, 1879 {MLOD} SB
Aqua 5-10
Green 100-125
Yellow Green 250-300
Olive Green 400-500

[470] (F-Skirt) W. BROOKFIELD (R-Skirt) NEW YORK SB
Aqua, Green Aqua 3-5
Light Blue, Light Green 5-10
Emerald Green, Lime Green 20-30

NO EMBOSSING

[010] [No embossing] {Large round dome} SB
Aqua, Dark Aqua 75-100
Dark Blue 100-125
Green 175-200

[020] [No embossing] {MLOD} SB
Mint Green 200-250
Yellow Green 300-350

[030] [No embossing] {Pennycuick style} SB
Dark Aqua 30-40
Green Aqua 40-50

NO NAME

[010] (F-Crown) 7 {Large number} {MLOD} SB
Green 350-400

[020] (F-Skirt) RD 149959 SB
Green Aqua 500-600
Jade Green Milk 1,000-1,250

PAT'D

[010] (F-Crown) PATd (R-Crown) [Number] SB
Green 400-500

PATENT - OTHER

[010] (F-Crown) ['PAT.JAN. 25, 1870' blotted out] SB
Aqua 15-20

CD 126.1

BROOKFIELD

[010] (F-Crown) W U T CO./CAUVETS PAT./ W. BROOKFIELD (R-Crown) [Number]/FEB. 22 70/NO 55 FULTON ST. N.Y. {MLOD} SB
Light Aqua, Light Blue 50-75

CD 126.3

A.U.

[010] (F-Crown) PATENT/DEC. 19. 1871 (F-Skirt) A.U. SB
Blue Aqua, Dark Aqua 50-75
Teal Green 500-600

[020] (F-Crown) PATENT/DEC. 19. 1871 (F-Skirt) A.U. (R-Crown) [Number] SB
Aqua, Dark Aqua, Green Aqua 50-75
Green 175-200

[030] (F-Crown) PATENT/DEC. 19. 1871 (F-Skirt) A.U. (R-Skirt) [Letter] SB
Aqua 50-75
Light Green 75-100
Green 175-200

CD 126.3

BROOKFIELD

[010] (F-Crown) (Arc)W. BROOKFIELD/45 CLIFF ST. N.Y. (R-Crown) PAT.JAN. 25, 1870 SB
Aqua 10-15

[020] (F-Crown) (Arc)W. BROOKFIELD/45 CLIFF ST. N.Y. (R-Crown) [Number]/PAT.JAN. 25, 1870 PAT.JAN. 14 1879 SB
Light Yellow Green 125-150

[030] (F-Crown) (Arc)W. BROOKFIELD/45 CLIFF ST./N.Y. (R-Crown) (Arc)CAUVET/PAT./JULY 25, 1865/JAN. 25, 1870/FEB. 22, 1870 SB
Light Aqua, Light Blue 10-15

[033] (F-Crown) (Arc)W. BROOKFIELD/45 CLIFF ST./N.Y. (R-Crown) (Arc)CAUVET/PAT./JULY 25, 1865/JAN. 25, 1870/FEB. 22, 1870 {MLOD} SB
Light Aqua 10-15

[037] (F-Crown) (Arc)W. BROOKFIELD/45 CLIFF ST/N.Y. (R-Crown) (Arc)DAUVET/PAT/JULY 25 1865/JAN 25 1870/FEB 22 1870 {Note spelling}{MLOD} SB
Light Aqua 10-15
Light Green 75-100

[040] (F-Crown) W. BROOKFIELD/45 CLIFF ST. N.Y. (R-Crown) PAT.JAN. 25, 1870 {MLOD} SB
Light Aqua 10-15
Light Green 100-125

[050] (F-Crown) W. BROOKFIELD/55 FULTON ST N.Y. (R-Crown) [Number]/PAT FEB. 22ND 1870/JAN. 14TH 1879 {MLOD} SB
Light Blue 10-15

[060] (F-Crown) W. BROOKFIELD/55 FULTON ST. N.Y. (R-Crown) PAT FEB 22 1870/JAN 25 1870/JAN 14 1879 {MLOD} SB
Light Aqua 10-15
Blue Gray 300-350

[070] (F-Crown) W. BROOKFIELD/55 FULTON ST. N.Y. (R-Crown) PAT FEB 22ND 1870/JAN 25TH 1870/JAN 14TH 1870 {Note 'JAN 14TH 1870' date} SB
Aqua, Blue Aqua 10-15

[080] (F-Crown) W. BROOKFIELD/55 FULTON ST. N.Y. (R-Crown) PAT.FEB 22ND 1870/JAN 25TH 1870/JAN 14TH 1879 {MLOD} SB
Light Blue 10-15
Apple Green, Lime Green 200-250

[090] (F-Crown) W. BROOKFIELD/55 FULTON ST. N.Y. (R-Crown) [Number]/PAT.FEB 22ND 1870/JAN 25TH 1870/JAN 14TH 1879 {MLOD} SB
Light Aqua 10-15
Lime Green 200-250
Green 350-400

[095] (F-Crown) [Backwards number]/(Arc)W. BROOKFIELD/45 CLIFF ST./N.Y. (R-Crown) (Arc)CAUVET/PAT./JULY 25, 1865/JAN. 25, 1870/FEB. 22, 1870 {MLOD} SB
Light Blue Aqua 10-15

[100] (F-Crown) [Number]/(Arc)W. BROOKFIELD/45 CLIFF ST./N.Y. (R-Crown) (Arc)CUAVET/PAT./JULY 25, 1865/JAN. 25, 1870/FEB. 22, 1870 {Note spelling} {MLOD} SB
Aqua, Blue Aqua, Light Aqua 10-15

[110] (F-Crown) [Number]/(Arc)W. BROOKFIELD/45 CLIFF ST./N.Y. (R-Crown) (Arc)DAUVET/PAT./JULY 25, 1865/JAN. 25, 1870/FEB. 22, 1870 {Note spelling}{MLOD} SB
Blue, Light Aqua 10-15

[115] (F-Crown) [Number]/(Arc)W. BROOKFIELD/45 CLIFF ST/N.Y. (R-Crown) CAUVET/PAT/JULY 25 1865/JAN 25 1870/FEB 22 1870 {MLOD} SB
Light Aqua 10-15

[120] (F-Crown) [Number]/W. BROOKFIELD/45 CLIFF ST. N.Y. (R-Crown) PAT.JAN 25 1870/PAT.JAN 14 1879 {MLOD} SB
Blue Aqua 10-15

[130] (F-Crown) [Number]/W. BROOKFIELD/45 CLIFF ST. N.Y. (R-Crown) PAT.JAN 25TH 1870 {MLOD} SB
Aqua, Blue Aqua, Light Aqua 10-15

[140] (F-Crown) [Number]/W. BROOKFIELD/55 FULTON ST. N.Y. (R-Crown) FEB. 22 1870/JAN. 25 1870/JAN 14 1879 {MLOD} SB
Light Aqua, Light Blue 10-15
Light Green 100-125

[150] (F-Crown) [Number]/W. BROOKFIELD/55 FULTON ST. N.Y. (R-Crown) FEB. 22, 1870/JAN. 25, 1870 SB
Aqua 10-15

[160] (F-Crown) [Number]/W. BROOKFIELD/55 FULTON ST. N.Y. (R-Crown) PAT.FEB 22TH 1870/JAN 25TH 1870/JAN 14TH 1879 SB
Yellow Green 500-600
Yellow Olive Green 1,500-1,750

CD 127

[170] (F-Crown) [Number]/W. BROOKFIELD/ 55 FULTON ST. N.Y. (R-Crown) PAT.FEB. 22, 1870/ " JAN. 25 " / " " 1A, 1879 {Note backwards number} {MLOD} SB
Aqua, Blue Aqua, Light Aqua, Light Blue 10-15
Light Green 100-125

[180] (F-Crown) [Number]/W. BROOKFIELD/ 55 FULTON ST. N.Y. (R-Crown) PAT.FEB. 22, 1870/ " JAN. 25, " / " " 14, 1870 {Note 'JAN 14, 1870' date} {MLOD} SB
Blue, Light Aqua 10-15

[190] (F-Crown) [Number]/W. BROOKFIELD/ 55 FULTON ST. N.Y. (R-Crown) PAT.FEB. 22, 1870/ " JAN. 25, " / " " 14, 1879 {MLOD} SB
Aqua, Blue Aqua, Green Aqua,
Light Blue Aqua 10-15
Light Green 100-125

[200] (F-Crown) [Number]/W. BROOKFIELD/ 55 FULTON ST. N.Y. (R-Crown) PAT.FEB. 22ND 1870/JAN 25TH 1870/JAN 14TH 1879 SB
Light Aqua 10-15

[210] (F-Crown) [Number]/W. BROOKFIELD/ 55 FULTON ST. N.Y. (R-Crown) PAT.FEB. 22ND 1870/JAN 25TH 1870/JAN 14TH 1879 {MLOD} SB
Light Aqua 10-15

NO EMBOSSING

[010] [No embossing] {Straight lower wire ridge variant} SB
Aqua 50-75

NO NAME

[010] (Dome) ['X' or '111'] SB
Aqua, Blue, Green Aqua, Light Aqua 20-30
Light Green 150-175
Light Yellow Green 300-350
Yellow Green 500-600
Olive Green 3,000-3,500
Yellow Olive Green 7,500-10,000

CD 126.4

AM.INSULATOR CO.

[010] (Base) AM INS CO PAT SEP 13 1881. BE
Aqua 20-30

[020] (Dome) [Number] (Base) AM INS CO./ PAT SEP 13 1881 BE
Aqua, Light Aqua, Light Blue 20-30

W.E.MFG.CO.

[010] (F-Crown) W.E.MFG.CO. (F-Skirt) PATENT/APPLIED FOR (R-Crown) W.U. SB
Aqua 15-20
Blue, Light Aqua 20-30
Lime Green 50-75
Cornflower Blue 800-1,000

[020] (F-Crown) W.E.MFG.CO. (F-Skirt) PATENT/APPLIED FOR/['DEC. 19. 1871' blotted out] (R-Crown) W.U. SB
Green Aqua, Light Green 15-20
Dark Aqua 20-30

[030] (F-Crown) W.E.MFG.CO. (F-Skirt) PATENT/DEC. 19. 1871 (R-Crown) W.U. SB
Aqua, Light Green Aqua 15-20
Dark Aqua, Ice Aqua, Light Blue 20-30
Light Lime Green 40-50
Off Clear 75-100
Teal Blue 250-300
Clear 300-350
Teal Green 350-400
Cornflower Blue 800-1,000
Royal Purple 10,000-15,000

[040] (F-Crown) W.E.MFG.CO. (F-Skirt) PATENT/DEC. 19. 1871 (R-Crown) W.U. (R-Skirt) [Letter] SB
Aqua, Blue Aqua, Green Aqua, Light Aqua .. 15-20
Lemon, Light Lime Green 40-50
Light Gray 175-200
Clear 300-350
Cornflower Blue 800-1,000

[050] (F-Crown) W.E.MFG.CO. (R-Crown) W.U. SB
Aqua 15-20
Lime Green 50-75

[060] (F-Crown) W.E.MFG.CO. (R-Crown) W.U. (R-Skirt) [Letter] SB
Blue 20-30
Lime Green 50-75

CD 127

BROOKE'S

[005] (Base) H. BROOKE'S PAT.JAN. 25TH 1870 BE
Light Aqua 75-100

45

CD 127

[010] (Dome) [Number] (Base) H. BROOKE'S PAT.JAN. 25TH 1870 BE
Aqua, Ice Blue, Light Aqua 75-100

BROOKFIELD

[010] (Dome) [Letter and number in a circle] (F-Crown) W. BROOKFIELD 55 FULTON ST. N.Y./CAUVET'S PAT. FEB. 22 '70 - PAT.JAN 25 1870 - SB
Light Blue 30-40

[015] (F-Crown) W. BROOKFIELD 55 FULTON ST   (R-Crown) W.U.T.Co./ CAUVETS PAT. FEB. 22, 1870. {MLOD} SB
Light Aqua 15-20

[020] (F-Crown) W. BROOKFIELD 55 FULTON ST. (R-Crown) CAUVETS PAT.APL. 4, 1870 {Note 'APL. 4, 1870' date} {MLOD} SB
Aqua, Blue 15-20

[030] (F-Crown) W. BROOKFIELD 55 FULTON ST. (R-Crown) PAT.APL. 4, 1871 {MLOD} SB
Blue 10-15
Green 400-500

[040] (F-Crown) W. BROOKFIELD 55 FULTON ST. (R-Crown) [Number]/PAT APRIL 4 1871 {MLOD} SB
Lime Green 250-300
Mint Green 300-350

[050] (F-Crown) W. BROOKFIELD 55 FULTON ST. (R-Crown) [Number]/PAT.APR. 4, 1871 {MLOD} SB
Aqua 10-15

[060] (F-Crown) W. BROOKFIELD/CAUVET'S PAT. (R-Crown) 55 FULTON [Large backwards '1'] S^T N.Y./FEB 22 1870 {The '1' also extends between 'FEB 22' and '1870'} {Small dome variant} {MLOD} SB
Aqua 250-300

[063] (F-Crown) W.U.T.Co./CAUVETS PAT./ W. BROOKFIELD (F-Skirt) [Letter] (R-Crown) ['†', letter or number]/FEB. 22 70/ NO 55 FULTON ST. N.Y. {MLOD} SB
Light Blue Aqua 15-20
Light Green 200-250

[067] (F-Crown) W.U.T.Co./CAUVETS PAT./ W. BROOKFIELD (F-Skirt) [Small letter] (R-Crown) [Number]/FEB. 22 70/NO 55 FULTON ST. {MLOD} SB
Aqua, Light Aqua 15-20

[070] (F-Crown) W.U.T.Co./CAUVETS PAT./ W. BROOKFIELD (R-Crown) [Number]/FEB. 22 70/NO 55 FULTON ST. {MLOD} SB
Aqua, Light Aqua 10-15

[080] (F-Crown) W.U.T.Co./CAUVETS PAT./ W. BROOKFIELD (R-Crown) ['†' or letter]/ FEB. 22 70/NO 55 FULTON ST. N.Y. {MLOD} SB
Aqua, Light Blue 10-15
Lime Green 250-300
Green 400-500

[090] (F-Crown) W.U.T.Co./CAUVET'S PAT/W. BROOKFIELD (R-Crown) FEB. 22, 1870/NO 55 FULTON ST N.Y. {MLOD} SB
Aqua 10-15

[095] (F-Crown) W.U.T.Co./CAUVET'S PAT/W. BROOKFIELD (R-Crown) [Number]/FEB 22 1870/NO 55 FULTON ST NY {MLOD} SB
Light Aqua 10-15

[100] (F-Crown) W.U.T.Co./CAUVET'S PAT/W. BROOKFIELD (R-Crown) [Number]/FEB 22 1870/NO 55 FULTON ST {MLOD} SB
Aqua 10-15

[110] (F-Crown) W.U.T.Co./CAUVET'S PAT/W. BROOKFIELD (R-Crown) [Number]/FEB. 22 1870/55 FULTON ST. {MLOD} SB
Aqua 10-15
Light Green 200-250

[120] (F-Crown) W.U.T.Co./CAUVET'S PAT/W. BROOKFIELD (R-Crown) [Number]/FEB. 22-70/NO 55 FULTON ST. N.Y. {MLOD} SB
Aqua, Ice Aqua, Light Blue 10-15
Light Yellow Green 350-400
Yellow Olive Green 2,000-2,500

[130] (F-Crown) [Number]/W. BROOKFIELD 55 FULTON ST. (R-Crown) PAT.APR. 4, 1871 {MLOD} SB
Aqua, Blue Aqua 10-15

[140] (F-Crown) [Number]/W. BROOKFIELD 55 FULTON ST. (R-Crown) PAT.APRIL 4, 1870 {Note 'APRIL 4, 1870' date} {MLOD} SB
Blue Aqua 10-15

[150] (F-Crown) [Number]/W. BROOKFIELD 55 FULTON ST. (R-Crown) PAT.APRIL 4, 1871 {MLOD} SB
Blue, Blue Aqua 10-15
Lime Green 300-350
Light Yellow Green 350-400

[160] (F-Crown) [Number]/W. BROOKFIELD 55 FULTON ST. (R-Crown) PATD APL 4 1871 {MLOD} SB
Blue Aqua 10-15

NO EMBOSSING

[010] [No embossing] SB
Green 250-300

CD 128

[020] [No embossing] {3-piece mold; mold line around crown} SB
Aqua, Dark Aqua 100-125

[030] [No embossing] {MLOD} SB
Aqua 20-30
Dark Aqua 50-75

[040] [No embossing] {Unembossed W/1} SB
Light Aqua, Light Blue 200-250
Dark Yellow Green, Emerald Green ... 1,500-1,750

NO NAME

[010] (Dome) ['W/1' in a circle] SB
Aqua, Light Blue 300-350
Dark Green 1,500-1,750
Dark Yellow Green, Emerald Green ... 1,750-2,000

W.U.

[010] (Dome) [Number] (F-Crown) L.A.C.S PAT.JULY 25TH 1865 (F-Skirt) W.U. PATTERN SB
Aqua 250-300
Blue Aqua 300-350

[020] (Dome) [Number] (F-Skirt) W.U.P. SB
Blue Aqua, Dark Aqua, Light Aqua 1,500-1,750
Dark Teal Green 2,500-3,000
Cobalt Blue, Dark Cobalt Blue 4,500-5,000

CD 127.4

NO EMBOSSING

[010] [No embossing] SB
Dark Aqua 350-400
Teal Aqua 400-500
Sage Aqua 500-600
Light Green 600-700
Light Purple 1,750-2,000

PATENT - DEC. 19, 1871

[010] (F-Skirt) PATENT/DEC. 19. 1871 SB
Blue Aqua, Dark Aqua, Green Aqua ... 1,250-1,500
Sage Green 1,500-1,750
Light Purple 4,000-4,500

CD 128

ARMSTRONG Ⓐ

[010] (F-Skirt) Ⓐrmstrong CSC (R-Skirt) MADE IN U.S.A. [Numbers and dots] SB
Clear x1
Blue Tint 1-2

[020] (F-Skirt) *Armstrong's* CSC (R-Skirt) MADE IN U.S.A. Ⓐ [Numbers and dots] SB
Clear, Straw x1
Purple Tint, Smoke 1-2

[030] (F-Skirt) *Armstrong's* CSC (R-Skirt) MADE IN U.S.A. [Numbers and dots] SB
Olive Green Tint 1-2

HEMINGRAY

[010] (F-Skirt) HEMINGRAY-CS/<Ⓗ> (R-Skirt) E-14B SB
Off Clear 10-15
Light Opalescent 50-75
Opalescent 125-150
Dark Opalescent 175-200
White Milk 400-500

[020] (F-Skirt) HEMINGRAY-CS/<Ⓗ> (R-Skirt) E-14B {Inside of skirt is threaded} SB
Clear 20-30

[030] (F-Skirt) HEMINGRAY-E.1. (R-Skirt) MADE IN U.S.A. {Inside of skirt is threaded} SB
Light Lemon, Off Clear 75-100
Ice Green, Lemon 100-125

[040] (F-Skirt) HEMINGRAY/[Numbers and dots] (R-Skirt) C.S.A. SB
Clear x1
Green Tint, Ice Green, Light Green 3-5
Light Lemon, Light Yellow 5-10
Dark Yellow 20-30
Light Lemon w/ Milk Swirls 50-75

[045] (F-Skirt) HEMINGRAY/[Numbers and dots] (R-Skirt) C.S.A. {Base has four indentations; experimental piece} SB
Clear 75-100

[050] (F-Skirt) HEMINGRAY/[Numbers and dots] (R-Skirt) C.S.C. CB
Clear x1
Green Tint, Ice Aqua 3-5

CD 128

[060] (F-Skirt) HEMINGRAY/[Numbers and dots] (R-Skirt) C.S.C. SB
Clear, Off Clear x1

[070] (F-Skirt) [Vertical bar]/HEMINGRAY/ [Numbers] (R-Skirt) [Vertical bar]/C.S.O. SB
Clear 5-10

[080] (F-Skirt) [Vertical bar]/HEMINGRAY/ [Numbers] (R-Skirt) [Vertical bar]/CSO {'O' is engraved over an 'A'} SB
Clear, Green Tint, Off Clear 5-10

[090] (F-Skirt) [Vertical bar]/HEMINGRAY/ [Number] (R-Skirt) C.S.C. SB
Light Green 3-5

[100] (F-Skirt) [Vertical bar]/HEMINGRAY/ [Number] (R-Skirt) C.S.O. SB
Ice Green 5-10

KERR

[010] (F-Skirt) KERR CSC (R-Skirt) MADE IN U.S.A. [Numbers and dots] SB
Blue Tint, Off Clear 3-5

NO EMBOSSING

[010] [No embossing] SB
Straw 50-75
Light Green 150-175

NO NAME

[020] (F-Skirt) MADE IN U.S.A./PAT. APPD. FOR [Letters] (R-Skirt) MADE IN U.S.A./PAT. APPD. FOR [Letters] {Pyrex product} SB
Clear 10-15

OWENS ILLINOIS ◁◎▷

[010] (F-Skirt) C.S./◁◎▷ (R-Skirt) E-14B {Inside of skirt is threaded} SB
Off Clear 10-15

PYREX

[003] (F-Skirt) PYREX [Letter]/T.M. REC. U.S. PAT. OFF. {Note spelling} (R-Skirt) C S A/MADE IN U.S.A. SB
Off Clear 3-5

[007] (F-Skirt) PYREX [Letter]/T.M. REG. U.S. PAT. OFF. (R-Skirt) C S A/MADE IN U.S.A. SB
Clear x1

[010] (F-Skirt) PYREX [Letter]/T.M. REG. U.S. PAT. OFF. (R-Skirt) C S A/MADE IN U.S.A. [Number] SB
Clear, Off Clear, Straw x1
Green Tint, Light Lemon 1-2

[020] (F-Skirt) PYREX [Letter]/T.M. REG. U.S. PAT. OFF. (R-Skirt) C S A/MADE NI U.S.A. [Number] {Note spelling} SB
Clear 5-10

[030] (F-Skirt) PYREX/REG. U.S. PAT. OFF. (R-Skirt) MADE IN U.S.A./PAT.APPD.FOR {Inside of skirt is threaded} SB
Clear, Straw x1

[040] (F-Skirt) PYREX/REG. U.S. PAT. OFF. (R-Skirt) [Letter] [Number] MADE IN U.S.A./ PAT.APPD.FOR {Inside of skirt is threaded} SB
Clear x1

[050] (F-Skirt) PYREX/REG. U.S. PAT. OFF. {1/2" shorter than standard size} {Inside of skirt is threaded} SB
Clear 5-10

[060] (F-Skirt) PYREX/REG. U.S. PAT. OFF./ MADE IN U.S.A. (R-Skirt) MADE IN U.S.A./ PAT.APPD.FOR {Inside of skirt is threaded} SB
Clear x1

[070] (F-Skirt) PYREX/REG. U.S. PAT. OFF./ MADE IN U.S.A. (R-Skirt) PAT.APPD FOR {Inside of skirt is threaded} SB
Clear x1

[080] (F-Skirt) PYREX/REG. U.S. PAT. OFF./ MADE IN U.S.A. (R-Skirt) [Letter] [Number] MADE IN U.S.A./PAT.APPD.FOR {Inside of skirt is threaded} SB
Clear x1

[090] (F-Skirt) PYREX/REG. U.S. PAT. OFF./ MADE IN U.S.A. (R-Skirt) [Number]/ PAT.APPD.FOR {Inside of skirt is threaded} SB
Clear x1

[100] (F-Skirt) PYREX/T.M. REG. U.S. PAT. OFF. (R-Skirt) MADE IN U.S.A./PAT.APPD FOR [Letter] {Inside of skirt is threaded} SB
Clear, Straw x1

[110] (F-Skirt) PYREX/T.M. REG. U.S. RAT. OFF. {Note spelling} (R-Skirt) MADE IN U.S.A./ PAT.APPD FOR [Letter] {Inside of skirt is threaded} SB
Clear 5-10

[120] (F-Skirt) RYREX/T.M. REG. U.S. PAT. OFF. {Note spelling} (R-Skirt) MADE IN U.S.A./ PAT.APPD FOR [Letter] {Inside of skirt is threaded} SB
Clear 20-30

CD 130.1

WHITALL TATUM ⓦ

[010] (F-Skirt) (Arc)WHITALL TATUM/C.S.C. (R-Skirt) (Arc)MADE IN U.S.A./[Numbers and dots]/Ⓐ SB
Clear 1-2

[020] (F-Skirt) (Arc)WHITALL TATUM/CSA (R-Skirt) (Arc)MADE IN U.S.A./[Numbers and dots]/Ⓐ {Base has four indentations; experimental piece} SB
Clear 100-125

[030] (F-Skirt) (Arc)WHITALL TATUM/[Number]/Ⓐ/[Number] (R-Skirt) C.S.C./MADE IN U.S.A. SB
Clear, Green Tint, Smoke 1-2

CD 128.4

HEMINGRAY

[010] (F-Skirt) HEMINGRAY/[Numbers and dots] (R-Skirt) SB {1" pinhole} SB
Aqua Tint, Clear 250-300

[020] (F-Skirt) HEMINGRAY/[Numbers] (R-Skirt) SB SB
Clear, Yellow Tint 100-125

CD 129

ARMSTRONG Ⓐ

[010] (F-Skirt) Ⓐrmstrong T.S. (R-Skirt) MADE IN U.S.A. Ⓐ [Numbers and dots] SB
Green Tint, Light Straw 3-5

[020] (F-Skirt) Ⓐrmstrong T.S. (R-Skirt) MADE IN U.S.A. [Numbers and dots] SB
Blue Tint, Clear, Green Tint, Off Clear 3-5

[030] (F-Skirt) *Armstrong's* T.S. (R-Skirt) MADE IN U.S.A. Ⓐ [Numbers and dots] SB
Clear, Green Tint, Olive Green Tint, Straw 3-5

HEMINGRAY

[005] (F-Skirt) HEMINGRAY (R-Skirt) T.S. CB
Clear 1-2

[010] (F-Skirt) HEMINGRAY/[Numbers and dots] (R-Skirt) T.S. CB
Clear, Light Straw x1
Blue Tint 1-2

KERR

[010] (F-Skirt) KERR T.S. (R-Skirt) MADE IN U.S.A. [Numbers and dots] SB
Off Clear 3-5

CD 130

CAL.ELEC.WORKS

[005] (F-Skirt) CAL.ELEC.WORKS/PATENT SB
Blue 300-350
Blue Aqua 350-400
Green Aqua, Ice Aqua 400-500
Green 3,500-4,000

[010] (F-Skirt) CAL.ELEC.WORKS/PATENT {MLOD} SB
Aqua, Blue, Light Blue 300-350
Blue Aqua 350-400
Ice Aqua 400-500
Dark Aqua 500-600
Light Green 2,000-2,500
Green, Yellow Green 3,500-4,000

CD 130.1

CAL.ELEC.WORKS

CD 130.1
$2^{1}/_{2} \times 4^{3}/_{4}$

[010] (F-Skirt) CAL.ELEC.WORKS/PATENT {MLOD} SB
Cobalt Blue 4,500-5,000
Aqua, Light Cobalt Blue 5,000-7,500

CD 130.2

CD 130.2

CD 130.2
SEILERS $2\,^5/_8 \times 4\,^1/_2$

[010] (F-Skirt) SEILERS PATENT/FEB. 6-1877 (R-Skirt) PATENT/DEC. 19. 1871 SB
Aqua, Light Green 7,500-10,000

CD 131

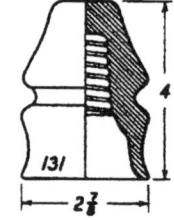

BROOKFIELD

[030] (F-Crown) (Arc)CAUVETS/PAT./JULY 25, 1865 (F-Skirt) W. BROOKFIELD/55 FULTON ST. N.Y. (R-Skirt) [Number] {MLOD} SB
Light Aqua 30-40

[010] (F-Crown) (Arc)CAUVETS/PAT./JULY 25, 1865 (F-Skirt) W. BROOKFIELD/55 FULTON ST. N.Y. (R-Skirt) [Number]/[Number] {MLOD} SB
Light Aqua 30-40

[020] (F-Crown) (Arc)CAUVETS/PAT./JULY 25, 1865 (F-Skirt) W. BROOKFIELD/55 FULTON ST. N.Y. {MLOD} SB
Aqua 30-40

[035] (F-Crown) (Arc)CAUVETS/PAT./JULY 25, 1865/JULY 25, 1865 {Note double date embossing: one small, one large} (F-Skirt) W. BROOKFIELD/55 FULTON ST. N.Y. (R-Skirt) [Number]/[Number] {MLOD} SB
Aqua 50-75

[040] (F-Crown) (Arc)CAUVET'S/PAT./JULY 25, 1865 (F-Skirt) W. BROOKFIELD/NO 55 FULTON ST. N.Y. (R-Crown) [Number] (R-Skirt) [Number] {MLOD} SB
Light Aqua 30-40
Light Blue 40-50

[050] (F-Crown) (Arc)W. BROOKFIELD/45 CLIFF ST./N.Y. (R-Crown) (Arc)CAUVET/JULY 25, 1865/PAT JAN " 1870/ " FEB. 22, " {MLOD} SB
Light Aqua 30-40
Ice Blue 50-75

[060] (F-Crown) (Arc)W. BROOKFIELD/45 CLIFF S^T/N.Y. (R-Crown) (Arc)CAUVET/PAT JULY 25 1865/ " JAN " 1870/ " FEB 22 " {MLOD} SB
Light Aqua 30-40
Light Green 175-200
Lime Green 200-250

[070] (F-Crown) [Number]/(Arc)CAUVETS/PAT./JULY 25, 1865 (F-Skirt) W. BROOKFIELD/55 FULTON ST. N.Y. (R-Skirt) [Number] {MLOD} SB
Aqua 30-40

[080] (F-Crown) [Number]/(Arc)CAUVETS/PAT./JULY 25, 1865 (F-Skirt) W. BROOKFIELD/55 FULTON ST. N.Y. {MLOD} SB
Aqua, Light Aqua 30-40

NO EMBOSSING

[010] [No embossing] {MLOD} SB
Aqua 50-75

TILLOTSON & CO.

[010] (F-Crown) (Arc)L.G. TILLOTSON/& CO./N.Y. (F-Skirt) BROOKE'S PAT/JAN 25TH 1870 SB
Light Aqua 300-350
Light Sage Green 700-800
Light Green 800-1,000
Gingerale, Pink 2,000-2,500

[020] (F-Crown) (Arc)L.G. TILLOTSON/& CO./N.Y. (F-Skirt) BROOKE'S PAT/JAN 25TH 1870 (R-Skirt) PAT.JAN 25, 1870/PAT.FEB 22, 1870 {MLOD} SB
Aqua, Light Aqua 300-350

[030] (F-Crown) (Arc)L.G. TILLOTSON/& CO./N.Y. (F-Skirt) BROOK'S PAT {Note spelling} SB
Dark Aqua 300-350

CD 131.4

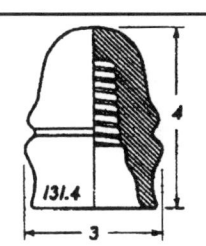

L.G.T. & CO.

[010] (F-Skirt) L.G.T.& C<u>o</u> (R-Skirt) PAT.JAN. 25, 1870/PAT.FEB. 22, 1870 {MLOD} SB
Aqua 100-125

[020] (F-Skirt) L.G.T.& C<u>o</u> {MLOD} SB
Aqua 150-175

NO EMBOSSING

[010] [No embossing] SB
Aqua, Blue Aqua, Dark Aqua 50-75
Dark Blue Aqua 100-125
Light Sage Green 175-200
Celery Green, Sage Green 200-250
Green 350-400
Light Purple 2,500-3,000
Dark Olive Green, Dark Yellow Green 3,500-4,000
Purple 5,000-7,500

PATENT - DEC. 19, 1871

[010] (F-Crown) PATENT/DEC. 19. 1871 (R-Crown) 1 SB
Aqua, Green Aqua, Light Aqua 75-100
Light Green 200-250
Sage Green 250-300
Green 500-600
Dark Green 1,000-1,250
Purple 5,000-7,500

CD 131.8

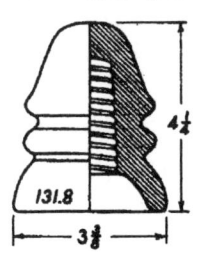

NO EMBOSSING

[010] [No embossing] SB
Aqua, Blue Aqua 2,000-2,500
Yellow Green 4,000-4,500

CD 132

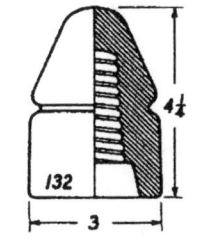

NO EMBOSSING

[010] [No embossing] SB
Light Green 125-150

CD 132.2

PAT APP FOR

[010] (F-Skirt) PAT APP FOR (R-Skirt) 2 SB
Brooke's Blue, Dark Aqua,
Dark Blue Aqua 400-500

PAT'D

[010] (F-Crown) 2 (Base) PAT^D BE
Blue Aqua, Green Aqua 1,000-1,250

PATENT - DEC. 19, 1871

[010] (F-Crown) PAT.DEC. 19. 1871 (R-Crown) 2 (R-Skirt) [Letter] SB
Aqua, Light Aqua, Light Green Aqua 50-75
Ice Green 100-125
Green 300-350
Teal Blue 600-700
Light Purple 1,500-1,750
Dark Yellow Green 2,000-2,500

[020] (F-Crown) PATENT/DEC. 19. 1871 (F-Skirt) [Letter] (R-Crown) 2 SB
Aqua, Green Aqua 15-20
Light Green 20-30

[030] (F-Crown) PATENT/DEC. 19. 1871 (R-Crown) 2 SB
Aqua, Green Aqua, Light Aqua 15-20
Dark Aqua 20-30
Steel Blue 40-50
Green 50-75
Sage Green 75-100
Light Smoke, Off Clear 100-125
Light Purple 150-175
Gray Blue, Purple 250-300
7-up Green, Emerald Green 1,000-1,250
Dark Olive Green 1,750-2,000
Cobalt Blue 5,000-7,500

[040] (F-Crown) PATENT/DEC. 19. 1871 (R-Crown) 2 (R-Skirt) [Letter] SB
Aqua, Blue Aqua, Green Aqua, Light Aqua .. 15-20
Dark Aqua, Ice Blue, Ice Green 20-30
Steel Blue 40-50
Off Clear, Smoke 100-125
Clear, Light Purple 150-175
Dark Blue 400-500
Cornflower Blue 500-600
Milky Light Green 700-800

CD 132.2

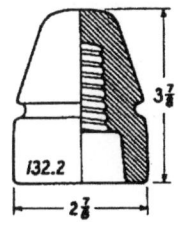

NO EMBOSSING

[010] [No embossing] SB
Dark Blue Aqua 100-125

51

CD 132.2

NO NAME

[010] (Dome) 2 SB
Aqua, Ice Blue, Light Aqua 75-100
Blue Aqua 125-150
Light Green 300-350
Blue, Lime Green 400-500
Teal Blue 800-1,000
Dark Green 2,000-2,500
Olive Green, Sapphire Blue,
Yellow Green 3,000-3,500
Cobalt Blue, Dark Cobalt Blue 3,500-4,000

PAISLEY

[010] (Dome) 1 (F-Skirt) S.T.PAISLEY/
MAKER/BEAVER FALLS, PA.
(R-Skirt) L.A.CAUVET'S/PAT'D/JULY 25TH
1865 SB
Aqua 1,750-2,000

[020] (F-Skirt) S.T.PAISLEY/MAKER/BEAVER
FALLS, PA. (R-Skirt) L.A.CAUVET'S/PAT'D/
JULY 25TH 1865 SB
Aqua 1,750-2,000

CD 132.4

PATENT - DEC. 19, 1871

[010] (F-Crown) PATENT/DEC. 19 1871 SB
Aqua, Dark Aqua, Light Aqua 500-600
Light Green 700-800

CD 133

B.G.M.CO.

[005] (F-Skirt) B.G.M.C. SB
Light Purple 100-125

[010] (F-Skirt) B.G.M.CO. SB
Light Purple 75-100
Purple 100-125

BROOKFIELD

[010] (Dome) [Number] (F-Skirt) BROOKFIELD
(R-Skirt) NEW YORK SB
Aqua, Green Aqua, Light Aqua x1
Blue Aqua 1-2
Yellow Green 10-15

[020] (Dome) [Number] (F-Skirt) BROOKFIELD
(R-Skirt) № 20 SB
Aqua, Dark Aqua 1-2
Green 5-10
Emerald Green, Olive Green, Yellow Green 10-15
Dark Olive Amber 100-125
Dark Yellow Amber 200-250

[030] (Dome) [Number] (F-Skirt) W.
BROOKFIELD (R-Skirt) NEW YORK SB
Blue Aqua, Light Aqua 1-2
Green 5-10
Yellow Green 15-20

[040] (F-Crown) (Arc)BROOKFIELD ['W'
blotted out]/['1 /55 FULTON ST.' blotted out]/
N.Y. (F-Skirt) ER (R-Crown) [Blotted out
embossing] {MLOD} SB
Aqua 15-20

[050] (F-Crown) (Arc)CAUVET'S/PAT/JULY
25, 1865 (F-Skirt) W. BROOKFIELD/№ 55
FULTON ST. (R-Skirt) 7 {MLOD} SB
Aqua 20-30

[060] (F-Crown) (Arc)W. BBOOKFIELD/45
CLIFF ST./N.Y. {Note spelling}
(R-Crown) PAT/JAN 25 1870 {MLOD} SB
Light Aqua 5-10
Yellow Green 175-200

[070] (F-Crown) (Arc)W. BROOKFIELD/45
CLIFF ST. N.Y. {MLOD} SB
Aqua 1-2
Green 30-40
Yellow Green 175-200

[080] (F-Crown) (Arc)W. BROOKFIELD/45
CLIFF ST. И.Y. {Note backwards letter}
{MLOD} SB
Aqua 5-10

[085] (F-Crown) (Arc)W. BROOKFIELD/45
CLIFF ST./N.Y. (R-Crown) PAT/JAN 25 1870
{MLOD} SB
Aqua 1-2

[090] (F-Crown) (Arc)W. BROOKFIELD/45
CLIFF ST./N.Y. (R-Crown) [Number] {MLOD}
SB
Purple 1,500-1,750

CD 133

[100] (F-Crown) (Arc)W. BROOKFIELD/45 CLIFF ST./N.Y. (R-Crown) [Number]/PAT/JAN 25 1870 {MLOD} SB
Aqua, Light Aqua 1-2
Blue 5-10
Green 30-40
Yellow Green 175-200
Olive Green 1,000-1,250
Purple 1,500-1,750

[110] (F-Crown) (Arc)W. BROOKFIELD/45 CLIFF ST./N.Y. {MLOD} SB
Light Aqua, Light Blue 1-2
Green 30-40
Yellow Green 175-200

[120] (F-Crown) (Arc)W. BROOKFIELD/45 CLIFF ST./И.Y. {Note backwards letter} (R-Crown) PAT./JAN. 25, 1870 {MLOD} SB
Aqua 5-10

[125] (F-Crown) (Arc)W. BROOKFIELD/45 ƆLIFF ST./N.Y. {Note backwards letter} (R-Crown) PAT/JAN 25 1870 {MLOD} SB
Light Aqua 10-15

[130] (F-Crown) (Arc)W. BROOKFIELD/45 CLIFF ST./NEW YORK {Note 'NEW YORK'} (R-Crown) [Number]/PAT/JAN 25, 1870 {MLOD} SB
Aqua 10-15
Olive Green 1,000-1,250

[135] (F-Crown) (Arc)W. BROOKFIELD/55 FULTON ST./N.Y. (R-Crown) (Arc)CAUVETS/PAT./JULY 25, 1865/JULY 25, 1865 {Note double date embossing; one small, one large} {MLOD} SB
Aqua 5-10

[140] (F-Crown) (Arc)W. BROOKFIELD/55 FULTON ST./N.Y. (R-Crown) (Arc)CAUVET'S/PAT/JULY 25 1865/PAT JAN 25 1870 {MLOD} SB
Light Aqua 3-5

[150] (F-Crown) (Arc)W. BROOKFIELD/55 FULTON ST./N.Y. (R-Crown) (Arc)CAUVET'S/PAT/JULY 25 1865/PAT JAN 25 1870/PAT FEB 22 1870 {MLOD} SB
Light Aqua 3-5

[160] (F-Crown) (Arc)W. BROOKFIELD/55 FULTON ST./N.Y. (R-Crown) (Arc)CAUVET'S/PAT./JULY 25, 1865 {MLOD} SB
Mustard Yellow 1,500-1,750

[170] (F-Crown) (Arc)W. BROOKFIELD/NEW YORK (R-Crown) (Arc)W. BROOKFIELD/NEW YORK {MLOD} SB
Aqua 20-30

[180] (F-Crown) (Arc)W. BROOKFIELD/NEW YORK (R-Crown) [Letter] {MLOD} SB
Aqua, Light Aqua, Light Blue 1-2
Ice Blue 3-5
Apple Green 40-50
Yellow Green 75-100
Yellow Olive Green 500-600

[190] (F-Crown) (Arc)W. BROOKFIELD/Ʌ5 CLIFF ST./N.Y. {Note backwards number} (R-Crown) (Arc)CAUVET'S/PAT./JULY 25, 1865/PAT.JAN. 25, 1870/PAT.FEB. 22, 1870 {MLOD} SB
Light Aqua 5-10

[200] (F-Crown) (Arc)W. BROOKFIELD/[Number]/55 FULTON ST. N.Y. (F-Skirt) E.R. (R-Crown) (Arc)CAUVET'S/PAT./JULY 25, 1865 (R-Skirt) PAT.JAN. 25, 1870 {MLOD} SB
Aqua 15-20

[210] (F-Crown) (Arc)W. BROOKFIELD/[Number]/55 FULTON ST. N.Y. (F-Skirt) E.R. (R-Crown) (Arc)CAUVET'S/PAT./JULY 25, 1865 {MLOD} SB
Aqua 15-20

[220] (F-Crown) (Arc)W. BROOKFIELD/[Number]/55 FULTON ST./N.Y. (F-Skirt) E.R. (R-Crown) (Arc)CAUVET'S/PAT./JULY 25, 1865 (R-Skirt) PAT.JAN. 25, 1870/PAT.FEB. 22, 1870 {MLOD} SB
Aqua, Blue, Light Aqua 15-20
Yellow Green 1,000-1,250

[223] (F-Crown) (Arc)W. BROOKFIELD/[Number]/55 FULTON ST./N.Y. (F-Skirt) ['E.R.' blotted out] (R-Crown) (Arc)CAUVET'S/PAT./JULY 25, 1865 (R-Skirt) PAT.JAN. 25, 1870/PAT.FEB. 22, 1870 {MLOD} SB
Light Aqua 10-15

[227] (F-Crown) (Arc)W. BROOKFIELD/[Number]/55 FULTON ST./N.Y. (F-Skirt) ['E.R.' blotted out] (R-Crown) (Arc)CAUVET'S/PAT./JULY 25, 1865 (R-Skirt) [Blotted out embossing] {MLOD} SB
Light Aqua 10-15

[230] (F-Crown) (Arc)W. BROOKFIELD/[Number]/55 FULTON ST./N.Y. (R-Crown) (Arc)CAUVET'S/PAT/JULY 25, 1865 {MLOD} SB
Blue Aqua, Light Aqua 10-15
Apple Green 175-200
Cornflower Blue, Dark Green 800-1,000
Sapphire Blue, Teal Blue 1,500-1,750

53

CD 133

[240] (F-Crown) (Arc)W. BROOKFILD/55 FULTON ST./N.Y. {Note spelling} (R-Crown) (Arc)CAUVET'S/PAT/JULY 25 1865/PAT JAN 25 1870/PAT FEB 22 1870 {MLOD} SB
Aqua 5-10

[250] (F-Crown) W. BROOKFIELD/NEW YORK (R-Crown) [Letter] SB
Aqua, Light Blue 1-2
Yellow Green 75-100

[260] (F-Crown) [Number]/(Arc)CAUVET'S/PAT/JULY 25, 1865 (F-Skirt) W. BROOKFIELD/NO 55 FULTON ST (R-Skirt) [Large '7'] {MLOD} SB
Aqua, Light Aqua 20-30

[270] (F-Crown) [Number]/(Arc)W. BOOKFIELD/45 CLIFF ST/N.Y. {Note spelling} {MLOD} SB
Green Aqua, Light Aqua 5-10

[280] (F-Crown) [Number]/(Arc)W. BOOKFIELD/45 CLIFF ST./N.Y. {Note spelling and backwards number} (R-Crown) (Arc)CAUVET'S/PAT./JULY 25, 1865/PAT.JAN. 25, 1870/PAT.FEB. 22, 1870 {MLOD} SB
Aqua 10-15

[285] (F-Crown) [Number]/(Arc)W. BROOKFIED/55 FULTON ST./N.Y. {Note spelling} (R-Crown) (Arc)CAUVET'S/PAT./JULY. 25, 1865/PAT.JAN. 25. 1870/PAT.FEB. 22. 1870 {MLOD} SB
Aqua 5-10

[290] (F-Crown) [Number]/(Arc)W. BROOKFIED/55 FULTON ST./N.Y. {Note spelling} (R-Crown) (Arc)CAUVET'S/PAT.JULY 25, 1865/PAT.JAN. 25, 1870/PAT.FEB. 22, 1870 {MLOD} SB
Aqua 5-10

[300] (F-Crown) [Number]/(Arc)W. BROOKFIELD/45 OLIFF ST {Note backwards letter} {Note no 'N.Y.'} (R-Crown) PAT./JAN. 25, 1870 {MLOD} SB
Aqua 5-10

[310] (F-Crown) [Number]/(Arc)W. BROOKFIELD/45 CLIFF ST {Note no 'N.Y.'} (R-Crown) PAT./JAN. 25, 1870 {MLOD} SB
Aqua, Light Aqua 5-10
Yellow Olive Green 1,000-1,250

[320] (F-Crown) [Number]/(Arc)W. BROOKFIELD/45 CLIFF ST. N.Y. (R-Crown) (Arc)CAUVET'S/PAT./JULY 25, 1865/JAN. 25, 1870/FEB. 22, 1870 {MLOD} SB
Aqua 3-5
Yellow Green 175-200

[330] (F-Crown) [Number]/(Arc)W. BROOKFIELD/45 CLIFF ST. N.Y. (R-Crown) PAT/JULY 25, 1865 {MLOD} SB
Aqua 1-2

[340] (F-Crown) [Number]/(Arc)W. BROOKFIELD/45 CLIFF ST./N.Y. (R-Crown) (Arc)CAUVET'S/PAT./JULY 25, 1865/JAN. 25, 1870/FEB. 22, 1870 {MLOD} SB
Aqua 3-5

[350] (F-Crown) [Number]/(Arc)W. BROOKFIELD/45 CLIFF ST./N.Y. (R-Crown) (Arc)CAUVET'S/PAT/JULY 25 1865/PAT JAN 25 1870/PAT FEB 22 1870 {MLOD} SB
Aqua 3-5
Yellow Olive Green 1,000-1,250

[360] (F-Crown) [Number]/(Arc)W. BROOKFIELD/45 CLIFF ST./N.Y. (R-Crown) PAT./JAN. 25 1870 {MLOD} SB
Aqua 1-2
Green 30-40
Yellow Green 175-200
Light Sapphire Blue 300-350
Light Yellow Olive Green 400-500
Olive Green 1,000-1,250
Purple 1,500-1,750

[370] (F-Crown) [Number]/(Arc)W. BROOKFIELD/45 CLIFF ST./N.Y. {MLOD} SB
Light Aqua 1-2
Ice Blue 3-5
Light Green 5-10

[380] (F-Crown) [Number]/(Arc)W. BROOKFIELD/45 CLIFF ST./И.Y. {Note backwards letter} (R-Crown) PAT./JAN. 25 1870. {MLOD} SB
Light Aqua 5-10

[385] (F-Crown) [Number]/(Arc)W. BROOKFIELD/45 CLIFF SI {Note no 'N.Y.'} (R-Crown) (Arc)PAT./JAN. 25, 1870 {MLOD} SB
Light Blue 3-5

CD 133

[390] (F-Crown) [Number]/(Arc)W. BROOKFIELD/55 FULTON ST./N.Y. (R-Crown) (Arc)CAUVET'S/PAT.JULY 25, 1865/PAT.JAN. 25, 1870/PAT.FEB. 22, 1870 {MLOD} SB
Aqua 3-5

[400] (F-Crown) [Number]/(Arc)W. BROOKFIELD/55 FULTON ST./N.Y. (R-Crown) (Arc)CAUVET'S/PAT/JULY 25 1865/PAT JAN 25 1870 {MLOD} SB
Light Blue 3-5

[405] (F-Crown) [Number]/(Arc)W. BROOKFIELD/55 FULTON ST./N.Y. (R-Crown) (Arc)CAUVET'S/PAT/JULY 25 1865/PAT JAN 25 1870/PAT FEB 22 1870 {MLOD} SB
Light Blue Aqua 3-5

[410] (F-Crown) [Number]/(Arc)W. BROOKFIELD/55 FULTON ST./N.Y. (R-Crown) (Arc)CAUVET'S/PAT/JULY 25 1865/PAT JAN 25 1870/TAT FEB 22 1870 {Note spelling} {MLOD} SB
Light Aqua 5-10

[420] (F-Crown) [Number]/(Arc)W. BROOKFIELD/NEW YORK SB
Light Blue 1-2

[430] (F-Crown) [Number]/(Arc)W. BROOKFIELD/NEW YORK (R-Crown) [Letter] {MLOD} SB
Aqua, Light Aqua, Light Blue 1-2

[440] (F-Crown) [Number]/(Arc)W. BROOKFIELD/NEW YORK {MLOD} SB
Aqua, Light Aqua 1-2
Green 20-30

[450] (F-Crown) [Number]/(Arc)W. BROOKFILD/45 CLIFF ST./N.Y. {Note spelling} (R-Crown) PAT.JAN. 25, 1870 {MLOD} SB
Aqua 5-10

[460] (F-Crown) [Number]/(Arc)W. BROOKFILD/55 FULTON ST./N.Y. {Note spelling} (R-Crown) (Arc)CAUVET'S/PAT.JULY 25, 1865/PAT.JAN. 25, 1870/PAT.FEB. 22, 1870 {MLOD} SB
Aqua 5-10

[465] (F-Skirt) BROOKFIELD SDP
Dark Aqua 10-15

[470] (F-Skirt) BROOKFIELD (R-Skirt) NEW YORK SB
Aqua, Light Aqua x1
Light Blue, Light Green 1-2
Green 5-10

[480] (F-Skirt) BROOKFIELD (R-Skirt) NEW YORK/['B.G.M.CO.' blotted out] SB
Light Aqua 5-10
Light Yellow Green 20-30

[490] (F-Skirt) BROOKFIELD (R-Skirt) №20 SB
Aqua, Dark Green Aqua 1-2
Green 5-10
Yellow Green 10-15
Dark Olive Green 20-30
Olive Green Amber 50-75

[500] (F-Skirt) BROOKFIELD (R-Skirt) №20 SDP
Dark Aqua 15-20
Olive Green, Yellow Green 75-100

[510] (F-Skirt) BROOKFIELD (R-Skirt) [Blotted out embossing] SB
Aqua 5-10

[520] (F-Skirt) W. BROOKFIELD (R-Skirt) NEW YORK SB
Aqua 1-2
Apple Green, Lime Green 15-20

CALIFORNIA

[010] (F-Skirt) CALIFORNIA SB
Aqua 5-10
Blue Aqua 10-15
Light Purple 75-100
Smoke, Yellow 100-125
Dark Rose, Purple 150-175
Burgundy, Dark Plum 200-250
Purple/Peach Two Tone 250-300
Peach 300-350

CITY FIRE ALARM

[010] (F-Skirt) CITY FIRE ALARM SB
Light Aqua 75-100

E.R.

[010] (F-Skirt) E R {Pennycuick mold} SB
Aqua, Light Aqua 30-40
Light Green 50-75

H.G.CO.

[005] (F-Skirt) H G CO/PATENT MAY 2 1893 {Note no periods} (R-Skirt) STANDARD SDP
Aqua 1-2

[010] (F-Skirt) H.G.CO. (R-Skirt) №7/STANDARD {Note no line under 'O'} SB
Aqua, Light Blue 5-10
Light Green 10-15
Light Steel Blue 15-20
Powder Blue 30-40
Clear 40-50
Green, Purple Tint 50-75
Orange Amber 7,500-10,000

CD 133

[020] (F-Skirt) H.G.CO. (R-Skirt) № 7 / ƧTANDARD {Note backwards letter} SB
Aqua, Blue Aqua 10-15
Ice Aqua, Light Green 15-20

[030] (F-Skirt) H.G.CO. (R-Skirt) STANDARD SB
Aqua 30-40

[040] (F-Skirt) H.G.CO./PATENT MAY 2 1893 (R-Skirt) № 20 SDP
Aqua 1-2
Lime Green 75-100

[050] (F-Skirt) H.G.CO./PATENT MAY 2 1893 (R-Skirt) № 20/['STANDARD' blotted out] SDP
Aqua 3-5

[060] (F-Skirt) H.G.CO./PATENT MAY 2 1893 (R-Skirt) STANDARD SB
Aqua 40-50

[070] (F-Skirt) H.G.CO./PATENT MAY 2 1893 (R-Skirt) STANDARD SDP
Aqua, Light Aqua 1-2
Light Green Aqua 3-5
Hemingray Blue, Light Green 10-15
Ice Green 20-30
Light Yellow Green 40-50
Sage Gray 75-100
Apple Green, Green, Lime Green 100-125
Lemon 125-150
Milky Aqua 250-300
Yellow Green 300-350
Milky Hemingray Blue, Milky Lime Green . 400-500
Jade Green Milk 800-1,000

[080] (F-Skirt) H.G.CO./PATENT MAY 2 1893 {'9' is engraved over an '8'} (R-Skirt) STANDARD SDP
Aqua 5-10

HEMINGRAY

[010] (F-Skirt) HEMINGRAY (R-Skirt) PATENT/MAY 2 1893 SDP
Aqua, Blue Aqua, Dark Aqua 1-2

[015] (F-Skirt) HEMINGRAY/№ 20 (R-Skirt) PATENT MAY 2 1893 RDP
Aqua 30-40

[020] (F-Skirt) HEMINGRAY/№ 20 (R-Skirt) PATENT MAY 2 1893 SDP
Aqua 20-30

[030] (F-Skirt) HEMINGRAY/PATENT MAY 2 1893 (R-Skirt) STANDARD SDP
Aqua, Blue Aqua 1-2
Light Green 10-15

NO EMBOSSING

[010] [No embossing] SB
Hemingray Blue 20-30

[020] [No embossing] SDP
Aqua 10-15
Blue Aqua 15-20

[025] [No embossing] {DEC 19, 1871 Patent style} SB
Celery Green, Dark Green 200-250

[030] [No embossing] {MLOD} SB
Light Aqua 10-15
Yellow Green 400-500
Dark Olive Amber 600-700

NO EMBOSSING - CANADA

[005] [No embossing] RB
Dark Green 800-1,000

[010] [No embossing] {Tall narrow dome} {MLOD} SB
Light Green 20-30
Blue, Light Aqua 30-40
Lemon, Powder Blue 50-75
Gray Purple, Purple 400-500
Dark Purple 500-600

NO NAME

[010] (Dome) [Number] (F-Skirt) № 20 SB
Aqua 1-2
Green 5-10
Yellow Green 15-20

[020] (F-Skirt) № 20 SB
Aqua, Green Aqua, Ice Blue, Light Blue 1-2
Light Green 3-5
Green 5-10
Emerald Green, Yellow Green 15-20
Olive Green 20-30

[030] (F-Skirt) № 20 SDP
Aqua 50-75

[040] (F-Skirt) Rd № 154745 SB
Aqua 800-1,000
Jade Green Milk 1,000-1,250

O.V.G.CO.

[010] (F-Skirt) O.V.G.CO. SB
Aqua 30-40

PATENT - DEC. 19, 1871

[010] (F-Crown) PATENT/DEC. 19. 1871 SB
Aqua, Ice Aqua 5-10
Light Green 20-30
Light Purple 1,000-1,250
Purple 1,500-1,750

CD 133.2

[020] (F-Crown) PATENT/DEC. 19. 1871
(R-Crown) 3 SB
Aqua, Green Aqua, Light Aqua 5-10
Sage Green 40-50
Light Gray 75-100
Milky Aqua 200-250

STAR ✶

[010] (F-Skirt) ✶ SB
Aqua, Blue, Blue Aqua, Light Blue,
Light Green 1-2
Green, Lime Green 5-10
Dark Green, Emerald Green, Yellow Green . 10-15
Olive Green, Yellow Olive Green 20-30
Aqua/Yellow Green Two Tone 30-40

TEL.FED.MEX.

[010] (F-Skirt) TEL.FED./MEX. (R-Skirt) JM.
{MLOD} SB
Dark Green Aqua 50-75
Teal Green 75-100

CD 133.1

BROOKE'S

CD 133.1
2 $\frac{1}{2}$ x 3 $\frac{3}{4}$

[010] (F-Skirt) HOMER BROOKE'S PAT
(R-Skirt) [Letter]/AUG 14 1883 SB
Aqua, Light Blue 30-40
Brooke's Blue 50-75

ELECTRICAL SUPPLY CO.

[010] (F-Crown) [Number] (F-Skirt) THE
ELECTRICAL SUPPLY Co./CHICAGO
{MLOD} SB
Aqua, Light Aqua, Light Blue 20-30
Light Green 100-125
Green 175-200
Yellow Green 200-250

[020] (F-Skirt) THE ELECTRICAL SUPPLY
Co./CHICAGO {MLOD} SB
Aqua, Green Aqua, Light Aqua, Light Blue .. 20-30
Light Green 100-125

[030] (F-Skirt) THE ELECTRICAL SUPPLY
CO./CHICAGO {MLOD} SB
Aqua 20-30
Light Green 100-125
Light Yellow Green 150-175

PAT APP FOR

[010] (F-Skirt) PAT APP FOR (R-Skirt) 3. SB
Dark Blue Aqua 30-40
Brooke's Blue 50-75
Teal Blue 100-125
Green 200-250
Emerald Green, Yellow Emerald Green 800-1,000

[020] (F-Skirt) PAT APP FOR (R-Skirt) 3/
[Letter] SB
Aqua 30-40
Brooke's Blue, Hemingray Blue 50-75

PATENT - DEC. 19, 1871

[010] (Dome) [Letter] (F-Skirt) PATENT/DEC.
19. 1871 (R-Skirt) 3 SB
Aqua, Light Aqua 3-5
Light Green 5-10
Apple Green, Green,
Light Hemingray Blue, Off Clear 30-40
Green w/ Milk Swirls, Purple Tint 75-100
Lavender 200-250

[020] (Dome) [Letter] (F-Skirt) PATENT/EC. 19.
1871 {Note spelling} (R-Skirt) 3 SB
Aqua, Light Aqua 10-15
Ice Green 15-20
Teal Blue 200-250

[030] (F-Skirt) PATENT/DEC. 19. 1871
(R-Skirt) 3/[Letter] SB
Light Green Aqua 3-5
Blue Aqua, Light Green 5-10
Apple Green, Green, Hemingray Blue,
Lime Green, Off Clear, Sky Blue 30-40
Clear, Purple Tint 75-100

TEL.FED.MEX.

[010] (F-Skirt) TEL.FED./MEX.
(R-Skirt) SAMUEL H$\underline{NOS}$/NEW YORK. SB
Aqua 40-50
Milky Green 200-250

[020] (F-Skirt) TEL.FED./MEXICO
(R-Skirt) MARTINEZ CHIBRAS/LIBRES SB
Emerald Green 500-600

CD 133.2

B.F.G.CO.

[010] (Dome) [Number] (F-Crown) (Arc)L.A.C.'S
PAT JULY 25$\underline{TH}$ 1865 (F-Skirt) B.F.G.Co. SB
Aqua, Blue Aqua 1,750-2,000

CD 133.2

[020] (Dome)$\mathbb{I}$ (F-Crown) (Arc)L.A.C.'S PAT JULY 25TH 1865 (F-Skirt) B.F.G.Co. SB
Aqua 1,750-2,000

NO EMBOSSING

[010] [No embossing] SB
Aqua 200-250

NO EMBOSSING - CANADA

[010] [No embossing] {Slope shoulder} {MLOD} {Comes with and without small button on dome} {Previously shown as CD 134 in the 1991 edition} SB
Blue Aqua, Dark Aqua,
Green Aqua, Light Blue 10-15
Blue 15-20
Milky Blue 200-250
Dark Green 250-300
Dark Yellow Green 300-350

NO NAME

[010] (Dome) M SB
Aqua, Blue Aqua 75-100

P & W

[010] (Dome) ['2' in a circle]
(F-Crown) (Arc)PATD JULY 25TH 1865/L.A.C. (F-Skirt) P & W SB
Aqua, Blue Aqua 75-100

[020] (F-Crown) (Arc)PATD JULY 25TH 1865/ L.A.C (F-Skirt) P & W SB
Aqua 75-100

[030] (F-Skirt) L.A.C/PATD JULY 25TH 1865/P & W SB
Aqua, Blue Aqua 100-125
Green 350-400

[040] (F-Skirt) PATD JULY 25TH 1865/P & W SB
Blue Aqua 100-125

CD 133.3

NO EMBOSSING

[010] [No embossing] SB
Light Aqua 30-40

NO NAME

[010] (F-Skirt) [Vertical bar] (R-Skirt) [Vertical bar] SB
Aqua 30-40
Green 50-75

[020] (F-Skirt) [Vertical bar] ϵ {Note backwards number} (R-Skirt) [Vertical bar] SB
Dark Aqua 30-40

[030] (F-Skirt) [Vertical bar] [Number] (R-Skirt) [Vertical bar] [Number] SB
Light Green 30-40

CD 133.4

NO EMBOSSING

[010] [No embossing] SB
Aqua 30-40
Light Blue 50-75
Blue, Light Green 75-100
Green, Sage Gray 150-175
Light Purple 300-350
Purple 350-400
Dark Blue, Gray Blue,
Purple/Peach Two Tone 400-500
Dark Plum, Dark Smoky Purple, Pink,
Purple/Sage Two Tone, Yellow Green 1,000-1,250
Emerald Green 2,000-2,500
Honey Amber 5,000-7,500

[020] [No embossing] {MLOD} SB
Lime Green 100-125
Green 175-200

[030] [No embossing] {West VA style} SB
Green Aqua 800-1,000
Off Clear, Teal Blue 1,250-1,500
Cobalt Blue 2,000-2,500
Purple 3,000-3,500

PAT'D

[010] (Base) PATD BE
Aqua, Green Aqua 50-75

[020] (Base) PAT'D BE
Aqua, Green Aqua 50-75
Green 300-350
Dark Green 700-800
Teal Green 800-1,000
Yellow Green 1,500-1,750
Purple/Sage Two Tone 1,750-2,000
Emerald Green 2,000-2,500
Cobalt Blue 5,000-7,500

PATENT - DEC. 19, 1871

[010] (F-Crown) PATENT/DEC. 19. 1871 (F-Skirt) [Letter] (R-Crown) 3 SB
Aqua, Green Aqua, Light Aqua 5-10
Dark Blue Aqua, Light Green 20-30
Green 40-50
Light Yellow Green 100-125
Light Citrine 175-200
Cornflower Blue, Light Sapphire Blue 350-400
Citrine 500-600

CD 134

[020] (F-Crown) PATENT/DEC. 19. 1871 (R-Crown) 3 SB
Aqua, Dark Aqua, Light Aqua 5-10
Dark Blue Aqua, Light Blue, Light Green 20-30
Teal Aqua 30-40
Green 40-50
Milky Aqua 200-250
Dark Green 500-600

[030] (F-Crown) PATENT/DEC. 19. 1871 (R-Crown) 3 (R-Skirt) [Letter] SB
Aqua, Blue Aqua, Dark Aqua, Light Aqua 5-10
Celery Green, Ice Green 15-20
Blue, Dark Blue Aqua, Light Green 20-30
Green 40-50
Light Purple 200-250
Dark Yellow Green 1,250-1,500
Olive Amber Blackglass 1,500-1,750

[040] (F-Crown) PATENT/DEC. 19. 1871 {Note 'E' is missing middle bar} (R-Crown) 3 (R-Skirt) [Letter] SB
Aqua 10-15
Ice Aqua 15-20
Ice Green 20-30

CD 133.5

TEL.FED.MEX.

CD 133.5
$2\ ^3/_4\ \times\ 4\ ^1/_4$

[010] (F-Skirt) TEL.FED./MEX (R-Skirt) JM {MLOD} SB
Teal Green 125-150
Teal Blue 175-200

[020] (F-Skirt) TEL.FED./MEX (R-Skirt) JM/ [Blotted out embossing] {MLOD} SB
Dark Green Aqua 75-100
Teal Green 125-150

CD 134

AM.INSULATOR CO.

[010] (Base) AM INS CO/PAT.SEP 13 1881 BE
Aqua, Blue Aqua, Light Aqua 15-20
Light Green 30-40

[020] (Base) AM INSЯ Co PATD SEPT. 13, '81. {Note backwards letter} BE
Aqua, Light Green Aqua 15-20
Light Green 30-40
Honey Amber, Olive Amber 4,000-4,500

[030] (Base) AM.INS.CO. PAT. SEPT 13. 81. BE
Light Aqua 15-20

[040] (Base) AM.INS.CO. PATD SEPT 13.'81. [Glass dots] BE
Aqua, Blue Aqua, Light Aqua,
Light Green Aqua 15-20
Light Green 30-40

[050] (Base) AM.INS.CO. PATD SEPT 13.'81./ ['OAKMAN' blotted out] [Glass dot] BE
Light Aqua 20-30

[060] (Base) AM.IИSʀ Co. PATD SEPT. 13, '81 {Note backwards letter} BE
Aqua, Ice Green, Light Aqua,
Light Green Aqua 15-20
Light Blue 20-30
Light Green 30-40
Green 200-250
Light Yellow Green 500-600
Jade Aqua Milk 1,000-1,250
Light Olive Green 1,250-1,500
Olive Green 2,000-2,500
Dark Yellow Olive Green 4,000-4,500

[070] (Base) AM.INSʀ CO. PATD SEPT. 13.'81./['OAKMAN' blotted out] [Glass dot] BE
Aqua 20-30

B.G.M.CO.

[010] (F-Skirt) B.G.M.CO. SB
Light Purple 100-125
Purple 150-175
Clear, Pink, Purple Tint 250-300
Light Lemon 350-400

BROOKFIELD

[010] (Dome) [Number] (F-Skirt) BROOKFIELD SB
Dark Green Aqua 1-2
Dark Green, Dark Yellow Green 3-5

[020] (Dome) [Number] (F-Skirt) BROOKFIELD SDP
Dark Aqua 3-5
Dark Green 5-10
Dark Yellow Green 20-30
Olive Amber 300-350

[030] (Dome) [Number] (F-Skirt) BROOKFIELD (R-Skirt) NEW YORK SB
Aqua, Green Aqua, Light Aqua x1

59

CD 134

[040] (Dome) [Number] (F-Skirt) BROOKFIFED {Note spelling} SB
Aqua, Dark Aqua 3-5

[045] (Dome) [Number] (F-Skirt) W. BROOKFIELD (R-Skirt) NEW YORK SB
Light Aqua 1-2

[050] (F-Crown) (Arc)W. BROOKFIELD/45 CLIFF ST. N.Y. (R-Crown) [Number] {MLOD} SB
Light Green Aqua 1-2

[060] (F-Crown) (Arc)W. BROOKFIELD/45 CLIFF ST. N.Y. (R-Crown) [Number]/PAT. JAN 25. 1870 {MLOD} SB
Aqua 1-2

[070] (F-Crown) (Arc)W. BROOKFIELD/45 CLIFF ST./N.Y. (R-Crown) [Number] {MLOD} SB
Ice Blue 3-5
Green 20-30
Apple Green 50-75
Yellow Olive Green 300-350

[080] (F-Crown) (Arc)W. BROOKFIELD/45 CLIFF ST./N.Y. {MLOD} SB
Aqua, Light Aqua 1-2
Green 20-30
Light Yellow Green 50-75
Yellow Green 200-250

[090] (F-Crown) (Arc)W. BROOKFIELD/45 CLIFF ST (R-Crown) [Number] {MLOD} SB
Light Green 10-15

[095] (F-Crown) (Arc)W. BROOKFIELD/45 CLIFF ST N.Y. (R-Crown) [Number] {MLOD} SB
Light Blue Aqua 1-2

[100] (F-Crown) (Arc)W. BROOKFIELD/45 CLIFF ST N.Y. {MLOD} SB
Aqua, Green Aqua 1-2
Green 20-30
Light Yellow Green 75-100
Yellow Green 200-250

[105] (F-Crown) (Arc)W. BROOKFIELD/45 CLIFF ST/N.Y. (R-Crown) [Number] {MLOD} SB
Light Aqua 1-2
Green 20-30

[110] (F-Crown) (Arc)W. BROOKFIELD/NEW YORK SB
Aqua, Green Aqua x1

[130] (F-Crown) (Arc)W. BROOKFIELD/NEW YORK (R-Crown) [Letter or number] {MLOD} SB
Aqua x1
Light Green 10-15

[120] (F-Crown) (Arc)W. BROOKFIELD/NEW YORK (R-Crown) [Letter] SB
Aqua, Green Aqua x1

[140] (F-Crown) (Arc)W. BROOKFIELD/NEW YORK {MLOD} SB
Aqua, Green Aqua x1

[145] (F-Crown) (Arc)W. BROOKFIELD/NEW YORK {MLOD} {This is a CD 162 with no inner skirt} SB
Light Blue 5-10

[150] (F-Crown) (Arc)WM BROOKFIELD/45 CLIFF ST. N.Y. (R-Crown) PAT.JAN. 25 1870 {MLOD} SB
Light Aqua 3-5
Light Green 15-20
Light Aqua w/ Milk Swirls 40-50

[160] (F-Crown) [Number]/(Arc)W. BROOKFIELD/45 CLIFF ST/N.Y. {MLOD} SB
Aqua 1-2
Light Green 10-15
Green 20-30

[170] (F-Skirt) BROOKFIELD SB
Aqua x1
Dark Green 5-10

[180] (F-Skirt) BROOKFIELD (R-Skirt) NEW YORK SB
Light Blue Aqua, Light Green 1-2
Green 3-5

[190] (F-Skirt) BROOKFIELD (R-Skirt) NEW YORK/['B.G.M.CO.' blotted out] SB
Aqua, Dark Aqua 5-10
Green 15-20

[200] (F-Skirt) BROOKFIELD (R-Skirt) № 36/ [Number] {Note '№ 36' style number} {This is a CD 162 with no inner skirt} SB
Dark Aqua 5-10
Blue, Blue Aqua 10-15

[210] (F-Skirt) BROOKFIELD (R-Skirt) № 41 SB
Aqua, Dark Aqua, Green Aqua 20-30
Green 30-40

[220] (F-Skirt) BROOKFIELD/[Number] (R-Skirt) № 36 {Note '№ 36' style number} {This is a CD 162 with no inner skirt} SB
Aqua, Blue Aqua, Dark Aqua 5-10

[225] (F-Skirt) BROOKFIFED {Note spelling} SB
Dark Green 10-15

[230] (F-Skirt) BROOИFIELD {Note spelling and backwards letter} (R-Skirt) № 41 SB
Aqua 30-40

60

CD 134

[240] (F-Skirt) W. BROOKFIELD (R-Skirt) NEW YORK SB
Aqua, Light Aqua, Light Blue Aqua 1-2
Green 10-15
Yellow Green 20-30

[250] (F-Skirt) W. BROOKFIELD (R-Skirt) NEW YORK/['B.G.M.CO.' blotted out] SB
Light Aqua 5-10

C.E.L.CO.

[010] (F-Skirt) C.E.L.CO. {MLOD} {Brookfield style} SB
Aqua, Blue Aqua 50-75
Green 175-200
Dark Yellow Green 800-1,000

[020] (F-Skirt) C.E.L.CO./[Blotted out embossing] {Pennycuick style} SB
Aqua, Blue Aqua, Green Aqua 50-75
Light Jade Aqua Milk 175-200
Milky Aqua 500-600
Yellow Olive Green 800-1,000

CALIFORNIA

[020] (F-Skirt) CALIFORNIA SB
Aqua, Blue Aqua 150-175

DIAMOND P ⟨P⟩

[010] (F-Skirt) ⟨P⟩ (R-Skirt) PAT.AUG 11. 85. SB
Aqua, Blue Aqua, Light Aqua 400-500
Light Apple Green,
Light Yellow Green, Lime Green 500-600
Green, Light Cornflower Blue, Steel Blue . 600-700
Apple Green, Light Green Slag,
Teal Aqua 800-1,000
Gray Green Vaseline,
Sage Green Vaseline,
Yellow Green Vaseline 1,250-1,500
Green/Sapphire Blue Two Tone,
Jade Green Milk 1,750-2,000
Sapphire Blue 4,000-4,500

ELECTRICAL SUPPLY CO.

[010] (F-Skirt) THE ELECTRICAL SUPPLY CO/CHICAGO (R-Skirt) 905 {MLOD} SB
Aqua, Light Aqua 75-100
Green 200-250

[020] (F-Skirt) THE ELECTRICAL SUPPLY CO/CHICAGO {MLOD} SB
Light Aqua 75-100

FALL RIVER

[010] (F-Skirt) FALL RIVER (R-Skirt) POLICE SIGNAL SB
Aqua 75-100

G.E.CO.

[010] (F-Crown) [Letter or number] (F-Skirt) G.E.CO./['T-H.E.CO.' blotted out] {MLOD} SB
Aqua, Dark Aqua, Light Aqua 5-10

[015] (F-Skirt) G.E.CO. SB
Aqua 5-10

[020] (F-Skirt) G.E.CO. {MLOD} {Brookfield style} SB
Aqua, Light Aqua 5-10
Green 20-30

[030] (F-Skirt) G.E.CO. {Pennycuick style} SB
Blue Aqua, Green Aqua, Light Aqua 5-10
Light Green 10-15
Green 20-30

GOOD

[010] (F-Skirt) GOOD SB
Aqua, Light Aqua 5-10
Blue Aqua 15-20
Light Sage Aqua 30-40

[020] (F-Skirt) GOOD/[1 or 2 dots] SB
Aqua, Light Aqua 5-10
Light Green 15-20

HEMINGRAY

[005] (F-Skirt) HEMIGRAY/PATENT MAY 2 1893/[Blotted out embossing] {Note spelling} (R-Skirt) STANDARD SDP
Aqua 10-15

[010] (F-Skirt) HEMINGRAY (R-Skirt) N⁰ 18 SB
Green 50-75

[020] (F-Skirt) HEMINGRAY (R-Skirt) N⁰ 18 SDP
Aqua, Blue Aqua, Light Aqua 3-5
Hemingray Blue 5-10

[030] (F-Skirt) HEMINGRAY (R-Skirt) PATENT/MAY 2 1893 SDP
Aqua, Green Aqua, Light Aqua 1-2
Hemingray Blue 5-10

[040] (F-Skirt) HEMINGRAY-18 (R-Skirt) MADE IN U.S.A. RDP
Aqua 5-10
Hemingray Blue 10-15

[045] (F-Skirt) HEMINGRAY-18 (R-Skirt) MADE IN U.S.A. SDP
Hemingray Blue 10-15

[050] (F-Skirt) HEMINGRAY-18/[Numbers and dots] (R-Skirt) MADE IN U.S.A./[Number] RDP
Clear 3-5
Ice Aqua 5-10

CD 134

[060] (F-Skirt) HEMINGRAY/MADE IN U.S.A. (R-Skirt) № 18 SDP
Aqua 5-10
Hemingray Blue 10-15

[070] (F-Skirt) HEMINGRAY/MADE IN U.S.A. (R-Skirt) № 18 {Crown variant - top looks like a CD 162} SDP
Hemingray Blue 75-100

[080] (F-Skirt) HEMINGRAY/PATENT MAY 2 1893 (R-Skirt) STANDARD SDP
Aqua 1-2

[090] (F-Skirt) HEMINGRAY/PATENT MAY 2 1893/[Blotted out embossing] (R-Skirt) STANDARD SDP
Aqua 1-2

[100] (F-Skirt) HEMINGRY {Note spelling} (R-Skirt) PATENT/MAY 2 1893 SDP
Aqua 10-15

K.C.G.W.

[010] (F-Skirt) [Glass dot]/K.C.G.W. (R-Skirt) [Glass dot]/FAIRMOUNT SB
Aqua, Green Aqua 15-20
Green, Ice Blue, Ice Green 20-30

[020] (F-Skirt) [Glass dot]/['K.C.G.W.' blotted out] (R-Skirt) [Glass dot]/['FAIRMOUNT' blotted out] SB
Aqua, Dark Aqua, Green Aqua, Ice Green ... 10-15

[030] (F-Skirt) ['K.C.G.W.' blotted out] (R-Skirt) ['FAIRMOUNT' blotted out] SB
Aqua, Light Aqua 10-15

NO EMBOSSING

[010] [No embossing] {Hemingray style} RDP
Aqua 20-30

[020] [No embossing] {Hemingray style} SDP
Aqua 20-30

[030] [No embossing] {K.C.G.W. product} SB
Blue Aqua 10-15
Ice Blue 15-20

[040] [No embossing] {Oakman mold; diamond plunger mark in the pinhole} SB
Aqua 15-20
Blue, Light Green 20-30

[050] [No embossing] {Oakman style; seen with and without a small diamond shape on the dome} SB
Aqua, Light Blue Aqua, Light Green Aqua ... 10-15
Light Green 20-30
Green 40-50
Apple Green 300-350
7-up Green, Emerald Green 1,500-1,750
Olive Amber 1,750-2,000
Yellow Olive Green 2,000-2,500
Root Beer Amber 2,500-3,000
Orange Amber 3,500-4,000

[060] [No embossing] {Paisley style} SB
Aqua 15-20

[070] [No embossing] {Pennycuick style} CD
Blue Aqua, Green Aqua 15-20
Milky Blue Aqua 40-50

[080] [No embossing] {Pennycuick style} SB
Aqua, Blue Aqua, Dark Aqua,
Green Aqua, Light Aqua 15-20
Dark Green Aqua 20-30
Light Green 30-40
Green 50-75
Blue 75-100
Lime Green 100-125
Dark Green, Forest Green,
Teal Aqua, Teal Green 200-250
Jade Blue Milk 300-350
Cornflower Blue, Dark Yellow Green ... 400-500
Olive Green 500-600
Dark Olive Amber, Dark Sapphire Blue . 800-1,000

NO EMBOSSING - CANADA

[010] [No embossing] RB
Light Aqua 30-40
Light Green, Sky Blue 40-50
Dark Green 350-400
Dark Yellow Green,
Depression Glass Green 500-600

NO NAME

[010] (F-Skirt) [Glass dot] (R-Skirt) [Glass dot] {K.C.G.W. product} SB
Aqua, Light Aqua 10-15
Light Green Aqua 15-20
Light Green 20-30

OAKMAN MFG.CO.

[010] (Base) OAKMAN PATD SEPT 13. 81. BE
Aqua 250-300
Emerald Green 2,000-2,500

PAT'D

[010] (Base) PATD ∧ {Note upside down letter} BE
Aqua 300-350

CD 134

PATENT - DEC. 19, 1871

[010] (Dome) [Letter] (F-Skirt) PATENT/DEC. 19. 1871 SB
Aqua, Light Green Aqua 3-5
Green Tint 5-10
Dark Aqua 10-15
Ice Blue 20-30
Smoky Gray 30-40
Clear 50-75
Gray Blue, Lime Green 75-100
Hemingray Blue, Lemon 100-125
Light Purple 200-250
Purple 400-500
Olive Green 1,250-1,500

[020] (Dome) [Letter] (F-Skirt) PATENT/DEC. 19. 1871 (R-Skirt) PATENT/MAY 2 1893 SDP
Light Aqua 5-10
Light Apple Green 125-150
Hemingray Blue 200-250
Off Clear 250-300
Gray Blue 400-500

[030] (F-Skirt) PATENT/DEC. 19. 1871 SB
Blue Aqua, Light Green 5-10
Ice Green, Light Blue 10-15
Ice Blue 20-30
Blue, Off Clear 30-40
Lime Green 75-100
Hemingray Blue, Lemon 100-125
Light Violet Blue, Purple 400-500
Depression Glass Green, Yellow Green ... 700-800
Dark Depression Glass Green 800-1,000
Olive Green 1,500-1,750
Peacock Blue 2,000-2,500

[040] (F-Skirt) PATENT/DEC. 19. 1871 (R-Skirt) PATENT MAY 2 1893 SDP
Aqua, Green Aqua 5-10
Light Green 20-30
Light Blue 30-40
Green Tint 40-50
Ice Aqua 100-125
Light Apple Green 125-150
Hemingray Blue 200-250
Lime Green, Off Clear, Sage Green 250-300
Clear 350-400
Dark Yellow 800-1,000
Gingerale 1,000-1,250
Peacock Blue 3,500-4,000

[050] (F-Skirt) PATENT/DEC. 19. 1871 (R-Skirt) PATENT/MAY 2 1893 SDP
Aqua, Light Aqua 5-10
Blue Aqua, Light Green 20-30
Blue 40-50
Light Apple Green 125-150
Depression Glass Green 700-800
Light Yellow Straw 800-1,000
Peacock Blue 3,500-4,000
Violet Blue 4,000-4,500

[060] (F-Skirt) PATENT/DEC. 19. 1871 (R-Skirt) [Letter] SB
Aqua, Green Aqua, Light Aqua 3-5
Light Green 5-10
Ice Green 10-15
Ice Blue 20-30
Blue, Sky Blue 30-40
Light Gray Green 40-50
Clear 50-75
Light Yellow Green, Lime Green,
Purple Tint 75-100
Light Purple 200-250

PATENT - OTHER

[010] (Base) PAT'D SEPT 13, '81 BE
Aqua 250-300
Light Purple 3,000-3,500

PETTINGELL ANDREWS

[010] (Skirt) PETTINGELL ANDREWS CO. BOSTON SB
Aqua, Blue Aqua, Light Green Aqua 175-200
Light Green 250-300
Blue, Light Lime Green 300-350
Teal Blue 350-400

STAR ✶

[010] (F-Skirt) ✶ SB
Aqua, Dark Aqua, Light Blue 3-5
Blue 5-10
Light Green 10-15
Green, Lime Green 20-30
Dark Green 40-50
Dark Yellow Green 75-100
Sapphire Blue 100-125

T-H.E.CO.

[010] (Dome) T-H.E.CO (F-Skirt) PATENT/ DEC. 19. 1871 SB
Aqua, Light Aqua 20-30
Dark Blue Aqua, Light Green,
Light Lime Green 40-50

[020] (F-Skirt) T-H.E.CO. SB
Aqua, Blue Aqua, Light Blue 5-10
Green 175-200

[030] (F-Skirt) T-H.E.CO. (R-Crown) [Number] {MLOD} SB
Aqua, Green Aqua, Light Blue Aqua 5-10
Blue, Light Green 15-20
Green 175-200
Yellow Green 800-1,000
Root Beer Amber 1,500-1,750

63

CD 134

W.F.G.CO.

[010] (F-Skirt) W.F.G.CO./DENVER, COLO. (R-Skirt) [0, 1 or 2 dots] SB
Aqua, Blue Aqua 5-10
Lavender, Pink, Purple,
Sage Green, Steel Blue 30-40
Dark Purple 40-50
Clear, Dark Steel Blue, Light Gray 75-100
Ice Green, Light Peach 100-125
Light Gingerale 175-200
Gingerale, Light Lime Green 200-250
Milky Purple, Purple/Straw Two Tone 300-350
Straw 400-500
Straw/Peach Two Tone 500-600

[020] (F-Skirt) W.F.G.CO./DENVER, COLO. (R-Skirt) [Blotted out embossing] SB
Lavender, Purple 30-40
Clear 75-100

W.G.M.CO.

[010] (F-Skirt) W.G.M.Co. SB
Light Purple, Purple 50-75
Clear 175-200
Straw 500-600

WESTINGHOUSE

[010] (F-Skirt) WESTINGHOUSE/NO.6 SB
Blue Aqua, Light Aqua 300-350
Light Green 500-600
Blue 800-1,000
Peacock Blue 2,500-3,000
Dark Peacock Blue 3,000-3,500

CD 134.4

NO EMBOSSING

[010] [No embossing] {American threads} SB
Aqua, Blue Aqua 75-100
Light Green 125-150
Emerald Green, Yellow Green 1,000-1,250
Olive Amber, Olive Green 3,000-3,500
Golden Amber,
Light Honey Amber, Mustard Yellow ... 5,000-7,500

CD 134.6

BROOKFIELD

[010] (F-Skirt) BROOKFIELD (R-Skirt) PHILADELPHIA/PAT'D APRIL 7, 1896 SB
Aqua 15,000-20,000

CD 135

CHICAGO INSULATING CO.

[010] (Base) CHICAGO INSULATING CO./ PAT OCT 16 1883. (F-Skirt) [Glass dots may be present] BE
Blue, Blue Aqua 75-100
Light Aqua 100-125
Milky Blue 600-700

[020] (Upper shoulder) CHI.INS.CO. PAT.OCT. 16. 1883 (F-Skirt) [Glass dots may be present] SB
Aqua, Blue 75-100

[030] (Upper shoulder) CHI.INS.CO. PATD OCT. 16. 1883. (F-Skirt) [Glass dots may be present] SB
Aqua, Blue 75-100
Milky Blue 600-700

CD 135.5

E.R.W.

[010] (F-Skirt) E.R.W. SB
Blue Aqua, Ice Blue, Light Aqua 1,000-1,250
Green Aqua 1,500-1,750
Lime Green 2,000-2,500

CD 136

B & O

[010] (F-Crown) B & O (R-Crown) PAT. JAN 25TH 1870 SB
Aqua, Light Aqua 15-20
Green, Lime Green 250-300
Yellow Olive Green 800-1,000

[020] (F-Crown) B & O (R-Crown) [Number] {MLOD} SB
Green 250-300

[030] (F-Crown) B & O {MLOD} SB
Green Aqua 15-20

[040] (F-Skirt) B & O (R-Skirt) PATENT/DEC. 19. 1871 SB
Blue Aqua 15-20

[050] (F-Skirt) B & O (R-Skirt) PATENT/PEC. 19. 1871 {Note spelling} SB
Light Aqua 20-30
Light Green 50-75

[060] (F-Skirt) B & O/[Letter] (R-Skirt) PATENT/DEC. 19. 1871 SB
Aqua, Green Aqua 15-20
Light Green 50-75
Apple Green, Green 250-300
Depression Glass Green 800-1,000

[070] (F-Skirt) B & O/[Letter] (R-Skirt) PATENT/PEC. 19. 1871 {Note spelling} SB
Aqua 20-30
Light Green 50-75

BROOKFIELD

[010] (F-Crown) (Arc)W. BROOKFIELD/45 CLIFF ST./N.Y. (F-Skirt) B & O (R-Crown) PAT/JAN 25TH 1870 {MLOD} SB
Light Blue 15-20

[020] (F-Crown) (Arc)W. BROOKFIELD/45 CLIFF ST./N.Y. (F-Skirt) B & O (R-Crown) [Letter]/PAT/JAN 25TH 1870 {MLOD} SB
Light Aqua 15-20
Lime Green 200-250

[030] (F-Crown) (Arc)W. BROOKFIELD/45 CLIFF ST./И.Y. {Note backwards letter} (F-Skirt) B & O (R-Crown) PAT/JAИ 25TH 1870 {Note backwards letter} {MLOD} SB
Aqua, Light Blue 15-20
Light Green 200-250

[040] (F-Crown) (Arc)W. BROOKFIELD/45 CLIFF ST./И.Y. {Note backwards letter} (F-Skirt) B & O (R-Crown) [Number]/PAT/JAИ 25TH 1870 {Note backwards letter} {MLOD} SB
Aqua 15-20
Apple Green 300-350

[050] (F-Crown) (Arc)W. BROOKFIELD/45 CLIFF ST./И.Y. {Note backwards letter} (F-Skirt) B & O (R-Crown) [Number]/PAT/JAN 25TH 1870 {MLOD} SB
Aqua, Light Aqua, Light Blue 15-20
Yellow Olive Green 1,500-1,750

[060] (F-Crown) [Number]/(Arc)W. BROOKFIELD/45 CLIFF ST./N.Y. (F-Skirt) B & O (R-Crown) PAT/JAN 25TH 1870 {MLOD} SB
Aqua 15-20

CD 136.4

NO EMBOSSING

[010] [No embossing] {Four segmented threads} SB
Aqua, Blue Aqua,
Dark Green Aqua, Light Aqua 600-700
Green 1,500-1,750
Emerald Green 2,000-2,500
Yellow Green 2,500-3,000
Yellow Olive Green 4,500-5,000

[020] [No embossing] {MLOD} SB
Aqua, Ice Aqua, Light Blue 50-75
Sky Blue 100-125
Dark Green Aqua 175-200
Sapphire Blue, Turquoise Blue 300-350
Dark Green, Dark Sapphire Blue,
Dark Teal Green 700-800
Dark Yellow Olive Green 5,000-7,500

NO NAME

[010] (F-Crown) [Backwards number] {Just above wire groove} SB
Aqua, Light Blue 50-75
Sky Blue 100-125
Dark Green Aqua 175-200
Sapphire Blue, Turquoise Blue 300-350
Green 500-600

CD 136.5

CD 136.5
3 x 4 1/8

BOSTON BOTTLE WORKS

[010] (Base of threads) BOSTON BOTTLE WORKS - PAT. OCT. 15 72 {Three segmented threads} {This is a CD 158.2 with no inner skirt} SB
Aqua 3,000-3,500

65

CD 136.7

CD 136.7

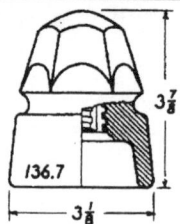

NO EMBOSSING

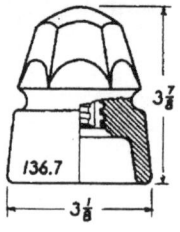

[010] [No embossing] {Three segmented threads} SB
Aqua 1,000-1,250

PATENT - OTHER

[010] (Base of threads) PAT.OCT. 15, 1872. {Three segmented threads} SB
Aqua, Blue Aqua, Ice Aqua 1,000-1,250
Celery Green 1,500-1,750
Off Clear 1,750-2,000
Apple Green 3,500-4,000

CD 137

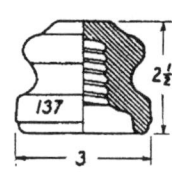

HEMINGRAY

[005] (F-Skirt) HEMINGRAY SB
7-up Green 300-350

[010] (F-Skirt) HEMINGRAY (R-Skirt) D-990 SB
Clear, Dark Aqua 5-10
Carnival 100-125

[020] (F-Skirt) HEMINGRAY [Numbers and dots] (R-Skirt) D-990 [Number] CB
Clear 5-10

[030] (F-Skirt) HEMINGRAY [Number] (R-Skirt) D-990 SB
Ice Blue 5-10

CD 137.5

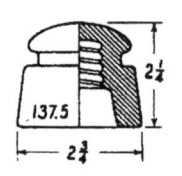

NO EMBOSSING

[010] [No embossing] {Hemingray product} SB
Clear 250-300

CD 138

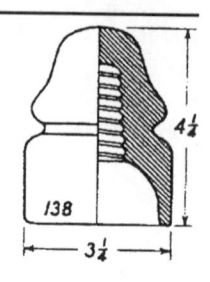

BROOKFIELD

[010] (F-Crown) (Arc)W. BROOKFIELD/45 CLIFF ST./N.Y. (F-Skirt) POSTAL TEL.Co. (R-Crown) PAT/JAN 25TH 1870 {Large embossing variation}{MLOD} SB
Light Aqua 20-30
Green 350-400

[020] (F-Crown) (Arc)W. BROOKFIELD/45 CLIFF ST./N.Y. (F-Skirt) POSTAL TEL.Co. (R-Crown) [Number]/PAT/JAN 25TH 1870 {MLOD} SB
Light Aqua, Light Blue 20-30
Light Green 125-150
Apple Green 350-400
Gray, Off Clear 800-1,000
Yellow Green 1,250-1,500
Olive Green 2,500-3,000

[030] (F-Crown) [Number]/(Arc)W. BROOKFIELD/45 CLIFF ST./N.Y. (F-Skirt) POSTAL TEL.Co. (R-Crown) PAT/JAN 25TH 1870 {MLOD} SB
Aqua 20-30
Gray, Off Clear 800-1,000

CD 138.2

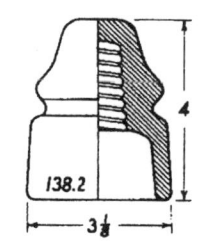

LAWRENCE GRAY

[010] (Base) MANUFACTURED BY LAWRENCE B. GRAY'S/PATENT PROCESS BE
Blue Gray 10,000-15,000

NATIONAL INSULATOR CO.

[010] (Base) NATIONAL INSULATOR CO./PAT.JAN.1 & OCT.7.1884 BE
Aqua, Light Aqua 50-75

NO EMBOSSING

[010] [No embossing] {Pennycuick style} SB
Dark Teal Aqua 800-1,000
Dark Olive Green 3,000-3,500

CD 140

STANDARD GLASS INSULATOR CO.

[010] (Base) THE STANDARD GLASS INSULATOR CO./BOSTON MASS. BE
Aqua, Light Aqua, Light Blue 50-75
Light Green 600-700

CD 138.9

PAT APP FOR

[010] (F-Skirt) PATENT/APPLIED FOR SB
Light Aqua .... 20,000+

CD 139

COMBINATION SAFETY

[010] (F-Skirt) COMBINATION/SAFETY (R-Skirt) PAT. APPLIED FOR SB
Aqua 15,000-20,000

[020] (F-Skirt) COMBINATION/SAFETY (R-Skirt) PAT. APPLIED FOR SDP
Aqua 15,000-20,000

[030] (F-Skirt) COMBINATION/SAFTY {Note spelling} (R-Skirt) PAT. APPLIED FOR SB
Aqua 15,000-20,000

CD 139.6

PYREX

CD 139.6
$4\,^{7}/_{8}$ x 3

[010] (F-Skirt) 660-PYREX-PAT.MAY-27-1919 (R-Skirt) CORNING MADE IN U.S.A. SB
Clear 5,000-7,500

CD 139.8

NO EMBOSSING

CD 139.8
$3\,^{3}/_{4}$ x 4

[010] [No embossing] SB
Root Beer Amber 5,000-7,500

CD 139.9

McLAUGHLIN

[010] (F-Skirt) McLAUGHLIN (R-Skirt) USLD {Peaked dome style} SB
Aqua 400-500

[020] (F-Skirt) McLAUGHLIN (R-Skirt) USLD {Rounded dome style} SB
Aqua 175-200

CD 140

JUMBO

[010] (F-Skirt) JUMBO SB
Aqua 250-300
Dark Aqua 400-500

NO EMBOSSING

[010] [No embossing] SB
Aqua, Blue 250-300
Turquoise Blue 350-400
Green 500-600
Dark Green 800-1,000
Lime Green 1,000-1,250
Yellow Green 1,750-2,000
Jade Aqua Milk 3,000-3,500
7-up Green, Olive Green 3,500-4,000
Clear 5,000-7,500
Light Purple, Olive Amber, Purple 7,500-10,000

CD 140

OAKMAN MFG.CO.

[010] (Base) OAKMAN M'F'G CO. BOSTON BE
Aqua 300-350

[020] (F-Skirt) JUMBO (Base) OAKMAN M'F'G' CO. BOSTON BE
Aqua, Blue 300-350

[030] (F-Skirt) JUMBO (R-Skirt) PATD JUNE 17, 1890, AUG 19, 1890 (Base) OAKMAN M'F'G.CO. BOSTON BE
Aqua, Light Green Aqua 300-350

CD 140.5

NO NAME

[010] (F-Skirt) [Glass dot] {MLOD} SB
Cobalt Blue, Cobalt Blue Blackglass .. 3,000-3,500
Light Cobalt Blue 4,000-4,500

CD 141

NO EMBOSSING

[010] [No embossing] SB
Dark Aqua 15-20
Green 20-30
Light Aqua 30-40
Emerald Green 75-100

CD 141.5

NO EMBOSSING

[010] [No embossing] SB
Aqua 2,500-3,000

CD 141.5
2 x 6 $^1/_2$

CD 141.6

PATENT - OTHER

CD 141.6
2 $^3/_8$ x 3 $^7/_8$

[010] (F-Skirt) PAT'D AUG 29-93 SB
Light Aqua 7,500-10,000

CD 141.7

TWIGGS

[010] (Dome) (Circle) W.R. TWIGGS PAT AUG 29 1905 SB
Crystal Clear .... 20,000+

CD 141.8

BUZBY

[010] (Crown behind wire groove) J.F.BUZBY/ PAT'D MAY 6/1890 SB
Aqua, Ice Aqua .... 20,000+

CD 141.9

EMMINGER'S

[010] (Base) EMMINGER'S PAT/FEBY 20TH 1872 BE
Light Aqua .... 20,000+

CD 143

CD 142

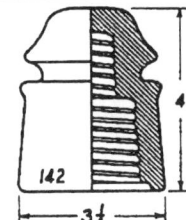

HEMINGRAY

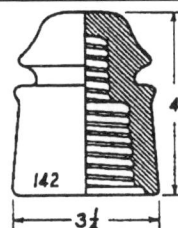

[010] (F-Skirt) TS-2/[Numbers and dots]/ HEMINGRAY (R-Skirt) W.U.T.CO. {Inside of skirt is threaded and fitted with a copper liner} RDP
Carnival (without copper liner) 30-40
Carnival (with copper liner) 40-50

[020] (F-Skirt) [Numbers and dots]/ HEMINGRAY (R-Skirt) W.U.T.CO./TS-2 {Inside of skirt is threaded and fitted with a copper liner} RDP
Carnival (without copper liner) 30-40
Carnival (with copper liner) 40-50
Ice Blue (without copper liner) 200-250

NO NAME

[010] (F-Skirt) TS-2 {Inside of skirt is threaded and fitted with a copper liner} {Hemingray product} RDP
Carnival (without copper liner) 20-30
Carnival (with copper liner) 30-40

CD 142.4

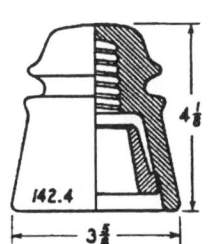

NO EMBOSSING

[010] [No embossing] {A glass insert is cemented inside the skirt} {Hemingray product} RDP
Carnival (without glass insert) 175-200
Carnival (with glass insert) 250-300

NO NAME

[010] (F-Skirt) TS 3 {Note backwards letter} {A glass insert is cemented inside the skirt} {Hemingray product} RDP
Carnival (without glass insert) 175-200
Carnival (with glass insert) 250-300

CD 143

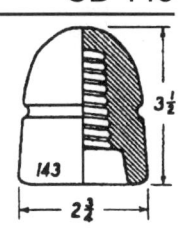

C.N.R.

[010] (F-Skirt) C N R (R-Skirt) STANDARD {'CNR' small embossing - 5/16"} SB
Aqua, Blue Aqua 5-10

[015] (F-Skirt) C.N.R. (R-Skirt) STANDARD {7/16" embossing - 2 periods} SB
Turquoise Blue 10-15

[020] (F-Skirt) C.N.R. (R-Skirt) STANDARD/ ['CANADIAN PACIFIC RY CO' blotted out] SB
Aqua 15-20

[030] (F-Skirt) C.N.R (R-Skirt) STANDARD/ ['CNR' blotted out] SB
Aqua 5-10
Light Gray, Steel Blue, Turquoise Blue 10-15
Gray Green 15-20
Light Purple 150-175
Purple 175-200
Dark Purple 200-250

[040] (F-Skirt) C.N.R. SB
Aqua 5-10
Light Gray, Steel Blue 10-15
Light Purple, Light Violet Blue 125-150
Purple 150-175
Dark Purple 200-250
Violet Blackglass 250-300

[050] (F-Skirt) C.N.R. (R-Skirt) STANDARD SB
Aqua 5-10
Steel Blue, Turquoise Blue 10-15
Light Purple 150-175
Dark Purple 200-250
Dark Violet Blue 250-300

[060] (F-Skirt) C.N.R. {Periods are shaped like prisms} SB
Aqua 5-10
Light Green 10-15
Green 20-30
Yellow Green 30-40
Blue, Olive Green 40-50

[070] (F-Skirt) C.N.R. {Periods are shaped like prisms} (R-Skirt) [Blotted out embossing] SB
Aqua 5-10
Yellow Green 30-40
Olive Green 40-50

69

CD 143

[080] (F-Skirt) C.N.R/['CNR' blotted out]
(R-Skirt) STANDARD {'CNR' small embossing} SB
Aqua 10-15
Steel Blue 15-20
Light Purple 150-175
Dark Purple 200-250

[090] (F-Skirt) C.N.R/['CNR' blotted out]
(R-Skirt) STANDARD/['CANADIAN PACIFIC RY CO' blotted out] SB
Aqua 30-40

[100] (F-Skirt) [Horizontal dashes]/['CNR' blotted out] SB
Aqua 20-30

[110] (F-Skirt) ['CNR' blotted out] SB
Aqua 20-30

[120] (F-Skirt) ['CNR' blotted out]
(R-Skirt) ['CPR' blotted out] SB
Aqua 20-30

[130] (F-Skirt) ['CNR' blotted out]
(R-Skirt) ['STANDARD' blotted out] {'CNR' small embossing} SB
Aqua 20-30

C.P.R.

[010] (F-Skirt) C.P.R. (R-Skirt) STANDARD SB
Aqua 5-10
Steel Blue 15-20
Light Purple 30-40

[020] (F-Skirt) C.P.R. (R-Skirt) [Horizontal dashes]/['CNR' blotted out] SB
Aqua 10-15
Blue Gray 15-20
Light Purple 30-40

[030] (F-Skirt) C.P.R. (R-Skirt) ['CANADIAN PACIFIC RY CO' blotted out] SB
Aqua 10-15
Blue Gray, Light Blue, Steel Blue 15-20
Light Purple 30-40

[040] (F-Skirt) C.P.R. (R-Skirt) ['CNR' blotted out] SB
Aqua 10-15
Blue Gray, Steel Blue 15-20
Light Purple 30-40

[050] (F-Skirt) C.P.R/['CNR' blotted out]
(R-Skirt) STANDARD/['CANADIAN PACIFIC RY CO' blotted out] SB
Aqua 10-15
Blue Gray, Steel Blue 15-20
Light Purple 30-40
Light Purple/Aqua Two Tone, Purple 40-50

[060] (F-Skirt) C.P.R/['CNR' blotted out]
(R-Skirt) STANDARD/['CNR' blotted out] SB
Aqua 10-15
Steel Blue 15-20
Light Purple 40-50

CANADIAN PACIFIC RY.CO.

[010] (F-Skirt) CANADIAN (R-Skirt) PACIFIC RY CO {MLOD} RB
Aqua, Light Aqua 1-2
Blue 5-10
Light Lime Green 20-30
Light Yellow Green 30-40
Light Olive Green 300-350
Dark Yellow Green 1,000-1,250

[020] (F-Skirt) CANADIAN (R-Skirt) PACIFIC RY CO {'RY' is engraved over an upside down 'RY'} {MLOD} RB
Aqua, Light Green Aqua 3-5
Blue, Light Green 5-10

[030] (F-Skirt) CANADIAN (R-Skirt) PACIFIC RY CO {'RY' is upside down} {MLOD} RB
Aqua, Blue Aqua 5-10
Blue, Light Green 10-15

[040] (F-Skirt) CANADIAN (R-Skirt) PACIFIC. RY.CO. {MLOD} RB
Aqua 3-5
Light Green 5-10
Dark Yellow Green 1,000-1,250

[050] (F-Skirt) CANADIAN PACIFIC RY CO (R-Skirt) ['STANDARD' blotted out] SB
Aqua 3-5
Gray, Steel Blue 5-10
Light Green 20-30
Light Purple 30-40
Purple 40-50

[055] (F-Skirt) CANADIAN PACIFIC RY CO (Top of pinhole) [Number] {MLOD} SB
Light Aqua 10-15

[060] (F-Skirt) CANADIAN PACIFIC RY CO {MLOD} SB
Aqua, Light Aqua 1-2
Light Green 10-15
Blue 20-30
Sage Green 40-50
Clear, Light Purple, Royal Purple 50-75
Purple 75-100

CD 143

[070] (F-Skirt) CANADIAN PACIFIC RY CO
{Note backwards letter} {MLOD} SB
Aqua, Light Aqua, Light Yellow Green 3-5
Light Green 5-10
Blue, Light Gray 20-30
Sage Green 40-50
Clear, Royal Purple 50-75
Olive Green 1,750-2,000
Yellow Olive Green 2,000-2,500

[080] (F-Skirt) CANADIAN PACIFIC RY CO
{Withycombe - 69 vertical ridges above and 9
left hand spiral ridges below wire groove} SB
Light Aqua 40-50
Light Purple 1,000-1,250

[090] (F-Skirt) CANADIAN PACIFIC RY {Slug
embossing} {MLOD} SB
Aqua, Dark Aqua 3-5
Light Green 5-10
Blue 10-15
Green 100-125
Yellow Olive Green 1,000-1,250

[100] (F-Skirt) CANADIAN PACIFIC RY.CO.
(R-Skirt) STANDARD RB
Aqua 3-5

[110] (F-Skirt) CANADIAN PACIFIC RY.CO.
(R-Skirt) STANDARD/['CANADIAN PACIFIC
RY CO' blotted out] SB
Aqua, Light Green 15-20

[120] (F-Skirt) CANADIAN PACIFIC RY.CO.
(R-Skirt) ['STANDARD' blotted out] SB
Aqua, Light Aqua 3-5
Light Gray, Steel Blue 5-10
Dark Purple,
Light Aqua w/ Milk Swirls, Light Purple 40-50
Royal Purple 50-75

[130] (F-Skirt) CANADIAN PACIFIC RY.CO./
['CNR' blotted out] (R-Skirt) STANDARD RB
Aqua 20-30

[140] (F-Skirt) CANADIAN PACIFIC RY.CO./
['CNR' blotted out] (R-Skirt) STANDARD/
['CANADIAN PACIFIC RY CO' blotted out] RB
Aqua 20-30

[150] (F-Skirt) CANADIAN. PACIFIC. R.Y.CO
{MLOD} SB
Aqua, Light Aqua 5-10
Light Green 10-15
Blue, Gray 15-20
Purple 50-75
Jade Green Milk 500-600

[160] (F-Skirt) CANADIAN.PACIFIC.RY.CO.
RB
Aqua 3-5

[170] (F-Skirt) CANADIAN.PACIFIC.RY.CO.
{Comes in many different 'period' patterns} SB
Aqua, Ice Blue, Light Aqua 1-2
Light Green 3-5
Blue, Light Gray, Steel Blue, Steel Gray 5-10
Light Aqua w/ Milk Swirls 30-40
Clear, Light Purple 40-50
Lavender w/ Purple Swirls, Purple,
Royal Purple 50-75
Light Gray w/ Milk Swirls 200-250
Light Yellow Amber 2,500-3,000
Yellow Amber 5,000-7,500

[180] (F-Skirt) ['CANADIAN PACIFIC RY CO'
blotted out] SB
Aqua, Light Blue 10-15

[190] (F-Skirt) ['CANADIAN PACIFIC RY CO'
blotted out] (R-Skirt) ['STANDARD' blotted out]
SB
Aqua 15-20

DWIGHT

[010] (Dome) [Number] (F-Skirt) ['G.N.W.'
blotted out] DWIGHT (R-Skirt) PATTERN GB
Aqua, Blue Aqua, Light Aqua 3-5
Light Green 30-40

[020] (F-Skirt) DWIGHT (R-Skirt) PATTERN
SB
Aqua 3-5
Light Green 30-40
Light Purple 350-400
Light Gray Purple, Purple 400-500

[030] (F-Skirt) G.N.W. DWIGHT
(R-Skirt) PATTERN GB
Aqua, Light Blue Aqua 100-125

[040] (F-Skirt) G.N.W. DWIGHT
(R-Skirt) PATTERN SB
Aqua, Green Aqua 125-150

[050] (F-Skirt) ['G.N.W.' blotted out] DWIGHT
(R-Skirt) PATTERN SB
Aqua 3-5
Green Aqua 10-15
Light Green 30-40
Steel Blue 100-125
Gray 125-150
Light Purple 350-400
Purple 400-500

G.N.R.

[010] (F-Skirt) G.N.R. (R-Skirt) STANDARD
SB
Aqua 15-20
Gray 75-100
Light Purple 200-250

71

CD 143

[020] (F-Skirt) G.N.R./['CNR' blotted out] (R-Skirt) STANDARD {'CNR' small embossing} SB
Aqua 15-20
Gray, Light Steel Blue 75-100
Light Purple 200-250

G.N.W.TEL.CO.

[010] (F-Skirt) G N W {Irregular spacing} {MLOD} SB
Aqua, Blue Aqua 10-15
Light Green 15-20
Gray, Smoke 200-250
Green 300-350
Purple 800-1,000

[020] (F-Skirt) G.N.W SB
Aqua 10-15
Light Green 15-20
Light Yellow Green 30-40

[030] (F-Skirt) G.N.W/[Raised slug plate] {MLOD} SB
Aqua, Light Green 40-50

G.P.R.

[010] (F-Skirt) G.P.R./['CNR' blotted out] (R-Skirt) STANDARD {'CNR' large or small embossing} SB
Aqua 10-15
Steel Blue 15-20
Light Purple 50-75
Purple 75-100

GREAT NORTHWESTERN

[010] (F-Skirt) GREAT NORTHWESTERN (R-Skirt) TELEGRAPH CO GB
Aqua 50-75
Light Green 75-100

[020] (F-Skirt) GREAT NORTHWESTERN (R-Skirt) TELEGRAPH.CO {MLOD} SB
Aqua 50-75
Light Green 75-100

MONTREAL

[010] (F-Skirt) MONTREAL TELEGRAPH (R-Skirt) CO {MLOD} SB
Aqua, Dark Aqua 15-20
Blue 20-30
Dark Sky Blue 400-500
Sapphire Blue 700-800
Green, Yellow Green 800-1,000
Dark Yellow Green 1,000-1,250
Emerald Green 1,250-1,500
Olive Green 1,500-1,750

[020] (F-Skirt) MONTREAL TELEGRAPH CO {MLOD} SB
Aqua, Green Aqua, Light Blue 15-20
Blue 20-30
Dark Sky Blue 400-500
Sapphire Blue 700-800
Green 800-1,000
Dark Yellow Green 1,000-1,250
Olive Green 1,500-1,750

NO EMBOSSING - CANADA

[005] [No embossing] {3-piece mold} GB
Blue Aqua 5-10

[010] [No embossing] {3-piece mold} SB
Purple 250-300

[020] [No embossing] {Double threaded} RB
Blue Aqua, Green 250-300

[030] [No embossing] {Double threaded} {MLOD} SB
Aqua 50-75
Blue Gray 75-100
Blue 100-125
Light Green 200-250
Green 500-600
Cornflower Blue 700-800
Olive Green 2,000-2,500
Yellow Olive Green 2,500-3,000
Mustard Yellow 5,000-7,500

[040] [No embossing] {MLOD} RB
Aqua, Light Aqua 3-5
Blue 5-10
Dark Sky Blue 20-30
Light Yellow Green 50-75
Ice Aqua 100-125
Clear, Gray, Smoky Gray 200-250
Blue Purple 300-350
Blue/Purple Two Tone, Royal Purple 400-500
Green 800-1,000
Emerald Green, Olive Green 1,000-1,250

[050] [No embossing] {MLOD} SB
Aqua, Blue 5-10
Light Green 10-15
Lime Green 100-125
Light Purple 200-250
Royal Purple 250-300
Dark Green 800-1,000

[060] [No embossing] {Tall variant} {MLOD} SB
Light Aqua 40-50
Light Green 200-250
Royal Purple 400-500
Purple 500-600
Light Purple 600-700

[065] [No embossing] {Threaded version of CD 743.1} SB
Aqua 250-300

72

CD 143.4

[070] [No embossing] {Threaded version of CD 743.2} SB
Blue Aqua, Light Aqua 400-500

[075] [No embossing] {Unembossed Montreal} {MLOD} SB
Blue 20-30

[080] [No embossing] {Whittle mold} {MLOD} SB
Aqua 10-15
Green 15-20
Blue 30-40
Gray, Smoke 75-100
Purple 150-175

[090] [No embossing] {Withycombe - 11 circular rings above and 10 circular rings below wire groove} {MLOD} SB
Aqua 1,000-1,250

[100] [No embossing] {Withycombe - 14 circular ridges above and 10 left hand spiral ridges below wire groove} {MLOD} SB
Light Green 200-250

[110] [No embossing] {Withycombe - 16 or 19 circular rings above and 8 right hand spiral ridges below wire groove} {MLOD} SB
Light Gray, Light Purple 200-250

[120] [No embossing] {Withycombe - 50, 52, 53 or 54 vertical ridges above and below wire groove} SB
Aqua 30-40
Blue 40-50
Light Green 50-75
Green 75-100

STANDARD

[010] (F-Skirt) STANDARD SB
Aqua, Light Aqua 3-5
Gray 20-30
Clear 30-40
Light Purple, Purple 40-50
Royal Purple 50-75
Light Aqua w/ Milk Swirls 75-100
Green 400-500

[020] (F-Skirt) STANDARD (R-Skirt) ['CNR' blotted out] SB
Aqua 10-15
Light Aqua w/ Milk Swirls 75-100
Green 400-500
Yellow Green 800-1,000
Olive Green Blackglass 3,500-4,000

[030] (F-Skirt) STANDARD (R-Skirt) ['CNR' blotted out] {'CNR' small embossing} SB
Aqua 10-15
Gray 20-30
Light Green 30-40
Light Purple, Royal Purple 50-75
Light Aqua w/ Milk Swirls 75-100

[040] (F-Skirt) STANDARD (R-Skirt) ['CPR' blotted out] SB
Light Blue 10-15

[050] (F-Skirt) STANDARD/['CANADIAN PACIFIC RY CO' blotted out] SB
Aqua, Light Aqua, Steel Gray 10-15
Light Gray Green 20-30
Light Aqua w/ Milk Swirls 75-100
Olive Green Blackglass 3,500-4,000

[060] (F-Skirt) STANDARD/['CANADIAN PACIFIC RY CO' blotted out] (R-Skirt) ['CNR' blotted out] SB
Aqua 10-15

[070] (F-Skirt) STANDARD/['CNR' blotted out] SB
Aqua, Steel Blue 10-15

[080] (F-Skirt) STANDARD/['CNR' blotted out] (R-Skirt) ['CNR' blotted out] SB
Aqua 10-15

CD 143.4

AM. INSULATOR CO.

[010] (Dome) [Letter] (Base) AM. INSULATOR CO. N.Y. DOUBLE PETTICOAT/PAT'D SEPT. 13, 1881, NOV. 13, 1883 {This is a CD 145 with no inner skirt} BE
Blue 150-175

[020] (Dome) [Letter] (Base) AM. INSULATOR CO. N.Y. DOUBLE PETTICOAT/PAT'D SEPT. 13, 1881, NOV. 13, 1883, FEB. 12, 1884 {This is a CD 145 with no inner skirt} BE
Blue 150-175

BROOKFIELD

[005] (F-Crown) (Arc)W. BROOKFIELD/45 CLIFF ST/N.Y. (R-Crown) [Number]/(Arc)PAT'D NOV. 13, 1883/FEB. 12, 1884 {This is a CD 145 with no inner skirt} SB
Aqua 75-100

[010] (F-Crown) [Number]/(Arc)W. BROOKFIELD/45 CLIFF ST/N.Y. (R-Crown) (Arc)PAT'D NOV. 13, 1883/FEB. 12, 1884 {This is a CD 145 with no inner skirt} SB
Aqua 75-100
Green 500-600

CD 143.4

[020] (F-Crown) [Number]/(Arc)W. BROOKFIELD/83 FULTON ST/N.Y. (R-Crown) (Arc)PATD.NOV. 13TH 1883/FEB 12TH 1884 {This is a CD 145 with no inner skirt} SB
Aqua 75-100

[030] (F-Skirt) W. BROOKFIELD (R-Skirt) NEW YORK {This is a CD 145 with no inner skirt} SB
Aqua 50-75

CD 143.5

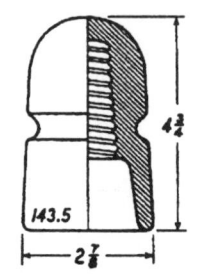

T-H.E.CO.

[010] (F-Crown) T-H.E.CO. {MLOD} {Comes in two dome shapes} SB
Aqua, Light Aqua 75-100
Light Green 125-150

CD 143.6

BOSTON BOTTLE WORKS
CD 143.6
3 x 4

[010] (Base of threads) BOSTON BOTTLE WORKS/PAT.OCT 15 '72 {Three segmented threads} SB
Aqua, Dark Green Aqua 2,500-3,000

CD 144

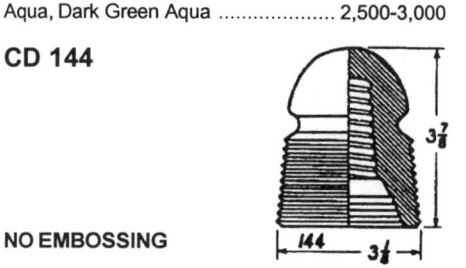

NO EMBOSSING

[010] [No embossing] {Foree Bain} {Inside of skirt has horizontal threaded ridges from the base to the pinhole} SB
Aqua, Blue Aqua 75-100
Light Green 125-150
Green 200-250

NO EMBOSSING - CANADA

[010] [No embossing] {Circular rings above and spiral ridges below wire groove; known as the "High groove Withycomb"} SB
Light Green 1,500-1,750

PATENT - OTHER

[010] (Dome) (Circle)PATENT.DEC. 23 1890. {Foree Bain} {Inside of skirt has horizontal threaded ridges from the base to the pinhole} SB
Aqua 150-175
Light Green 200-250
Lime Green 300-350
Depression Glass Green 600-700

CD 144.5

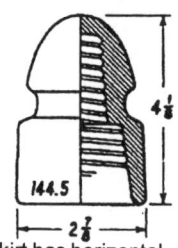

NO NAME

[010] (Dome) 111 {Inside of skirt has horizontal threaded ridges from the base to the pinhole; unembossed Oakman} SB
Light Aqua 350-400

PATENT - OTHER

[010] (Dome) 111 (Base) PAT'D AUG. 19, 1890 {Inside of skirt has horizontal threaded ridges from the base to the pinhole} BE
Aqua, Light Aqua 350-400

CD 145

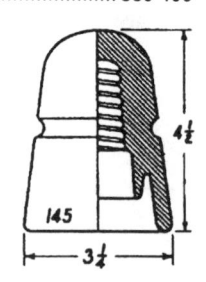

AM.INSULATOR CO.

[010] (Base) AM.INSOLATOR CO. N.Y. DOUBLE PETTICOAT/PAT'D SEPT. 13, 1881, NOV. 13, 188Ɛ, NOV. 188Ɛ {Note spelling; backwards numbers; no day in third date} BE
Light Blue Aqua, Light Green,
Light Green Aqua 50-75

CD 145

[020] (Dome) [Letter] (Base) AM. INSULATOR CO. N.Y. DOUBLE PETTICOAT/PAT'D SEPT. 13, 1881, NOV. 13, 1883, FEB. 12, 1884 BE
Aqua, Blue Aqua, Green Aqua,
Light Aqua, Light Blue 10-15
Blue 15-20
Apple Green,
Blue/Green Two Tone, Green 350-400
Yellow Green 500-600
Yellow Olive Green 1,000-1,250

[030] (Dome) [Letter] (Base) AM. INSULATOR CO. N.Y. DOUBLE PETTICOAT/PAT'D SEPT. 13, 1881, NOV. 13, 1883, JAN. 29, 1884 {Note 'JAN. 29, 1884' date} BE
Aqua, Light Aqua, Light Blue Aqua 10-15
Light Green 50-75
Lime Green 150-175
Green 350-400
Yellow Green 500-600
Yellow Olive Green 1,000-1,250

[040] (Dome) [Letter] (Base) AM. INSULATOR CO. N.Y. DOUBLE PETTICOAT/PAT'D SEPT. 13, 1881. NOV. 13, 1883. BE
Aqua, Green Aqua, Light Aqua, Light Blue .. 10-15
Blue 15-20
Apple Green, Green,
Light Jade Aqua Milk 300-350
Yellow Green 500-600
Yellow Olive Green 1,000-1,250
Yellow Olive Amber 1,750-2,000
Golden Amber 4,000-4,500

B

[010] (Dome) [Letter or number] (F-Skirt) B SB
Aqua, Blue, Green Aqua, Light Aqua,
Light Blue, Light Green x1
Emerald Green, Green 1-2
Teal Aqua 3-5
Yellow Green 5-10
Dark Olive Green 15-20

[020] (Dome) [Letter or number] (F-Skirt) B (R-Skirt) B SB
Aqua, Light Aqua, Light Blue, Light Green x1
Emerald Green 5-10
7-up Green, Off Clear 10-15
Blue, Gray, Light Purple 15-20
Purple 20-30

[030] (Dome) [Number] (F-Skirt) B SDP
Aqua 3-5
Dark Teal Green 10-15

[035] (Dome) [Number] (F-Skirt) B (R-Skirt) [Number] SB
Light Aqua 3-5
Light Green Aqua 5-10

[040] (F-Skirt) B SB
Aqua, Green Aqua, Light Aqua, Light Blue x1
Dark Green, Emerald Green, Green 1-2
Light Apple Green 3-5
Dark Yellow Green 5-10
Dark Olive Green 15-20
Apple Green 20-30
Olive Amber 100-125

[045] (F-Skirt) B SDP
Teal Aqua 5-10

[050] (F-Skirt) B (R-Skirt) B SB
Aqua x1
Green 1-2
Yellow Green 5-10

[060] (F-Skirt) B (R-Skirt) B SDP
Aqua, Blue 3-5
Green 5-10
Dark Yellow Amber 250-300

[070] (F-Skirt) B (R-Skirt) № 44/[Number] RB
Aqua, Blue Aqua, Green Aqua 3-5

[080] (F-Skirt) B (R-Skirt) [Number] SB
Light Aqua 3-5
Light Green 5-10

[090] (F-Skirt) B/['B.G.M.CO.' blotted out] (R-Skirt) B SB
Light Aqua 5-10

B.G.M.CO.

[010] (F-Skirt) B.G.M.CO. SB
Light Purple 200-250
Purple 250-300
Green Tint, Off Clear 300-350
Dark Purple 350-400
Lemon, Pink 400-500

BROOKFIELD

[010] (Dome) [Letter or number] (F-Crown) (Arc)W. BROOKFIELD/NEW YORK (R-Crown) PAT.NOV. 13, 1883/FEB. 12 1884 SB
Aqua x1

[020] (Dome) [Letter or number] (F-Skirt) BROOKFIELD SB
Aqua x1
Green 1-2
Lime Green 3-5
Dark Yellow Green 5-10
Teal Aqua 15-20
Chartreuse 125-150
Olive Amber 150-175

75

CD 145

[030] (Dome) [Number] (F-Skirt) BROOKFIELD (R-Skirt) NEW YORK SB
Aqua, Green Aqua, Light Aqua, Light Blue x1
Green, Light Green 1-2
Dark Green 3-5
Yellow Green 5-10

[040] (Dome) [Number] (F-Skirt) BROOKFIELD (R-Skirt) NEW YORK SDP
Green 5-10

[050] (Dome) [Number] (F-Skirt) BROOKFIELD (R-Skirt) NEW YORK {Note backwards letter} SB
Aqua 10-15

[120] (F-Crown) (Arc)W. BROOKFIELD/45 CLIFF ST (R-Crown) [Number]/(Arc)PAT'D NOV. 13TH 1883/FEB. 12TH 1884 {'4' in '45 CLIFF ST' is upside down and backwards} {MLOD} SB
Aqua, Green Aqua 10-15

[060] (F-Crown) (Arc)W. BROOKFIELD/45 CLIFF ST/N.Y. (F-Skirt) ER (R-Crown) [Number]/(Arc)PAT'D NOV. 13TH 1883/FEB. 12TH 1884 {MLOD} SB
Aqua, Light Aqua, Light Blue 15-20
Green 100-125
Yellow Green 400-500

[070] (F-Crown) (Arc)W. BROOKFIELD/45 CLIFF ST/N.Y. (R-Crown) (Arc)PAT.NOV. 13, 1883/FAB. 12, 1884 {Note spelling} {MLOD} SB
Aqua 3-5
Green 75-100
Yellow Olive Green 400-500

[080] (F-Crown) (Arc)W. BROOKFIELD/45 CLIFF ST/N.Y. (R-Crown) PAT/JAN. 25TH 1870 {MLOD} SB
Light Aqua 1-2
Purple 250-300
Blue Gray, Lavender, Yellow Green 300-450
Yellow Olive Green 600-700

[085] (F-Crown) (Arc)W. BROOKFIELD/45 CLIFF ST/N.Y. (R-Crown) [Number] {MLOD} SB
Aqua, Light Blue Aqua 1-2

[090] (F-Crown) (Arc)W. BROOKFIELD/45 CLIFF ST/N.Y. (R-Crown) [Number]/(Arc)PAT'D NOV. 13TH 1883/FEB. 12TH 1884 {MLOD} SB
Aqua, Blue Aqua 1-2
Light Green 20-30
Milky Aqua 150-175
Yellow Green 300-350
Yellow Olive Green 400-500

[100] (F-Crown) (Arc)W. BROOKFIELD/45 CLIFF ST/N.Y. (R-Crown) [Number]/PAT/JAN. 25TH 1870 {MLOD} SB
Aqua, Light Aqua 1-2
Light Blue 15-20
Light Green 20-30
Light Purple,
Light Sapphire Blue, Purple 250-300
Blue Gray, Yellow Green 300-350

[110] (F-Crown) (Arc)W. BROOKFIELD/45 CLIFF ST/N.Y. (R-Crown) [Number]/[Letter] {MLOD} SB
Aqua 1-2
Light Purple 250-300

[130] (F-Crown) (Arc)W. BROOKFIELD/ 83.FULTON.ST/N.Y. (R-Crown) [Number]/ (Arc)/PAT'D NOV 13TH 1883/FEB 12TH 1884 {MLOD} SB
Aqua 3-5
Yellow Green 300-350

[140] (F-Crown) (Arc)W. BROOKFIELD/NEW YORK (R-Crown) PAT.NOV. 13TH 1883/" FEB. 12TH 1881 {Note 'FEB. 12TH 1881' date} SB
Aqua, Green Aqua 3-5

[150] (F-Crown) (Arc)W. BROOKFIELD/NEW YORK (R-Crown) PAT.NOV. 13TH 1883/" FEB. 12TH 1884 {MLOD} SB
Aqua 1-2
Blue Aqua 5-10
Light Blue 15-20
Light Yellow Green 175-200

[160] (F-Crown) (Arc)W. BROOKFIELD/NEW YORK (R-Crown) [Letter]/PAT.NOV. 13TH 1883/" FEB. 12TH 1881 {Note 'FEB. 12TH 1881' date} SB
Aqua, Green Aqua 3-5

[170] (F-Crown) (Arc)W. BROOKFIELD/NEW YORK (R-Crown) [Letter]/PAT.NOV. 13TH 1883/" FEB. 12TH 1884 {MLOD} SB
Aqua, Light Blue Aqua 1-2
Dark Green Aqua 5-10
Green 75-100

[180] (F-Crown) (Arc)W. BROOKFIELD/NEW YORK (R-Crown) [Number]/PAT.NOV. 13, 1883/FEB. 12 1884 {MLOD} SB
Aqua 1-2
Green 75-100

[190] (F-Crown) W. BROOKFIELD/45 CLIFF ST/N.Y. (R-Crown) [Number]/(Arc)PAT'D NOV. 13TH 1883/FEB. 12TH 1884 {MLOD} SB
Aqua 1-2
Ice Blue 5-10
Yellow Green 300-350
Yellow Olive Green 400-500

CD 145

[200] (F-Crown) W<u>M</u> BROOKFIELD/45 CLIFF S<u>T</u>/N.Y. (R-Crown) [Number]/PAT./JAN. 25<u>TH</u> 1870 {MLOD} SB
Light Green 30-40

[205] (F-Crown) [Backwards number]/[Number]/(Arc)W. BROOKFIELD/45 CLIFF S<u>T</u>/N.Y. {MLOD} SB
Aqua 1-2
Yellow Green 175-200

[210] (F-Crown) [Number]/(Arc)W. BROOKFIELD/45 CLIFF S<u>T</u>/N.Y. (R-Crown) (Arc)PAT'D NOV. 13<u>TH</u> 1883/FEB. 12<u>TH</u> 1884 {MLOD} SB
Aqua, Light Green Aqua 1-2
Ice Blue 5-10
Light Green 20-30
Milky Light Aqua 100-125
Yellow Green 300-350

[220] (F-Crown) [Number]/(Arc)W. BROOKFIELD/45 CLIFF S<u>T</u>/N.Y. (R-Crown) [Number]/PAT./JAN. 25TH 1870 {MLOD} SB
Aqua, Light Aqua 1-2
Light Sapphire Blue, Purple 250-300

[230] (F-Crown) [Number]/(Arc)W. BROOKFIELD/45 CLIFF S<u>T</u>/N.Y. {MLOD} SB
Aqua, Light Aqua, Light Blue,
Light Green Aqua 1-2
Green 75-100
Light Sapphire Blue, Purple 250-300
Gray Blue, Yellow Green 300-350

[235] (F-Crown) [Number]/(Arc)W. BROOKFIELD/83. FULTON. ST/N.Y (R-Crown) (Arc)PATD. NOV. 13. 1883/EEB. 12<u>TH</u> 1884 {Note 'EEB. 12<u>TH</u> 1884' date} {MLOD} SB
Aqua 10-15

[240] (F-Crown) [Number]/(Arc)W. BROOKFIELD/83. FULTON. ST/N.Y. (R-Crown) (Arc)PATD. NOV. 13TH 1883/FEB 12TH 1884 {MLOD} SB
Aqua 1-2

[250] (F-Crown) [Number]/(Arc)W. BROOKFIELD/NEW YORK (R-Crown) PAT NOV. 13<u>TH</u> 1883/ " FEB. 12<u>TH</u> 1884 {MLOD} SB
Aqua, Light Aqua, Light Green Aqua 1-2
Green 30-40

[253] (F-Crown) [Number]/[Number]/(Arc)W. BROOKFIELD/45 CLIFF S<u>T</u>/N.Y. (R-Crown) (Arc)PAT'D NOV. 13<u>TH</u> 1883/FEB. 12<u>TH</u> 1884 {MLOD} SB
Aqua 1-2

[257] (F-Crown) [Number]/[Number]/(Arc)W. BROOKFIELD/45 CLIFF S<u>T</u>/N.Y. {MLOD} SB
Light Aqua 1-2

[260] (F-Skirt) BROOKFIELD (R-Skirt) NEW YORK/['B.G.M.CO.' blotted out] SB
Aqua 5-10

[270] (F-Skirt) W. BROOKFIELD (R-Skirt) NEW YORK SB
Light Aqua x1
Yellow Green 50-75

[280] (F-Skirt) W. BROOKFIELD (R-Skirt) NEW YORK/[Blotted out embossing] SB
Aqua 1-2

CALIFORNIA

[010] (F-Skirt) CALIFORNIA SB
Sage Green, Smoke 10-15
Green 15-20
Light Purple 20-30
Pink, Purple 30-40
Burgundy 40-50
Purple/Sage Two Tone 300-350

E.D.R.

[010] (F-Skirt) E.D.R./['G.N.W. TEL.CO.' blotted out] SB
Light Aqua, Light Blue 15-20

G.N.W.TEL.CO.

[020] (F-Skirt) G.N.W. TEL Co./['G.T.P. TEL.CO.' blotted out] {Note small 'o'} SB
Light Aqua 10-15

[010] (F-Skirt) G.N.W. TEL.CO. SB
Light Aqua, Light Blue 5-10
Ice Blue, Light Green, Steel Blue 10-15
Light Purple 40-50
Aqua w/ Milk Swirls, Dark Purple,
Royal Purple 50-75

[030] (F-Skirt) ['G.N.W. TEL.CO.' blotted out] SB
Straw 200-250

G.T.P.TEL.CO.

[010] (Dome) [Number] (F-Skirt) G.T.P. (R-Skirt) B SB
Aqua, Dark Aqua 10-15
Green 15-20
Dark Yellow Green, Emerald Green 30-40

[020] (F-Skirt) G.T.P. SB
Light Aqua, Light Blue 5-10
Blue Gray, Gray, Steel Blue 10-15
Light Lavender 15-20
Light Green 20-30

CD 145

[040] (F-Skirt) G.T.P. TEL Co. {Note small 'o'} SB
Light Aqua 10-15

[030] (F-Skirt) G.T.P. TEL.CO. SB
Blue, Blue Aqua, Green Aqua, Ice Blue,
Light Aqua, Steel Blue 5-10
Blue Gray 10-15
Light Gray 15-20
Clear, Gray, Light Purple 20-30

[050] (F-Skirt) G.T.P./['G.N.W. TEL.CO.' blotted out] SB
Aqua 5-10

[060] (F-Skirt) G.T.P./['G.T.P. TEL.CO.' blotted out] SB
Light Aqua, Steel Blue 5-10
Blue Gray 10-15
Light Purple 20-30
Delft Blue 30-40

[070] (F-Skirt) G.T.P./['T.C.R.' blotted out] SB
Light Aqua, Light Blue Aqua 5-10
Blue Gray 10-15
Light Gray, Light Lavender 15-20

[080] (F-Skirt) G.T.P.TEL.CO./['G.N.W. TEL.CO.' blotted out] SB
Green Aqua, Light Aqua, Light Blue 5-10
Blue 10-15
Blue Gray 15-20
Light Gray Lavender, Light Purple 20-30

H.B.R.

[010] (F-Skirt) H.B.R. SB
Aqua, Ice Blue 20-30

[020] (F-Skirt) H.B.R./[Blotted out embossing] SB
Aqua, Ice Blue, Light Aqua 20-30

H.G.CO.

[010] (Dome) H (F-Skirt) H G CO {Note no periods} SB
Aqua 1-2
Light Jade Green Milk 15-20

[020] (Dome) H (F-Skirt) H.G.CO. SB
Aqua, Dark Aqua, Green Aqua, Light Aqua x1
Hemingray Blue, Light Green 1-2
Jade Green Milk 15-20
Green 20-30
Milky Aqua 50-75
Jade Blue Milk 200-250

[030] (Dome) H (F-Skirt) H.G.CO. (R-Skirt) ['H.G.CO.' blotted out] SB
Aqua, Blue Aqua, Dark Aqua 10-15
Light Jade Green Milk 20-30

[040] (Dome) H (F-Skirt) ['[Vertical bar]' blotted out]/H.G.CO. (R-Skirt) ['[Vertical bar]' blotted out] SB
Green 10-15

[050] (Dome) [Letter] (F-Skirt) H.G.CO. (R-Skirt) PETTICOAT SB
Green Tint, Ice Green, Light Aqua 3-5
Light Green, Off Clear, Yellow Green Tint 5-10
Light Apple Green 30-40
Clear, Purple Tint 50-75
Purple 400-500
Yellow Olive Green 1,500-1,750

[060] (Dome) [Letter] (F-Skirt) H.G.CO. (R-Skirt) PETTICOAT {Narrow style} SB
Aqua 30-40
Yellow Olive Green 1,500-1,750

[070] (Dome) [Letter] (F-Skirt) H.G.CO. (R-Skirt) PETTICOAT/[Letter] SB
Aqua, Light Aqua 1-2
Blue Aqua, Ice Aqua, Ice Green,
Light Blue Aqua 3-5
Light Green 5-10
Ice Blue, Light Blue 10-15
Light Steel Blue 15-20
Light Cornflower Blue 50-75
Blue, Lime Green 75-100
Dark Aqua, Light Purple 200-250
Dark Lavender, Jade Blue Milk 800-1,000
Sapphire Blue 2,000-2,500

[080] (Dome) [Letter] (F-Skirt) H.G.CO./[Letter] (R-Skirt) PETTICOAT SB
Green Aqua, Light Aqua 1-2
Light Green 5-10

[090] (F-Skirt) H.G.Co SB
Aqua, Hemingray Blue 10-15

[095] (F-Skirt) H.G.CO. SB
Dark Aqua 1-2

[100] (F-Skirt) H.G.CO. (R-Skirt) PETTICOAT SB
Aqua, Green Aqua, Green Tint 1-2
Blue Aqua, Ice Aqua, Ice Green,
Light Blue Aqua 3-5
Ice Blue 10-15
Blue, Clear, Light Cornflower Blue,
Light Yellow Green, Lime Green 50-75
Green, Lavender 200-250
Light Olive Green 400-500
Apple Green 500-600
Dark Yellow Green,
Depression Glass Green 700-800
Straw, Teal Green 800-1,000
Orange Amber 1,000-1,250
Violet Blue 1,750-2,000

78

CD 145

[110] (F-Skirt) H.G.CO. (R-Skirt) PETTICOAT {Narrow style} SB
Aqua, Light Blue, Light Green 30-40
Light Purple, Purple 800-1,000
Dark Purple 1,000-1,250

[120] (F-Skirt) H.G.CO. (R-Skirt) PETTICOAT/[Letter] SB
Light Aqua 1-2
Blue Aqua, Green Tint, Ice Aqua,
Light Blue Aqua, Light Green Aqua 3-5
Light Green 5-10
Ice Blue 10-15
Light Steel Blue 15-20
Sky Blue 50-75
Blue, Light Lavender, Lime Green, Pink 75-100
Light Purple 200-250
Milky Aqua 350-400
Purple 400-500
Apple Green, Jade Aqua Milk 500-600
Yellow Green 600-700
Jade Blue Milk 800-1,000
Honey Amber, Orange Amber 1,000-1,250
Sapphire Blue, Yellow 2,000-2,500
Amber Olive Green 3,000-3,500

[130] (F-Skirt) H.G.CO./[Letter] (R-Skirt) PETTICOAT SB
Aqua, Light Aqua 1-2
Blue Aqua, Ice Aqua 3-5
Light Green 5-10
Ice Blue, Light Blue 10-15
Light Steel Blue 15-20
Steel Blue 20-30
Sky Blue 40-50
Light Cornflower Blue, Light Yellow Green .. 50-75
Blue, Light Lavender 75-100
Dark Aqua, Green 200-250
Purple 400-500
7-up Green 500-600
Yellow Green 600-700
Depression Glass Green 700-800
Emerald Green 800-1,000
Golden Amber, Honey Amber,
Orange Amber 1,000-1,250
Yellow Amber 1,250-1,500
Yellow Olive Amber 2,000-2,500

[140] (F-Skirt) [Letter]/H.G.CO. (R-Skirt) PETTICOAT SB
Aqua, Light Aqua, Light Green Aqua 1-2
Aqua Tint, Blue Aqua, Green Aqua, Ice Aqua,
Ice Green, Light Blue Aqua 3-5
Blue Tint, Light Green, Off Clear 5-10
Ice Blue, Light Blue,
Light Sky Blue, Sage Green 10-15
Light Steel Blue 15-20
Steel Blue 20-30
Light Lime Green 30-40
Dark Blue Aqua, Sky Blue 40-50
Clear, Light Cornflower Blue, Powder Blue .. 50-75
Blue, Light Lavender 75-100
Aqua w/ Purple Swirls 100-125
Cornflower Blue, Dark Aqua, Lavender,
Light Purple, Light Smoky Purple 200-250
Hemingray Blue 300-350
Apple Green, Jade Aqua Milk,
Jade Green Milk 500-600
Yellow Green 600-700
Christmas Tree Green,
Forest Green, Teal Blue 800-1,000
Dark Orange Amber,
Golden Amber, Honey Amber 1,000-1,250
Yellow Amber 1,250-1,500
Sapphire Blue, Yellow 2,000-2,500
White Milk 10,000-15,000

HAWLEY ⌗

[010] (Dome) [Letter or number] (F-Skirt) ⌗/HAWLEY PA./U.S.A. SB
Aqua, Blue Aqua, Green Aqua, Light Aqua 3-5
Milky Blue Aqua 20-30

[020] (Dome) [Letter or number] (F-Skirt) ⌗/HAWLEY PA./U.S.A. (R-Skirt) [Letter or number] SB
Green Aqua 3-5

[030] (Dome) [Letter or number] (F-Skirt) ⌗/HAWLEY, PA./U.S.A. {Note backwards letter} SB
Aqua 3-5
Milky Aqua 20-30

[040] (Dome) [Letter or number] (F-Skirt) ⌗/HAWLEY, PA./U.S.A. {Note backwards letter} (R-Skirt) [Letter or number] SB
Aqua, Green Aqua, Light Aqua 3-5
Ice Green 5-10
Light Aqua w/ Milk Swirls 20-30

[045] (F-Skirt) ⌗/HAWLEY, PA./U.S.A. SB
Aqua 3-5
Milky Aqua 20-30

[050] (F-Skirt) ⌗/HAWLEY, PA./U.S.A. (R-Skirt) [Letter or number] SB
Aqua, Light Green Aqua 3-5
Light Blue 5-10
Light Aqua w/ Milk Swirls 20-30

CD 145

[060] (F-Skirt) ⌀/HAWLEY, PA./U.S.A.
{Postal style} SB
Aqua, Light Aqua 15-20

[070] (F-Skirt) ⌀/HAWLEY, PA./U.S.A.
{Note backwards letter} (R-Skirt) [Letter or
number] SB
Aqua 3-5

HEMINGRAY

[010] (F-Skirt) HEMINGRAY SB
Aqua, Blue Aqua 1-2

[020] (F-Skirt) HEMINGRAY SDP
Aqua, Blue Aqua 1-2
Hemingray Blue 3-5
Yellow Green 30-40

[030] (F-Skirt) HEMINGRAY
(R-Skirt) DOUBLE/PETTICOAT SB
Aqua 1-2

[040] (F-Skirt) HEMINGRAY
(R-Skirt) DOUBLE/PETTICOAT SDP
Aqua 5-10
Hemingray Blue 15-20

[050] (F-Skirt) HEMINGRAY (R-Skirt) № 21
SB
Aqua 3-5

[060] (F-Skirt) HEMINGRAY (R-Skirt) № 21
SDP
Aqua, Blue Aqua, Green Aqua 1-2
Hemingray Blue 3-5
Green 10-15
Forest Green, Yellow Green 20-30
Olive Green 30-40

[070] (F-Skirt) HEMINGRAY-21
(R-Skirt) MADE IN U.S.A. RDP
Aqua 1-2
Clear, Hemingray Blue 3-5

[080] (F-Skirt) HEMINGRAY-21
(R-Skirt) MADE IN U.S.A. SDP
Blue Aqua 1-2
Hemingray Blue 3-5

[090] (F-Skirt) HEMINGRAY-21/[Numbers and
dots] (R-Skirt) MADE IN U.S.A./[Number] RDP
Clear 3-5

[100] (F-Skirt) HEMINGRAY/MADE IN U.S.A.
(R-Skirt) № 21 SDP
Aqua 1-2
Hemingray Blue 3-5

K.C.G.W.

[010] (F-Skirt) K.C.G.W. SB
Aqua, Light Aqua, Light Green 20-30
Green 30-40
Light Yellow Green 40-50

LYNCHBURG Ⓛ

[010] (Dome) [Letter or number]
(F-Skirt) LYNCHBURG/Ⓛ (R-Skirt) No.-43/
MADE IN [Number]/U.S.A. SDP
Aqua, Blue Aqua 10-15

[020] (Dome) [Letter or number]
(F-Skirt) LYNCHBURG/Ⓛ (R-Skirt) No.43/
MADE IN/U.S.A. SDP
Aqua, Blue Aqua, Light Aqua 10-15
Celery Green 15-20

[030] (Dome) [Letter or number]
(F-Skirt) LYNCHBURG/Ⓛ (R-Skirt) ['B' blotted
out]/No.-43/MADE IN [Number]/U.S.A. SDP
Light Aqua 10-15

N.E.G.M.CO.

[010] (F-Skirt) N.E.G.M.CO. SB
Aqua, Light Aqua 1-2
Blue Aqua 3-5
Green, Ice Blue 15-20
Emerald Green, Lime Green, Yellow Green 50-75
Olive Green 175-200
Olive Amber, Sapphire Blue 300-350

NO EMBOSSING

[010] [No embossing] {American style} SB
Blue Aqua, Light Aqua 20-30

[020] [No embossing] {Grand Canyon style} SB
Blue Aqua, Light Aqua 15-20
Ice Aqua, Off Clear 20-30
Gray, Steel Blue 40-50
Blue Tint, Clear, Pink 50-75
Light Lavender 100-125
Lime Green 150-175
Steel Blue/Lavender Two Tone 175-200
Lavender 250-300
Dark Lavender 400-500
Cornflower Blue 800-1,000

[030] [No embossing] {Pennycuick style} SB
Aqua, Blue Aqua, Dark Aqua,
Green Aqua, Ice Aqua 15-20
Gray Green 50-75
Apple Green 150-175
Milky Aqua 200-250
Yellow Green 300-350

NO EMBOSSING - CANADA

[010] [No embossing] {Canadian style} SB
Straw 5-10
Peach 10-15
Pink 20-30
Light Green, Light Yellow Green,
Lime Green 40-50
Yellow 50-75
Light Green w/ Milk Swirls 75-100

CD 145.6

[020] [No embossing] {Canadian style} {MLOD} SB
Blue Aqua 50-75
Green 200-250

[030] [No embossing] {Canadian style} {Shallow slug plate around half of skirt} SB
Straw 5-10
Peach 10-15
Pink 20-30
Light Green, Light Yellow Green,
Lime Green 40-50
Yellow 50-75

NO NAME

[010] (Dome) H (F-Skirt) [Vertical bar] (R-Skirt) [Vertical bar] {Hemingray product} SB
Aqua 1-2
Hemingray Blue, Light Green 3-5
Emerald Green 300-350

[015] (Dome) H (F-Skirt) [Vertical bar]/ ['[Vertical bar]' blotted out] (R-Skirt) [Vertical bar] {Hemingray product} SB
Aqua 3-5

[020] (Dome) H {Hemingray product} SB
Aqua, Green Aqua, Light Green 3-5

[030] (Dome) [Letter] {American style} SB
Aqua, Ice Blue 20-30

[040] (F-Skirt) [Vertical bar] SB
Straw 30-40

O.V.G.CO.

[010] (F-Skirt) O.V.G.CO. SB
Aqua, Celery Green, Light Green 5-10
Blue 20-30
Green 40-50

POSTAL

[010] (F-Skirt) POSTAL SB
Aqua 3-5
Green, Sage Green, Sapphire Blue 10-15
Burgundy, Purple 20-30
Yellow Green 30-40
Light Gray 75-100
Olive Green 175-200
Purple/Sage Two Tone 200-250

STANDARD GLASS INSULATOR CO.

[010] (Base) THE STANDARD GLASS INSULATOR CO./BOSTON MASS. BE
Blue Aqua, Light Aqua 175-200

STAR ✯

[010] (F-Skirt) ✯ SB
Aqua, Blue Aqua, Light Blue, Light Green 1-2
Green 3-5
Dark Green, Emerald Green 5-10
7-up Green, Lime Green, Yellow Green 10-15
Olive Green, Teal Green 20-30
Olive Amber 125-150

[020] (F-Skirt) ✯ (R-Skirt) ✯ {Pointed dome} SB
Aqua, Blue Aqua, Light Blue 3-5
Green 20-30
Dark Green 50-75
Yellow Green 75-100
Dark Yellow Green, Emerald Green 100-125

[030] (F-Skirt) ✯ {Postal style} SB
Aqua, Blue Aqua, Dark Aqua 1-2
Green Aqua, Hemingray Blue 3-5
Blue 5-10
Green 15-20
Dark Green, Emerald Green 40-50
Dark Yellow Green, Olive Green 75-100

T.C.R.

[010] (F-Skirt) T.C.R. SB
Blue Aqua, Ice Blue, Light Aqua 10-15
Purple Tint 175-200
Light Purple 200-250

[020] (F-Skirt) T.C.R./[Blotted out embossing] SB
Aqua 10-15

W.G.M.CO.

[010] (F-Skirt) W.G.M.Co. SB
Lavender, Purple 50-75
Royal Purple 75-100
Gray 100-125
Purple Blackglass 125-150
Straw 500-600

[020] (F-Skirt) W.G.M.Co. {Postal style} SB
Purple 75-100
Gray 125-150

CD 145.6

BOSTON BOTTLE WORKS

[010] (Base of inner skirt) BOSTON BOTTLE WORKS/PATENT APPLIED FOR {Four segmented threads} {MLOD} SB
Aqua, Blue Aqua 1,750-2,000
Dark Olive Green,
Root Beer Amber 10,000-15,000

81

CD 145.6

[020] (Base of inner skirt) BOSTON BOTTLE WORKS/PATENT APPLIED FOR {Three segmented threads} {MLOD} SB
Aqua 3,000-3,500

[030] (Base of inner skirt) BOSTON BOTTLE WORKS/PATENT OCT 15 72 SB
Aqua 2,000-2,500

NO EMBOSSING

[010] [No embossing] {Four segmented threads} SB
Light Blue Aqua 1,750-2,000

CD 146.4

NO EMBOSSING

[010] [No embossing] {Pennycuick style} {This is a CD 104 with an inner skirt} SB
Aqua, Dark Aqua 50-75
Green 100-125
Emerald Green 125-150

CD 146.5

NO EMBOSSING

[010] [No embossing] SB
Light Green 400-500

NO NAME

[010] (F-Skirt) [Vertical bar] SB
Light Green 350-400

CD 147

HEMINGRAY

[010] (F-Skirt) HEMINGRAY (R-Skirt) PATENTED OCT. 8 1907. SB
Aqua, Light Aqua x1

[020] (F-Skirt) HEMINGRAY (R-Skirt) PATENTƎD OCT. 8 1907. {Note backwards letter} SB
Aqua 5-10

NO EMBOSSING

[010] [No embossing] SB
Aqua 200-250

PAT APP FOR

[010] (F-Skirt) PAT. APPL'D FOR SB
Aqua 200-250

PATENT - OTHER

[005] (Dome) [Number] (F-Skirt) PATENTED OCT 8 1907 SB
Aqua x1

[010] (Dome) [Number] (F-Skirt) PATENT'D OCT 8 1907 SB
Aqua 5-10

[020] (F-Skirt) PATENTED SB
Aqua 200-250

[030] (F-Skirt) PATENTED (R-Skirt) OCT. 8TH 1907. SB
Aqua x1
Green 20-30
Dark Yellow Green 75-100

[040] (F-Skirt) PATENTED OCT 8 1907. SB
Aqua, Light Aqua x1
Green 20-30
Emerald Green 40-50
Milky Aqua 125-150

[050] (F-Skirt) PATNTED OCT. 8 1907. {Note spelling} SB
Aqua 5-10

CD 149

BROOKFIELD

[010] (F-Skirt) BROOKFIELD (R-Skirt) NEW YORK SB
Aqua, Dark Aqua 75-100
Dark Green 125-150

NO EMBOSSING

[010] [No embossing] CD
Dark Aqua 20-30
Yellow Olive Green 1,750-2,000

CD 151

[020] [No embossing] SB
Aqua, Blue Aqua, Green Aqua,
Light Green Aqua 15-20
Yellow Green 1,250-1,500
Sapphire Blue 2,000-2,500

CD 150

BARCLAY

[010] (F-Skirt) BARCLAY (R-Skirt) PATENTED OCT. 8 1907. SDP
Aqua 3,000-3,500

BROOKFIELD

[010] (F-Skirt) BROOKFIELD/PAT.OCT. 8-1907 SB
Aqua 150-175

CD 151

BROOKFIELD

[010] (Dome) [Number] (F-Skirt) W. BROOKFIELD (R-Skirt) NEW YORK SB
Aqua, Blue Aqua, Light Aqua 1-2
Light Blue 3-5
Light Green 5-10

[020] (F-Crown) [Number]/(Arc)W. BROOKFIELD/NEW YORK {MLOD} SB
Aqua, Blue Aqua, Light Aqua 1-2
Light Green 10-15

[030] (F-Skirt) W. BROOKFIELD (R-Skirt) NEW YORK SB
Aqua 1-2
Apple Green 100-125

H.G.CO.

[010] (Crown) N.A.T.Co. (F-Skirt) H.G.CO./ PAT^D MAY 2 1893 (R-Skirt) PETTICOAT SDP
Aqua 15-20
Peacock Blue 500-600
Electric Blue 600-700
Dark Electric Blue 800-1,000

[020] (Crown) N.A.T.Co. (F-Skirt) H.G.CO./ PAT^D MAY 2 1893 (R-Skirt) PETTICOAT
{Sharp drip points appear only on the outer skirt; the inner skirt is smooth} SDP
Aqua 20-30
Peacock Blue 600-700

[030] (Crown) N.A.T.Co. (F-Skirt) H.G.CO./ PAT^D MAY 2ND 1893 (R-Skirt) PETTICOAT SDP
Aqua 15-20
Green 200-250
Peacock Blue 500-600
Electric Blue 600-700
Light Electric Blue 800-1,000
Cobalt Blue 1,750-2,000

[040] (Crown) N.A.T.Co. (F-Skirt) H.G.CO./ PAT^D MAY 2ND 1893 (R-Skirt) PETTICOAT
{Sharp drip points appear only on the outer skirt; the inner skirt is smooth} SDP
Peacock Blue 600-700
Dark Peacock Blue 700-800

[050] (Crown) N.A.T.Co. (F-Skirt) H.G.CO./ PAT^D MAY 2ND 1893 (R-Skirt) PETTICOAT/ ['H.G.CO.' blotted out] SDP
Peacock Blue 500-600

[060] (Crown) N.A.T.Co. (F-Skirt) H.G.CO./ PAT'D MAY 2 1893 (R-Skirt) PETTICOAT SDP
Aqua 15-20

[070] (F-Skirt) H.G.CO. (R-Skirt) PETTICOAT SB
Aqua, Light Aqua 3-5
Light Green 5-10
Ice Aqua, Ice Blue 10-15
Blue Aqua, Light Blue, Light Steel Blue 20-30
Green Tint 75-100
Smoke 100-125
Hemingray Blue,
Light Cornflower Blue, Purple Tint 150-175
Cornflower Blue 250-300
Dark Cornflower Blue 350-400
Apple Green 400-500
Clear 500-600
Depression Glass Green,
Green, Teal Blue 800-1,000
Sapphire Blue 1,000-1,250
Jade Green Milk 1,250-1,500
Forest Green 3,500-4,000
Olive Green 4,000-4,500
Dark Yellow Green,
Yellow Olive Green 5,000-7,500
Orange Brown Amber, Purple 10,000-15,000

[080] (F-Skirt) H.G.CO. (R-Skirt) PETTICOAT SDP
Aqua 15-20

83

CD 151

[090] (F-Skirt) H.G.CO./PATENT MAY 2 1893 (R-Skirt) PETTICOAT SDP
Aqua 1-2
Milky Aqua 1,000-1,250

[100] (F-Skirt) H.G.CO./PATENT MAY 2 1893 (R-Skirt) PETTICOAT {Sharp drip points appear only on the outer skirt; the inner skirt is smooth} SDP
Aqua 5-10

[110] (F-Skirt) H.G.CO./PATENT MAY 2ND 1893 (R-Skirt) PETTICOAT SDP
Aqua 1-2
Light Green 5-10

[120] (F-Skirt) H.G.CO./PATD MAY 2 1893 (R-Skirt) PETTICOAT SDP
Aqua 1-2
Ice Blue 20-30
Dark Blue Aqua 40-50
Hemingray Blue 50-75
Purple 10,000-15,000

[140] (F-Skirt) H.G.CO./PATD MAY 2ND 1893 (R-Skirt) PETTICOAT SDP
Aqua, Blue Aqua 1-2
Light Aqua 5-10
Light Green 10-15
Ice Aqua 20-30
Hemingray Blue 75-100
Green, Lime Green 300-350
Ice Aqua w/ Purple Swirls,
Peacock Blue 800-1,000
Gray Purple 1,000-1,250
Dark Yellow Green,
Yellow Olive Amber 5,000-7,500

[150] (F-Skirt) H.G.CO./PATD MAY 2ND 1893 (R-Skirt) PETTICOAT {Sharp drip points appear only on the outer skirt; the inner skirt is smooth} SDP
Aqua, Light Aqua 5-10

[160] (F-Skirt) H.G.CO./PAT'D MAY 2 1893 (R-Skirt) PETTICOAT SDP
Aqua 1-2

[165] (F-Skirt) H.G.CO./['PATENT MAY 2 1893' blotted out] (R-Skirt) PETTICOAT SDP
Aqua 3-5

[170] (F-Skirt) H.G.CO./['PATD MAY 2 1893' blotted out] (R-Skirt) PETTICOAT SB
Aqua 10-15

[180] (F-Skirt) H.G.CO./['PAT'D MAY 2 1893' blotted out] (R-Skirt) PETTICOAT SDP
Aqua, Green Aqua 3-5

NO EMBOSSING

[010] [No embossing] SB
Dark Green Aqua, Ice Aqua 50-75
Dark Green 75-100
Emerald Green 200-250
Root Beer Amber 5,000-7,500

NO NAME

[010] (F-Skirt) 7 (R-Skirt) 7 SB
Dark Aqua, Dark Green, Dark Green Aqua,
Green, Light Aqua 50-75

[020] (F-Skirt) [Vertical bar] (R-Skirt) [Vertical bar] SB
Dark Green Aqua, Green Aqua,
Light Green 50-75
Dark Green, Emerald Green 75-100

[030] (F-Skirt) [Vertical bar] 8 (R-Skirt) [Vertical bar] 8 SB
Celery Green 75-100

CD 152

B

[010] (F-Skirt) B SB
Aqua 1-2
Green 3-5
Emerald Green 5-10
Dark Olive Amber, Dark Olive Green 15-20
Root Beer Amber 75-100

BROOKFIELD

[010] (Dome) [Number] (F-Skirt) BROOKFIELD SB
Blue Aqua, Green Aqua x1
Dark Green 1-2
Light Blue 3-5
Blue, Dark Yellow Green 5-10
Olive Green 10-15

[013] (Dome) [Number] (F-Skirt) BROOKFIELD SDP
Aqua 1-2
Blue Aqua 3-5

[017] (Dome) [Number] (F-Skirt) BROOKFIELD {Australian threads} SB
Dark Aqua 175-200

CD 152

[020] (F-Skirt) BBOOKFIELD {Note spelling} SB
Aqua 3-5
Green 5-10
Emerald Green 10-15
Olive Green 15-20
Dark Teal Green 40-50

[030] (F-Skirt) BBOOKFIELD {Note spelling} SDP
Aqua 5-10
Blue 10-15
Emerald Green 15-20
Olive Green 20-30
Olive Amber 75-100

[040] (F-Skirt) BROOKFIELD SB
Aqua, Dark Aqua, Dark Blue Aqua,
Green Aqua x1
Emerald Green, Green 1-2
Blue 3-5
Olive Green 10-15
Dark Teal Green, Olive Amber 20-30

[050] (F-Skirt) BROOKFIELD SDP
Aqua, Dark Blue Aqua, Green Aqua ... 1-2
Green 3-5
Emerald Green 5-10
Olive Green 15-20
Olive Amber 30-40

[055] (F-Skirt) BROOKFIELD/[Number] SB
Aqua 1-2
Green 3-5

[060] (F-Skirt) BROOKFIELD/[Number] (R-Skirt) No 48 SB
Aqua, Green Aqua 1-2
Dark Blue Aqua 3-5

[070] (F-Skirt) BROOKFIELD/[Number] (R-Skirt) Иo 48 {Note backwards letter} SB
Aqua, Blue Aqua, Green Aqua 3-5
Green 5-10

CALIFORNIA

[010] (F-Skirt) CALIFORNIA SB
Sage Green 3-5
Light Green 5-10
Aqua, Blue, Light Purple, Smoke 10-15
Clear, Peach 50-75
Straw, Yellow 125-150
Purple/Sage Two Tone 175-200

DIAMOND ◇

[010] (F-Skirt) ◇ SB
Ice Green, Light Green 3-5
Aqua, Blue, Light Sage Green, Straw ... 5-10

HEMINGRAY

[010] (Dome) [Number] (F-Skirt) HEMINGRAY (R-Skirt) №40 SDP
Aqua, Light Green Aqua x1
Light Blue Aqua 1-2
Green, Hemingray Blue 3-5
Dark Yellow Green 10-15

[020] (Dome) [Number] (F-Skirt) HEMINGRAY/ MADE IN U.S.A. (R-Skirt) №40 SDP
Aqua 3-5
Olive Green 10-15

[030] (F-Skirt) HEMINGRAY (R-Skirt) №40 SB
Aqua 75-100

[040] (F-Skirt) HEMINGRAY (R-Skirt) №40 SDP
Aqua, Dark Aqua x1
Light Aqua, Light Green 1-2
Emerald Green, Green 3-5
Blue, Dark Yellow Green,
Hemingray Blue, Light Sage Green ... 5-10
Olive Green 10-15
Aqua w/ Amber Swirls, Dark Olive Green ... 15-20
Sage Green 20-30
Olive Amber 30-40
Milky Aqua 75-100
Jade Aqua Milk 350-400

[050] (F-Skirt) HEMINGRAY (R-Skirt) № : 40 or № 40: {Comes with several punctuation variations, with dots or dashes on front or rear skirt} SDP
Aqua 1-2

[060] (F-Skirt) HEMINGRAY/MADE IN U.S.A. (R-Skirt) №40 SDP
Aqua 3-5

[070] (F-Skirt) HEMINGRAY/№40 (R-Skirt) HEMINGRAY SDP
Aqua 1-2
Hemingray Blue 5-10

[080] (F-Skirt) HEMINGRAY/№40 (R-Skirt) PATENTED/MAY 2 1893 SDP
Aqua x1
Hemingray Blue 5-10
Green 10-15
Milky Aqua 75-100

[090] (F-Skirt) HEMINGRAY/№40 (R-Skirt) PATENTED/MAY 2 1893 {Inner skirt is extended about 1/2"} SB
Aqua 250-300

CD 152

[100] (F-Skirt) HEMINGRAY/·N⁰· 40· or N⁰ ·· 40 (R-Skirt) PATENTED/MAY 2 1893 {Comes with several punctuation variations, with dots or dashes on front or rear skirt} SDP
Aqua x1
Milky Aqua 75-100

McLAUGHLIN

[010] (F-Skirt) McLAUGHLIN (R-Skirt) N⁰ 40 SDP
Aqua 175-200
Green 400-500

NO EMBOSSING

[010] [No embossing] SB
Aqua 20-30
Green 30-40

NO NAME

[020] (Dome) [Number] (F-Skirt) N⁰ 40 SB
Aqua, Blue Aqua, Dark Aqua 5-10
Blue, Green 10-15

[010] (Dome) [Number] (F-Skirt) N⁰ 48 SB
Aqua 3-5
Blue Aqua 5-10
Blue, Green 10-15

[030] (Dome) [Number] (F-Skirt) N⁰ 48 SDP
Aqua 5-10
Blue, Green 10-15

CD 153

BROOKFIELD

[010] (F-Skirt) BROOKFIELD SB
Aqua, Dark Green Aqua 40-50
Green 100-125

GAYNER

[005] (F-Skirt) GAYNER (R-Skirt) NO. 48-400/[Number] RDP
Aqua 10-15

[010] (F-Skirt) GAYNER (R-Skirt) NO. 48-400/[Number] SB
Aqua 10-15

[020] (F-Skirt) GAYNER (R-Skirt) NO. 48-400/[Number] SDP
Blue Aqua, Light Green Aqua 3-5
Light Green 5-10

NO NAME

[005] (F-Skirt) NO.48-40 SB
Aqua, Light Aqua 3-5

[010] (F-Skirt) NO.48-40/[Number] SB
Aqua, Blue Aqua, Green Aqua, Light Aqua 3-5

WHITALL TATUM ▽

[010] (F-Skirt) WHITALL TATUM CO. N⁰ 1 (R-Skirt) MADE IN U.S.A. SB
Ice Aqua, Light Blue Aqua 175-200

CD 154

A.A. (AA)

[010] (F-Skirt) (AA) RDP
Aqua 40-50
Forest Green, Green 75-100
Yellow Green 100-125

[020] (F-Skirt) (AA) SDP
Green Aqua 50-75

C.N.R.

[010] (F-Skirt) C.N.R. (R-Skirt) ◇ SB
Aqua, Pink, Straw 400-500
Peach 500-600

CIA TELEFONICA

[010] (F-Skirt) CIA.TELEFONICA Y TELEGRAFICA MEXICANA RDP
Light Aqua 175-200

CRISOL TEXCOCO ▽

[010] (F-Skirt) CRISOL/▽/TEXCOCO SB
Emerald Green 125-150

[020] (F-Skirt) VID/CRISOL/▽/TEXCOCO (R-Skirt) [Blotted out embossing] SDP
Green 125-150

DIAMOND ◇

[010] (F-Skirt) ◇ RDP
Light Green, Straw 3-5
Peach 5-10

[020] (F-Skirt) ◇ (R-Skirt) ◇ RDP
Clear, Green, Straw 5-10

[025] (F-Skirt) ◇ (R-Skirt) ◇ SB
Pink 10-15

CD 154

[030] (F-Skirt) ◇ (R-Skirt) ◇ {Threaded inner skirt to base} RDP
Straw 5-10
Peach 10-15

[040] (F-Skirt) ◇ (R-Skirt) ◇ /['C.N.R.' blotted out] SB
Light Pink 30-40

[050] (F-Skirt) ◇ RDP
Dark Straw, Peach, Straw 10-15

[060] (F-Skirt) ◇ SB
Ice Yellow Green, Light Lime Green,
Peach, Straw 5-10
Aqua 15-20

[065] (F-Skirt) ◇ (R-Skirt) ◇ SB
Green Aqua 10-15

[070] (F-Skirt) ◇ RDP
Straw 5-10
Peach 10-15

[080] (F-Skirt) ◇ SB
Light Green, Straw 3-5
Blue Aqua, Ice Yellow Green,
Light Yellow Green, Peach, Pink 5-10

[090] (F-Skirt) ◇ (R-Skirt) ◇ SB
Light Pink 10-15

[100] (F-Skirt) ◇ (R-Skirt) ◇ /['C.N.R.' blotted out] SB
Light Green, Light Pink, Straw 30-40

DOMINION ◇

[010] (F-Skirt) DOMINION-42 (R-Skirt) ◇ RDP
Green Aqua, Light Green,
Light Sage Green, Light Straw 3-5
Dark Straw, Light Peach, Pink 10-15
Gingerale, Lemon 30-40
Clear, Dark Red Amber,
Honey Amber, Orange Amber 50-75
Yellow Amber 75-100
Light Yellow Amber 100-125
Honey Amber/Yellow Amber Two Tone,
Light Yellow 150-175
Light Cornflower Blue 300-350
Cornflower Blue 700-800
Dark Cornflower Blue 800-1,000

[015] (F-Skirt) DOMINION-42 (R-Skirt) ◇ SDP
Light Straw 15-20

[020] (F-Skirt) DOMINION-42 (R-Skirt) ◇ [Number] {Number may appear on any side of '◇'} RDP
Ice Green, Light Green, Light Straw, Straw 3-5
Peach, Sage Green 10-15
Light Aqua 15-20
Green 20-30
Clear, Dark Red Amber, Honey Amber,
Orange Amber, Red Amber 50-75
Gray Green, Yellow Amber 75-100
Light Yellow Amber 100-125
Light Yellow 150-175
Turquoise Blue 200-250
Light Cornflower Blue 300-350
Cornflower Blue 700-800

[030] (F-Skirt) DOMINION-42 (R-Skirt) ◇ / [Number] SDP
Straw 10-15

GAYNER

[010] (F-Skirt) GAYNER (R-Skirt) MADE IN NO.44 U.S.A. /[Number] SB
Aqua 1-2
Light Aqua 3-5

[020] (F-Skirt) GAYNER (R-Skirt) MADE IN NO.44 U.S.A./[Number] SDP
Aqua 1-2
Light Green 3-5

[030] (F-Skirt) GAYNER (R-Skirt) NO.44/ [Number] SB
Blue Aqua 3-5

HEMINGRAY

[010] (Dome) 1 (F-Skirt) HEMINGRAY-42 (R-Skirt) MADE IN U.S.A. SB
Hemingray Blue 20-30

[020] (Dome) 1 (F-Skirt) HEMINGRAY-42 (R-Skirt) MADE IN U.S.A. SDP
Hemingray Blue 1-2

[030] (F-Skirt) HEMIGRAY-42 {Note spelling} (R-Skirt) MADE IN U.S.A. SDP
Hemingray Blue 5-10

[040] (F-Skirt) HEMINGRAY {Note no '-42'} (R-Skirt) MADE IN U.S.A. RDP
Blue Aqua 5-10

[050] (F-Skirt) HEMINGRAY-42 RDP
Hemingray Blue 10-15

[053] (F-Skirt) HEMINGRAY-42 (R-Skirt) MADE IN U S A {Note no periods} RDP
Blue Aqua 5-10

87

CD 154

[057] (F-Skirt) HEMINGRAY-42
(R-Skirt) MADE IN U S A {Note no periods} SB
Hemingray Blue 20-30

[060] (F-Skirt) HEMINGRAY-42
(R-Skirt) MADE IN U S A {Note no periods} SDP
Hemingray Blue 5-10

[070] (F-Skirt) HEMINGRAY-42
(R-Skirt) MADE IN U.S.A. RDP
Aqua, Blue Aqua x1
Hemingray Blue, Ice Blue, Ice Green,
Light Blue, Light Green 1-2
Green 10-15
Dark Green 20-30
Aqua/Hemingray Blue Two Tone,
Blue/Green Two Tone 75-100
Clear/Blue Two Tone 150-175
Carnival 400-500

[080] (F-Skirt) HEMINGRAY-42
(R-Skirt) MADE IN U.S.A. SB
Hemingray Blue 20-30

[090] (F-Skirt) HEMINGRAY-42
(R-Skirt) MADE IN U.S.A. SDP
Aqua x1
Hemingray Blue 1-2

[095] (F-Skirt) HEMINGRAY-42 (R-Skirt) MADE IN U.S.A. {Note both 'E's are missing middle bar} RDP
Blue Aqua 3-5

[100] (F-Skirt) HEMINGRAY-42
(R-Skirt) MADE IN U.S.A./[Numbers and dots] RDP
Clear, Ice Blue, Straw, Yellow Tint x1

[110] (F-Skirt) HEMINGRAY-42/[Blotted out embossing] (R-Skirt) MADE IN U.S.A. SDP
Hemingray Blue 3-5

[120] (F-Skirt) HEMINGRAY-42/[Number and dots] (R-Skirt) MADE IN U.S.A./['HEMINGRAY-42' blotted out]/[Number] RDP
Clear 1-2

[130] (F-Skirt) HEMINGRAY-42/[Numbers and dots] (R-Skirt) MADE IN U.S.A. RDP
Clear, Green Tint, Ice Blue x1
Carnival 400-500

[140] (F-Skirt) HEMINGRAY-42/[Numbers and dots] (R-Skirt) MADE IN U.S.A./[Numbers] RDP
Clear, Ice Blue, Light Straw, Off Clear x1
Ice Yellow Green 1-2
White Milk 600-700
Opalescent 1,000-1,250

[143] (F-Skirt) HEMINGRAY-42/['MADE IN U.S.A.' blotted out] (R-Skirt) MADE IN U.S.A. RDP
Green Aqua 1-2

[147] (F-Skirt) HEMINGRAY-4S {Note backwards number} (R-Skirt) MADE IN U.S.A. RDP
Hemingray Blue 5-10

[150] (F-Skirt) HEMINGRAY-4S {Note backwards number} (R-Skirt) MADE IN U.S.A. SDP
Aqua 5-10

[160] (F-Skirt) MR/HEMINGRAY-42/[Numbers and dots] (R-Skirt) MADE IN U.S.A./[Number] RDP
Clear 500-600

[170] (F-Skirt) ['HEMINGRAY-42' blotted out] (R-Skirt) MADE IN U.S.A. {Many varieties of blot outs on the front skirt} RDP
Hemingray Blue 5-10

[180] (F-Skirt) ['HEMINGRAY-42' blotted out] (R-Skirt) MADE IN U.S.A. {Many varieties of blot outs on the front skirt} SDP
Hemingray Blue 5-10

LYNCHBURG Ⓛ

[010] (Dome) Ⓛ (F-Skirt) LYNCHBURG (R-Skirt) MADE IN NO.44 U.S.A./[Numbers] RDP
Aqua 3-5
Blue 10-15
Green 15-20

[020] (Dome) Ⓛ (F-Skirt) LYNCHBURG (R-Skirt) MADE IN NO.44 U.S.A./[Number] SDP
Blue Aqua, Light Aqua 3-5
Blue 15-20

[030] (Dome) Ⓛ (F-Skirt) LYNCHBURG/ ['GAYNER' blotted out] (R-Skirt) MADE IN NO.44 U.S.A./[Numbers] SDP
Light Blue Aqua 5-10
Light Blue Aqua w/ Milk Swirls 50-75

[033] (F-Skirt) LYNCHBURG (R-Skirt) MADE IN NO.44 U.S.A. SDP
Green Aqua 5-10
Sage Green 15-20
Olive Green 30-40

[035] (F-Skirt) LYNCHBURG (R-Skirt) MADE IN NO.44 U.S.A./[Number] SDP
Blue Aqua 5-10

CD 154

[037] (F-Skirt) LYNCHBURG (R-Skirt) MADE IN NO.ᔕᔕ U.S.A. {Note backwards number} SDP
Aqua 5-10

[040] (F-Skirt) LYNCHBURG (R-Skirt) [Number]/MADE IN NO.44 U.S.A. RDP
Lime Green 10-15
Sage Aqua, Sage Green, Smoke 15-20
Pink Tint 20-30
Yellow Green 30-40

[050] (F-Skirt) LYNCHBURG (R-Skirt) [Number]/MADE IN NO.44 U.S.A. SDP
Aqua, Light Aqua 3-5
Blue, Green Aqua 10-15
Green, Ice Green 15-20
Ice Blue, Smoky Straw, Yellow Green 20-30
Dark Green 30-40
Olive Green 40-50
Aqua w/ Milk Swirls 50-75

[053] (F-Skirt) Ⓛ/LYNCHBURG (R-Skirt) MADE IN NO.44 U.S.A. RDP
Sage Green 15-20
Yellow Green 30-40

[057] (F-Skirt) Ⓛ/LYNCHBURG (R-Skirt) MADE IN NO.44 [Number] U.S.A. SDP
Aqua 5-10

[060] (F-Skirt) Ⓛ/LYNCHBURG (R-Skirt) MADE IN NO.ᔕᔕ U.S.A. {Note backwards number} RDP
Aqua 10-15

[070] (F-Skirt) Ⓛ/LYNCHBURG (R-Skirt) [Number]/MADE IN NO.44 U.S.A. RDP
Green Aqua 3-5
Ice Green 10-15
Green, Ice Blue, Sage Aqua 15-20
Gingerale, Pink Tint, Smoky Straw 20-30

[080] (F-Skirt) Ⓛ/LYNCHBURG (R-Skirt) [Number]/MADE IN NO.44 U.S.A. {'[Number]' may appear above or below 'MADE IN NO. 44 U.S.A.'} SDP
Light Aqua 3-5
Blue Aqua 5-10
Blue 10-15
Green, Ice Green 15-20
Gingerale, Ice Blue, Light Blue, Smoke, Straw 20-30
7-up Green, Yellow Green 30-40
Light Aqua w/ Milk Swirls, Teal Blue 50-75

[090] (F-Skirt) Ⓛ/[Number] LYNCHBURG (R-Skirt) MADE IN NO.44 U.S.A. RDP
Sage Green 10-15
Light Green, Sage Aqua 15-20
Pink Tint, Smoky Straw, Yellow Green 20-30
Clear, Olive Green 30-40

[100] (F-Skirt) Ⓛ/[Number] LYNCHBURG (R-Skirt) MADE IN NO.44 U.S.A. SDP
Blue Aqua, Light Aqua 5-10
Ice Blue, Straw 20-30

MAYDWELL

[010] (F-Skirt) MAYDWELL-42 (R-Skirt) U.S.A. RDP
Straw 1-2
Ice Green 3-5

[020] (F-Skirt) MAYDWELL-42 (R-Skirt) U.S.A. SDP
Off Clear, Straw 1-2
Clear, Green Straw, Green Tint, Ice Green, Yellow Straw 3-5
Gingerale, Ice Yellow Green, Light Peach 5-10
Pink 10-15

[030] (F-Skirt) MAYDWELL-42 (R-Skirt) U.S.A./ ['MᶜLAUGHLIN' blotted out] RDP
Straw 1-2

[040] (F-Skirt) MAYDWELL-42/['MᶜLAUGHLIN' blotted out] (R-Skirt) U.S.A. SDP
Straw 1-2
Clear 3-5
Light Yellow Green 5-10
Pink 10-15

MᶜLAUGHLIN

[010] (Dome) [Number] (F-Skirt) MᶜLAUGHLIN (R-Skirt) № 42 RDP
Aqua, Dark Aqua, Ice Green, Light Aqua, Light Green 1-2
Blue Gray 5-10
Green 10-15
Apple Green, Lime Green 30-40

[020] (Dome) [Number] (F-Skirt) MᶜLAUGHLIN (R-Skirt) № 4Ƨ {Note backwards number} RDP
Aqua, Light Aqua, Light Green 3-5
Apple Green 30-40
Lime Green 40-50
Emerald Green 50-75

[023] (Dome) [Number] (F-Skirt) MᶜLAUGHLIN - 42 SDP
Light Aqua, Mint Green 100-125

[027] (F-Skirt) 6/MᶜLAUGHLIN-42 RDP
Aqua, Green, Light Green, Sage Green 10-15
Apple Green, Lime Green 30-40
Emerald Green, Olive Green 50-75

89

CD 154

[060] (F-Skirt) M̲c̲LAUGHLIN (R-Skirt) NO. 42 SB
Aqua, Sage Green 5-10
Blue Aqua 10-15
Forest Green 30-40
Delft Blue 40-50

[030] (F-Skirt) M̲c̲LAUGHLIN (R-Skirt) № 40 {Note '№ 40' style number} RDP
Blue 40-50

[040] (F-Skirt) M̲c̲LAUGHLIN (R-Skirt) № 40 {Note '№ 40' style number} SB
Aqua, Green Aqua, Sage Green, Steel Blue 10-15
Steel Gray 15-20

[050] (F-Skirt) M̲c̲LAUGHLIN (R-Skirt) № 42 RDP
Light Aqua, Light Blue, Light Green,
Sage Green 1-2
Light Steel Blue 5-10
Blue Gray, Dark Steel Blue 10-15
Delft Blue, Light Cornflower Blue 40-50

[065] (F-Skirt) M̲c̲LAUGHLIN (R-Skirt) № 4ᔕ {Note backwards number} RDP
Aqua 3-5

[070] (F-Skirt) M̲c̲LAUGHLIN - (R-Skirt) № 40 {Note dash and '№ 40' style number} SB
Green Aqua 15-20

[080] (F-Skirt) M̲c̲LAUGHLIN 42 SB
Aqua, Blue, Light Green, Sage Green 5-10

[090] (F-Skirt) M̲c̲LAUGHLIN-42 FDP
Aqua, Light Green 1-2
Green 5-10
Yellow Green 30-40
Delft Blue, Lime Green 40-50
Emerald Green, Olive Green 50-75

[100] (F-Skirt) M̲c̲LAUGHLIN-42 RDP
Apple Green 40-50
Emerald Green 50-75

[110] (F-Skirt) M̲c̲LAUGHLIN-42 (R-Skirt) MADE IN U.S.A. RDP
Green Aqua, Light Aqua,
Light Green, Sage Green 20-30
Blue Gray, Green, Off Clear 30-40

NO EMBOSSING

[010] [No embossing] {Gayner/Lynchburg product} RDP
Gingerale 50-75

NO NAME

[010] (F-Skirt) MADE IN NO.44 U.S.A./ [Number] {Gayner/Lynchburg product} SDP
Light Aqua 20-30
Gingerale 30-40

[020] (F-Skirt) MADE IN U.S.A. (R-Skirt) MADE IN U.S.A. {Hemingray product} RDP
Aqua 30-40

[030] (F-Skirt) MADE IN U.S.A. (R-Skirt) MADE IN U.S.A. {Hemingray product} SDP
Hemingray Blue 30-40

WHITALL TATUM ▽

[003] (F-Skirt) WHITALL TATUM CO. № 1 (R-Skirt) MADE IN U.S.A. SB
Aqua x1
Purple 20-30

[007] (F-Skirt) WHITALL TATUM CO. № 1 (R-Skirt) MADE IN U.S.A. ▽ SB
Peach 5-10

[010] (F-Skirt) WHITALL TATUM CO. № 1 (R-Skirt) MADE IN U.S.A./[Number] SB
Aqua, Light Aqua, Light Green Aqua x1
Light Sage Green, Pink 10-15
Purple 20-30

[020] (F-Skirt) WHITALL TATUM CO. № 1 (R-Skirt) [Number] MADE IN U.S.A. SB
Aqua, Blue Aqua, Light Aqua x1
Light Pink 10-15

[025] (F-Skirt) WHITALL TATUM CO. № 1 (R-Skirt) [Number] MADE IN U.S.A. ▽ SB
Clear, Off Clear x1

[030] (F-Skirt) WHITALL TATUM CO. № 1 (R-Skirt) [Number]/MADE IN U.S.A. ▽ SB
Light Aqua, Straw x1
Pink 10-15

[040] (F-Skirt) WHITALL TATUM CO. № 1/ [Number] (R-Skirt) MADE IN U.S.A. SB
Blue Aqua x1

[050] (F-Skirt) WHITALL TATUM CO. № 1/ [Number] (R-Skirt) MADE IN U.S.A. ▽ SB
Light Aqua, Light Green x1
Clear, Ice Aqua, Straw 1-2
Peach 5-10
Pink 10-15
Carnival, Light Aqua w/ Carnival 350-400
Dark Red Amber 2,000-2,500

[060] (F-Skirt) WHITALL TATUM CO. ▽ (R-Skirt) № 1. MADE IN U.S.A. SB
Light Aqua x1

[070] (F-Skirt) WHITALL TATUM CO. ▽ (R-Skirt) № 1. MADE IN U.S.A./[Number] SB
Aqua x1

[080] (F-Skirt) WHITALL TATUM CO. ▽ (R-Skirt) [Number]/№ 1 MADE IN U.S.A. SB
Light Aqua x1
Light Lavender 20-30

CD 155

[090] (F-Skirt) WHITALL TATUM № 1
(R-Skirt) MADE IN U.S.A. 〈W〉 SB
Clear x1
Pink Tint 10-15

CD 154.5

DERFLINGHER

[010] (F-Skirt) DERFLIGHER-T.N.-1 {Note spelling} RDP
Aqua 20-30
Green 40-50

[020] (F-Skirt) DERFLINGHER-T.N.-1 RDP
Aqua, Dark Green Aqua, Green,
Green Aqua 15-20
Dark Green 20-30

[030] (F-Skirt) DERFLINGHER-T.N.-1 SDP
Yellow Green 175-200

NO EMBOSSING - MEXICO

[010] [No embossing] RDP
Aqua, Green Aqua, Sage Green 20-30
Yellow Green 100-125
Yellow Olive Green 200-250

CD 155

ARMSTRONG Ⓐ

[010] (F-Skirt) Ⓐrmstrong DP 1 (R-Skirt) MADE IN U.S.A. Ⓐ [Numbers and dots] SB
Off Clear, Straw x1

[020] (F-Skirt) Ⓐrmstrong DP 1 (R-Skirt) MADE IN U.S.A. [Numbers and dots] SB
Clear x1
Blue Tint, Light Smoky Straw 1-2

[030] (F-Skirt) *Armstrong's* DP 1
(R-Skirt) MADE IN U.S.A. Ⓐ [Numbers and dots] SB
Clear, Light Smoke, Pink Tint, Straw x1
Green Smoke, Light Green 1-2
Light Olive Green 3-5

[040] (F-Skirt) 〈KPP〉 [Number] Armstrong D P 1 (R-Skirt) MADE IN U.S.A. [Numbers and dots] SB
Green Smoke, Off Clear 3-5

CRIMSA

[010] (F-Skirt) CRIMSA 45
(R-Skirt) [Letters]-[Number] SB
Clear 10-15
Ice Green 30-40

CRISOL TEXCOCO 〈T〉

[010] (F-Skirt) CRISOL 〈T〉 /TEXCOCO
(R-Skirt) 45 SB
Emerald Green 75-100

DIAMOND ◇

[010] (F-Skirt) ◇/[Number] SB
Off Clear 3-5

DOMINION ◇

[010] (F-Skirt) DOMINION 42 (R-Skirt) ◇ [Number]/[Number] SB
Clear, Off Clear, Straw 1-2
Ice Aqua, Ice Blue 3-5
Light Green Aqua 5-10

[013] (F-Skirt) DOMINION 42 (R-Skirt) ◇/ [Number] {Number may appear on any side of '◇'} SB
Clear, Straw 1-2

[017] (F-Skirt) DOMINION-42 (R-Skirt) ◇ SB
Light Pink 3-5

[020] (F-Skirt) DOMINION-42 (R-Skirt) ◇ [Number] (Top of pinhole) [Number] SB
Light Green 3-5

[030] (F-Skirt) DOMINION-42 (R-Skirt) ◇/ [Number and dots] SB
Clear, Light Straw 1-2
Light Green, Light Peach, Light Sage Green ... 3-5
Light Aqua 5-10

[040] (F-Skirt) DOMINION-42 (R-Skirt) ◇/ [Number] {Number may appear on any side of '◇'} SB
Light Peach 3-5

HEMINGRAY

[010] (F-Skirt) HEMINGRAY {Note no '-45'}
(R-Skirt) MADE IN U.S.A./[Numbers and dots] SB
Clear 5-10

[015] (F-Skirt) HEMINGRAY-42 {Note '-42'}
(R-Skirt) MADE IN U.S.A. RDP
Ice Blue 200-250

[020] (F-Skirt) HEMINGRAY-45
(R-Skirt) MADE IN U.A.S./[Numbers and dots] {Note spelling} SB
Clear 5-10

91

CD 155

[040] (F-Skirt) HEMINGRAY-45
(R-Skirt) MADE IN U.S.A./[Numbers and dots]
CB
Clear, Ice Blue, Ice Green, Straw x1
Light Lemon, Pink 1-2

[030] (F-Skirt) HEMINGRAY-45
(R-Skirt) MADE IN U.S.A./[Numbers and dots]
SB
Clear, Ice Green x1

[045] (F-Skirt) HEMINGRAY-45
(R-Skirt) MADE IN U.S.A./[Numbers and dots]
{'U' is engraved over an 'S'} CB
Clear x1

[050] (F-Skirt) HEMINGRAY-45
(R-Skirt) [Numbers and dots]/MADE IN U.S.A.
SB
Clear, Ice Blue, Off Clear 10-15

[060] (F-Skirt) HEMINGRAY-45
(R-Skirt) [Numbers] SB
Blue Tint 10-15

[070] (F-Skirt) H∃MINGRAY-45 {Note backwards letter} (R-Skirt) MADE IN U.S.A./
[Numbers and dots] SB
Clear 5-10

KERR

[010] (F-Skirt) KERR DP 1 (R-Skirt) MADE IN U.S.A. [Numbers and dots] SB
Blue Tint, Clear, Off Clear 1-2
Clear (Irridized) 125-150
Clear w/ Cobalt Blue Blobs 500-600

[020] (F-Skirt) KERR DP 1 (R-Skirt) MADE IN U.S.A. [Numbers and dots] {Flat dome variant} SB
Clear 100-125

NO EMBOSSING

[010] [No embossing] {Flat dome variant} {Kerr product} SB
Clear 100-125

NO EMBOSSING - MEXICO

[010] [No embossing] SB
Green Aqua 10-15
Emerald Green 50-75
Dark Yellow Green, Olive Green 75-100
Olive Amber 200-250

NO NAME - MEXICO

[010] (F-Skirt) 3 SB
Light Aqua 20-30
Emerald Green 100-125

[020] (F-Skirt) C 45 SB
Clear 30-40

P.S.S.A. ⓢ

[010] (F-Skirt)ⓢ/PSSA 42 RDP
Light Olive Green 30-40
Purple 400-500

[020] (F-Skirt) ⓢ/PSSA 42
(R-Skirt) DCE-2930 RDP
Ice Green, Smoky Sage Green 30-40

R.Y T. ⓡ

[010] (F-Skirt) ⓡ (R-Skirt) 45 SB
Aqua 15-20
Green 30-40

[020] (F-Skirt) ['ⓡ' upside down in a circle]
(R-Skirt) ['▽' entwined upside down in a triangle] SB
Green Aqua, Light Sage Green 15-20
Lime Green 20-30
Emerald Green 50-75
Ice Blue, Yellow Green 75-100
Cornflower Blue, Honey Amber 1,250-1,500

WHITALL TATUM ▽

[005] (F-Skirt) (Arc)WHITALL TATUM/№ 1
(R-Skirt) (Arc)MADE IN U.S.A./[Numbers]/Ⓐ
SB
Clear, Light Straw, Off Clear, Pink Tint x1

[010] (F-Skirt) (Arc)WHITALL TATUM/№ 1
(R-Skirt) (Arc)MADE IN U.S.A./[Numbers]/Ⓐ/
[Number] SB
Blue Tint, Clear, Off Clear, Pink Tint,
Smoky Straw, Straw x1
Smoky Olive Green 1-2

[020] (F-Skirt) WHITALL TATUM CO. № 1/
[Number] (R-Skirt) MADE IN U.S.A. ▽ SB
Clear, Off Clear, Straw x1

[030] (F-Skirt) WHITALL TATUM № 1/
[Number] (R-Skirt) MADE IN U.S.A. ▽ SB
Clear, Green Tint, Straw x1

[040] (F-Skirt) WHITALL TATUM № 1/
[Number] (R-Skirt) MADE IN U.S.A. ▽/
[Number] SB
Clear, Light Straw, Straw x1

CD 157

CD 155.6

HEMINGRAY

CD 155.6
$3 \, ^3/_8$ x 4

[010] (F-Skirt) HEMINGRAY SB
Aqua 7,500-10,000
Emerald Green 10,000-15,000

CD 156

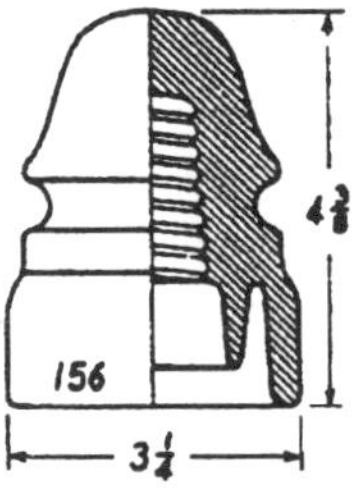

AM. INSULATOR CO.

[010] (Base) AM INSULATOR CO/PAT SEP 13 1881. BE
Aqua, Blue, Blue Aqua, Green Aqua,
Light Aqua, Light Green 30-40

POSTAL

[010] (Base) POSTAL TELEGRAPH CO PAT SEP 1Ɛ 1881 {Note backwards number} BE
Aqua, Light Green 30-40

CD 156.1

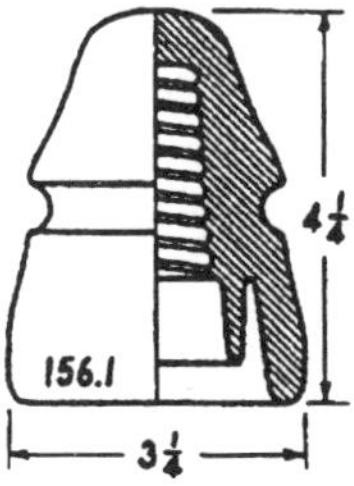

AM. INSULATOR CO.

[010] (Base) AM INSULATOR CO/PAT SEP 13 1881. BE
Blue Aqua, Ice Blue, Ice Green, Light Aqua,
Light Blue Aqua, Light Green 20-30

[020] (Base) AM. INSULATOR CO. N.Y. DOUBLE PETTICOAT PAT'D SEPT. 13. 1881 NOV. 13. 1883 BE
Aqua, Blue Aqua, Light Aqua,
Light Green Aqua 20-30
Green 350-400
Lime Green 400-500
Apple Green 500-600
Yellow Green 1,500-1,750
Olive Green 2,500-3,000

NO EMBOSSING

[010] [No embossing] SB
Light Blue 30-40

PATENT - OTHER

[010] (Base) PAT SEP 13 1881. BE
Blue Aqua, Light Aqua, Light Green 30-40

[020] (Base) PAT SEP {Note no day or year in date} BE
Light Aqua 50-75

POSTAL

[010] (Base) POSTAL TELEGRAPH CO. PAT SEP 1Ɛ 1881 {Note backwards number} BE
Aqua, Light Green 30-40

CD 156.2

CD 156.2
3 x $3 \, ^7/_8$

NO EMBOSSING

[010] [No embossing] SB
Blue Aqua 2,000-2,500

PATENT - OTHER

[010] (Base of inner skirt) PAT OCT 15 1872 SB
Aqua, Blue Aqua 2,000-2,500

CD 157

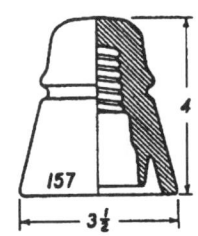

BROOKFIELD

[010] (F-Skirt) BROOKFIELD SB
Aqua, Dark Aqua 15-20

HEMINGRAY

[010] (F-Skirt) HEMINGRAY/N⁰ 38 (R-Skirt) PATENTED/MAY 2 1893 SDP
Aqua, Dark Aqua 10-15

CD 157.5

CD 157.5

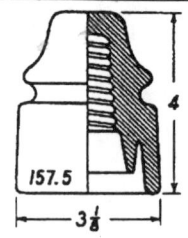

STANDARD GLASS INSULATOR CO.

[010] (Base) THE STANDARD GLASS INSULATOR CO./BOSTON MASS BE
Blue Aqua, Light Aqua 50-75
Light Green 600-700

CD 158

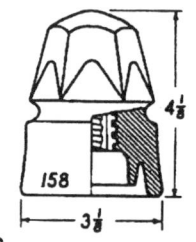

BOSTON BOTTLE WORKS

[010] (Base of inner skirt) BOSTON BOTTLE WORKS/PATENT APPLIED FOR {Four segmented threads} {Narrow wire groove} {MLOD} SB
Aqua, Light Blue Aqua 600-700
Green Aqua 700-800

[020] (Base of inner skirt) BOSTON BOTTLE WORKS/PATENT APPLIED FOR {Four segmented threads} {Wide wire groove} {MLOD} SB
Aqua, Blue Aqua 500-600
Dark Aqua, Green Aqua 700-800
Light Green, Light Lime Green 1,250-1,500
Dark Red Amber,
Dark Yellow Amber 7,500-10,000

NO EMBOSSING

[010] [No embossing] {Four segmented threads} {Wide wire groove} {MLOD} SB
Aqua 500-600

PATENT - OTHER

[010] (Base of inner skirt) PAT OCT 15 1872 {Three segmented threads} {This is a CD 136.7 with an inner skirt} SB
Aqua 1,500-1,750

CD 158.1

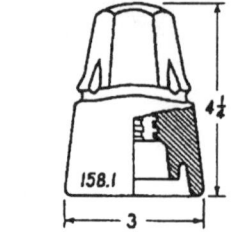

CHESTER

[010] (Base of inner skirt) CHESTER.104.CENTRE ST. N.Y./PATENTED {Four segmented threads} {MLOD} SB
Aqua 3,000-3,500
Blue Aqua 3,500-4,000

NO EMBOSSING

[010] [No embossing] {Four segmented threads} {MLOD} SB
Blue 2,500-3,000

CD 158.2

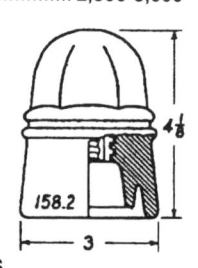

BOSTON BOTTLE WORKS

[010] (Base of inner skirt) BOSTON BOTTLE WORKS/-PATENT APPLIED FOR- {MLOD} {Four segmented threads} SB
Aqua, Blue Aqua 800-1,000
Light Yellow Green 1,750-2,000
Lime Green 2,000-2,500

[020] (Base of inner skirt) BOSTON BOTTLE WORKS/PATENT APLD FOR {MLOD} {Four segmented threads} SB
Aqua, Blue, Blue Aqua 800-1,000
Dark Green Aqua 1,000-1,250
Green 1,750-2,000
Honey Amber, Olive Amber,
Root Beer Amber 7,500-10,000

[030] (Base of inner skirt) BOSTON BOTTLE WORKS/PATENT OCT 15 72 {MLOD} {Three segmented threads} SB
Aqua 1,250-1,500
Yellow Green 4,000-4,500

[040] (Base of inner skirt) BOSTON BOTTLE WORKS/PATENT.OCT.15.72 {MLOD} {Four segmented threads} SB
Aqua, Light Aqua 800-1,000
Dark Green 3,000-3,500

[050] (Base of inner skirt) BOSTON BOTTLE WORKS/PATENTED OCT 15 72 {MLOD} {Four segmented threads} SB
Aqua 800-1,000

[060] (Crown) BOSTON BOTTLE WORKS PATENTED OCT 15 72 {MLOD} {Four segmented threads} SB
Aqua 1,000-1,250
Dark Green 2,500-3,000

[070] (Crown) [Blotted out embossing] (Base of inner skirt) BOSTON BOTTLE WORKS/ PATENT.OCT. 15. 72 {Four segmented threads} {MLOD} SB
Dark Green 2,500-3,000

NO EMBOSSING

[010] [No embossing] {Four segmented threads} SB
Aqua, Dark Green Aqua 700-800

CD 158.9

BOSTON BOTTLE WORKS

[010] (Base of inner skirt) BOSTON BOTTLE WORKS/-PATENT APPLIED FOR- (Top of threaded projection) {May have an embossed triangle} {Four segmented threads} SB
Aqua, Blue Aqua, Dark Aqua,
Ice Aqua, Ice Green,
Light Aqua, Light Celery Green 5,000-7,500
Dark Green 10,000-15,000
Root Beer Amber .... 20,000+

CD 159

GREGORY

[010] (F-Skirt) PAT.APR. 7TH 1903/F.W. GREGORY SB
Aqua, Green Aqua 1,750-2,000

CD 160

ARMSTRONG Ⓐ

[010] (F-Skirt) *Armstrong's* NO.14 (R-Skirt) MADE IN U.S.A. Ⓐ [Numbers and dots] SB
Clear 10-15

B

[010] (F-Skirt) B SB
Aqua 3-5

[020] (F-Skirt) B SDP
Aqua 5-10

[030] (F-Skirt) B (R-Skirt) № 32 SB
Aqua 3-5
Blue Aqua 5-10
Blue, Green 15-20
Olive Green 40-50
Olive Amber 75-100

BROOKFIELD

[010] (Dome) [Number] (F-Skirt) BROOKFIELD (R-Skirt) NEW YORK SB
Aqua, Light Aqua, Light Green Aqua 1-2
Green 3-5
Lime Green 10-15
Yellow Green 15-20

[015] (Dome) [Number] (F-Skirt) BROOKFIELD (R-Skirt) NEW YORK SDP
Green 10-15

[020] (F-Crown) (Arc)WM BROOKFIELD/NEW YORK (F-Skirt) PAT'D NOV. 13TH, 1883 {MLOD} SB
Aqua, Light Aqua 3-5

[030] (F-Skirt) BROOKFIELD SDP
Aqua 3-5

[040] (F-Skirt) BROOKFIELD (R-Skirt) NEW YORK SB
Aqua, Blue Aqua, Dark Aqua,
Green Aqua, Light Aqua 1-2
Green 3-5
Emerald Green 5-10
Yellow Green 15-20
Olive Green 20-30
Olive Amber 50-75

CD 160

[050] (F-Skirt) BROOKFIELD (R-Skirt) NEW YORK SDP
Aqua, Dark Aqua 3-5
Green 10-15
Emerald Green 20-30
Olive Green 30-40
Milky Yellow Green 150-175

[060] (F-Skirt) BROOKFIELD/N.Y. (R-Skirt) PAT'D NOV. 13TH 1883 SB
Aqua, Light Aqua 3-5
Light Green 20-30

[070] (F-Skirt) W. BROOKFIELD/N.Y. (R-Skirt) PAT'D.NOV. 13TH 1883 SB
Aqua, Light Aqua, Light Blue 3-5
Light Green 20-30
Yellow Green 50-75
Emerald Green 150-175

CALIFORNIA

[010] (F-Skirt) CALIFORNIA SB
Gray, Light Green, Light Purple,
Off Clear, Smoke 10-15
Smoky Purple, Smoky Straw 15-20
Blue, Blue Aqua, Green, Purple 20-30
Burgundy 40-50
Purple/Sage Two Tone, Yellow 125-150

GAYNER

[010] (F-Skirt) GAYNER (R-Skirt) NO. 140/ [Number] SDP
Aqua, Light Aqua, Light Blue Aqua 15-20

H.G.CO.

[005] (F-Skirt) H G CO/PATENT MAY 2 1893 {Note no periods} (R-Skirt) PETTICOAT SDP
Aqua 3-5

[010] (F-Skirt) H.G.CO. (R-Skirt) PETTICOAT RDP
Lime Green 100-125

[020] (F-Skirt) H.G.CO. (R-Skirt) PETTICOAT SB
Aqua, Blue Aqua, Green Aqua 20-30
Green Tint, Light Blue 30-40
Light Apple Green, Lime Green 50-75
Blue, Clear 75-100

[030] (F-Skirt) H.G.CO. (R-Skirt) PETTICOAT SDP
Aqua, Blue Aqua 40-50
Clear, Light Green 75-100

[120] (F-Skirt) H.G.CO./A (R-Skirt) PETTICOAT SB
Aqua, Light Aqua 20-30
Ice Steel Blue, Light Blue,
Light Green, Off Clear 30-40
Clear, Light Lemon 75-100
Cornflower Blue 250-300

[130] (F-Skirt) H.G.CO./A (R-Skirt) PETTICOAT SDP
Blue Aqua, Light Aqua, Light Green Aqua ... 40-50
Light Green 75-100

[040] (F-Skirt) H.G.CO./PATENT MAY 2 1893 (R-Skirt) PETTICOAT RDP
Aqua, Green Aqua 15-20
Hemingray Blue 20-30

[050] (F-Skirt) H.G.CO./PATENT MAY 2 1893 (R-Skirt) PETTICOAT SDP
Aqua x1
Hemingray Blue 3-5
Light Green 20-30
Ice Aqua, Ice Blue 75-100
Lime Green 100-125
Milky Aqua 200-250
Light Jade Green Milk 250-300
Jade Blue Milk 700-800
Depression Glass Green 800-1,000

[060] (F-Skirt) H.G.CO./PATENT MAY 2 1893 (R-Skirt) PETTICOAT {Smooth base mold; transition style embossing} SDP
Aqua 15-20
Dark Olive Green,
Yellow Olive Green 1,500-1,750

[070] (F-Skirt) H.G.CO./PATENT MAY 2 1893 (R-Skirt) PETTICOAT {'C' is engraved over an 'O'} SDP
Aqua 3-5

[080] (F-Skirt) H.G.CO./PATENT MAY 2 1893 (R-Skirt) PETTICOAT {'E' is engraved over an '8'} SDP
Aqua 3-5

[085] (F-Skirt) H.G.CO./PATENT MAY 2 1893 (R-Skirt) PETTICOAT {'E' is engraved over an 'A'} SDP
Aqua 3-5
Light Jade Green Milk 250-300

[090] (F-Skirt) H.G.CO./PATENT MAY 2 1893 (R-Skirt) PETTICOET {Note spelling} SDP
Aqua, Green Aqua 5-10
Hemingray Blue 10-15

CD 160

[100] (F-Skirt) H.G.CO./PATENT MAY 2 1893
{'P' is engraved over an 'M'}
(R-Skirt) PEITTICOAT {Note spelling} SDP
Aqua 3-5
Light Green 20-30
Green 75-100
Light Jade Green Milk 250-300

[110] (F-Skirt) H.G.CO./PATTENT MAY 2 1893
{Note spelling} (R-Skirt) PETTICOAT SDP
Aqua 5-10

[140] (F-Skirt) H.G.CO./['S.S & Co. MFG'S./
CHICAGO' blotted out]/Nº 40
(R-Skirt) DOUBLE/PETTICOAT SB
Aqua, Light Aqua 200-250

HAWLEY ⌸

[010] (F-Skirt) HAWLEY P.A./U.S.A.
(R-Skirt) ⌸ SB
Aqua, Light Aqua 125-150

HEMINGRAY

[005] (Dome) [Number]
(F-Skirt) HEMINGRAY-14/[Number]
(R-Skirt) MADE IN U.S.A. RDP
Blue Tint 1-2

[010] (F-Skirt) HEMINGRAY (R-Skirt) Nº 14
RDP
Aqua 1-2

[020] (F-Skirt) HEMINGRAY (R-Skirt) Nº 14
SDP
Aqua, Blue Aqua, Dark Aqua 1-2

[030] (F-Skirt) HEMINGRAY-14
(R-Skirt) MADE IN U.S.A. RDP
Aqua, Blue Aqua, Blue Tint, Dark Aqua 1-2
Green Tint, Hemingray Blue, Ice Green 3-5
Light Green 5-10

[040] (F-Skirt) HEMINGRAY-14
(R-Skirt) MADE IN U.S.A. SDP
Aqua 1-2
Hemingray Blue 3-5

[050] (F-Skirt) HEMINGRAY-14
(R-Skirt) MADE IN U.S.A./[Numbers and dots]
RDP
Clear, Light Straw 1-2

[060] (F-Skirt) HEMINGRAY-14
(R-Skirt) MADE IN U.S.A./['Nº 14' blotted out]
SDP
Blue Aqua 1-2
Hemingray Blue 3-5

[070] (F-Skirt) HEMINGRAY-14/PETTICOAT
(R-Skirt) MADE IN U.S.A./PATENT MAY 2
1893 SDP
Aqua 5-10
Hemingray Blue 10-15

[080] (F-Skirt) HEMINGRAY-14/PETTICOAT
(R-Skirt) MADE IN U.S.A./[Blotted out
embossing]/PATENT MAY 2 1893 SDP
Hemingray Blue 10-15

[090] (F-Skirt) HEMINGRAY-14/[Numbers and
dots] (R-Skirt) MADE IN U.S.A. RDP
Clear, Ice Aqua 1-2
Green Tint 3-5

[100] (F-Skirt) HEMINGRAY-14/[Numbers and
dots] (R-Skirt) MADE IN U.S.A./[Number] RDP
Clear 1-2
Ice Blue 3-5

[110] (F-Skirt) HEMINGRAY-14/[Number]
(R-Skirt) MADE IN U.S.A./['HEMINGRAY'
blotted out] RDP
Ice Blue 3-5

[120] (F-Skirt) HEMINGRAY-14/['H.G.CO.'
blotted out]/PATENT MAY 2 1893
(R-Skirt) MADE IN U.S.A./PETTICOAT SDP
Hemingray Blue 10-15

[130] (F-Skirt) HEMINGRAY-14/
['HEMINGRAY-14' blotted out] (R-Skirt) MADE
IN U.S.A./['MADE IN U.S.A.' blotted out]/
[Numbers and dots] {Smaller embossing
replaces large} RDP
Blue Tint, Clear, Light Straw 1-2

[140] (F-Skirt) HEMINGRAY/PATENT MAY 2
1893 (R-Skirt) MADE IN U.S.A./PETTICOAT
SDP
Blue Aqua 5-10
Hemingray Blue 10-15

[150] (F-Skirt) HEMINGRAY/['H.G.CO./
PATENT MAY 2 1893' blotted out]
(R-Skirt) MADE IN U.S.A./['PETTICOAT'
blotted out] SDP
Aqua 3-5

[160] (F-Skirt) HEMINGRAY/['H.G.CO.' blotted
out] (R-Skirt) MADE IN U.S.A. SDP
Hemingray Blue 10-15

[170] (F-Skirt) HEWINGRAY {Note spelling}
(R-Skirt) Nº 14 SDP
Aqua, Blue Aqua 5-10

[180] (F-Skirt) HEWINGRAY-14 {Note spelling}
(R-Skirt) MADE IN U.S.A./['Nº 14' blotted out]
SDP
Blue Aqua, Hemingray Blue 5-10

K.C.G.W.

[010] (F-Skirt) K C G W (R-Skirt) [Number] SB
Aqua, Ice Green, Light Green 75-100
Dark Green Aqua 200-250

LYNCHBURG Ⓛ

[010] (F-Skirt) Ⓛ/LYNCHBURG
(R-Skirt) NO.32/MADE IN [Number] U.S.A. RDP
Aqua, Light Aqua 15-20

[020] (F-Skirt) Ⓛ/LYNCHBURG
(R-Skirt) NO.32/MADE IN [Number] U.S.A. SDP
Aqua, Green Aqua, Light Aqua 15-20
Light Green 75-100

MAYDWELL

[010] (F-Skirt) MAYDWELL-14 (R-Skirt) U.S.A. FDP
Gingerale 5-10

[020] (F-Skirt) MAYDWELL-14 (R-Skirt) U.S.A. RDP
Clear, Smoke, Straw, Yellow Tint 3-5
Gingerale, Purple Tint 5-10

[030] (F-Skirt) MAYDWELL-14 (R-Skirt) U.S.A. SB
Clear, Straw 5-10

[040] (F-Skirt) MAYDWELL-14 (R-Skirt) U.S.A. SDP
Pink, Smoke, Straw, Yellow Tint 3-5

[050] (F-Skirt) MAYDWELL-14/['MᶜLAUGHLIN' blotted out] (R-Skirt) U.S.A. RDP
Light Smoke 3-5

MᶜLAUGHLIN

[010] (F-Skirt) MᶜLAUGHLIN (R-Skirt) №14 FDP
Ice Green 5-10
Amber Blackglass,
Olive Amber Blackglass,
Olive Green Blackglass 15-20

[020] (F-Skirt) MᶜLAUGHLIN (R-Skirt) №14 RDP
Aqua, Light Aqua, Light Blue, Light Green,
Sage Green, Steel Blue 5-10
Gray Green, Ice Blue, Light Sky Blue 10-15
Amber Blackglass, Mint Green,
Olive Amber Blackglass, Olive Blackglass .. 15-20
Olive Green 20-30
Apple Green 50-75

[030] (F-Skirt) MᶜLAUGHLIN (R-Skirt) №14 SDP
Green Aqua, Light Aqua, Light Blue,
Sage Green 5-10

NO NAME

[010] (F-Skirt) [Glass dot] (R-Skirt) [Glass dot] SB
Aqua 50-75

[020] (F-Skirt) [Glass dot] (R-Skirt) [Number] {Tall narrow dome} SB
Aqua 50-75

[030] (F-Skirt) [Number] (R-Skirt) [Blotted out embossing] SB
Light Green 75-100

[040] (F-Skirt) [Number] (R-Skirt) [Number] SB
Aqua, Light Green 75-100

[050] (F-Skirt) [Vertical bar] [Vertical bar] (R-Skirt) [Number] {K.C.G.W. product} SB
Aqua, Ice Green, Light Green 50-75
Green 125-150

O.V.G.CO.

[010] (F-Skirt) O.V.G.CO.
(R-Skirt) PETTICOAT PONY SB
Aqua, Light Blue Aqua 250-300

S.S.& CO.

[010] (F-Skirt) S.S & Co. MFG'S./CHICAGO/ Nº 40 (R-Skirt) DOUBLE/PETTICOAT SB
Clear, Green Tint, Off Clear 175-200
Ice Aqua 200-250
Ice Blue, Ice Green 250-300

STAR ✷

[010] (F-Skirt) ✷ SB
Aqua, Light Aqua 1-2
Blue Aqua, Dark Aqua, Green Aqua 3-5
Green, Light Blue 5-10
Blue 10-15
Lime Green, Yellow Green 15-20
Dark Yellow Green 20-30
Olive Green 30-40
Light Yellow Olive Green 75-100
Olive Amber 150-175

STERLING £

[010] (F-Skirt) STERLING (R-Skirt) £ SB
Aqua, Blue Aqua 15-20

WHITALL TATUM ▽

[010] (F-Skirt) WHITALL TATUM CO. Nº 14/ [Number] (R-Skirt) MADE IN U.S.A. ▽ SB
Clear, Straw 3-5

[020] (F-Skirt) WHITALL TATUM Nº 14/ [Number] (R-Skirt) MADE IN U.S.A. Ⓐ/ [Number and dots] SB
Clear, Pink Tint, Straw 3-5

CD 161

[030] (F-Skirt) WHITALL TATUM № 14 / [Number] (R-Skirt) MADE IN U.S.A. W SB
Clear, Straw 3-5

CD 160.4

AM.INSULATOR CO.

CD 160.4
2 3/4 x 3 3/4

[010] (Base) AM.INSULATOR CO.N.Y. DOUBLE PETTICOAT/PAT'D SEPT. 13, 1881 NOV. 13, 1883 {This insulator has the base of a CD 160.7 and the dome of a CD 134} {Previously shown as CD 162 in the 1995 edition} BE
Light Blue Aqua 500-600

CD 160.6

AM.TEL.& TEL.CO.

[010] (F-Crown) [Number] (F-Skirt) AM.TEL & TEL.CO. (R-Skirt) PATD.NOV. 13TH 1883. {MLOD} SB
Light Aqua 300-350

[020] (F-Skirt) AM.TEL & TEL.CO. (R-Crown) [Number] (R-Skirt) PATD.NOV. 13TH 1883. {MLOD} SB
Light Yellow Green 2,000-2,500

CD 160.7

AM.INSULATOR CO.

[010] (Base) AM INSULATOR CO.N.Y. DOUBLE PETTICOAT/PAT'D SEPT. 13, 1881, NOV. 13, 1883, FEB. 12, 1884 BE
Aqua, Blue Aqua 175-200
Green Aqua 200-250
Light Green 300-350
Lime Green 1,250-1,500
Apple Green 1,500-1,750
Olive Green, Yellow Olive Green 2,500-3,000

[015] (Base) AM INSULATOR CO.N.Y. DOUBLE PETTICOAT/PAT'D SEPT. 13, 1881, NOV. 13, 1883, FEB. 15, 1884 {Note 'FEB 15, 1884' date} BE
Lime Green 1,250-1,500

[020] (Base) AM.INSULATOR CO.N.Y. DOUBLE PETTICOAT/PAT'D SEPT. 13, 1881 NOV. 13, 1883 BE
Aqua, Blue Aqua, Light Blue 175-200
Green 700-800
Apple Green 1,500-1,750
7-up Green 5,000-7,500
Dark Orange Amber 15,000-20,000

CD 161

CALIFORNIA

[010] (Dome) A (F-Skirt) CALIFORNIA SB
Light Sage Green, Sage Green 3-5
Light Green, Smoke 5-10
Light Purple 10-15
Aqua, Blue Aqua 15-20
Green, Off Clear, Purple 20-30
Burgundy, Dark Purple, Plum 30-40
Peach, Smoky Straw 50-75
Yellow 200-250
Purple/Sage Two Tone,
Purple/Yellow Two Tone 250-300

[020] (F-Skirt) CALIFORNIA SB
Light Sage Green, Sage Green 3-5
Light Green, Smoke 5-10
Light Purple 10-15
Aqua, Blue Aqua 15-20
Green, Light Yellow Green,
Off Clear, Purple 20-30
Dark Purple, Plum 30-40
Smoky Straw 50-75
Peach, Purple/Green Two Tone, Yellow ... 200-250

[030] (F-Skirt) CALIFORNIA SDP
Light Purple, Sage Green,
Smoky Peach 1,750-2,000

STAR ✶

[010] (F-Skirt) ✶ SB
Aqua, Green Aqua 3-5
Dark Green, Lime Green 10-15
Yellow Green 15-20

[020] (F-Skirt) ✶ SDP
Aqua, Dark Aqua 10-15

99

CD 161.2

CD 161.2

BROOKFIELD

[010] (F-Skirt) BROOKFIELD SB
Aqua, Dark Aqua 10-15
Dark Green 20-30
Dark Yellow Green 50-75

[020] (F-Skirt) [Blotted out embossing]/
BROOKFIELD SB
Aqua 10-15

CD 162

B.G.M.CO.

[010] (F-Skirt) B.G.M.CO. SB
Light Purple 200-250
Purple 250-300
Clear, Pink 350-400

BROOKFIELD

[010] (Dome) [Letter or number]
(F-Skirt) BROOKFIELD SB
Green Aqua x1
Green 1-2

[020] (Dome) [Letter or number]
(F-Skirt) BROOKFIELD SDP
Aqua x1
Green 1-2

[030] (Dome) [Letter or number]
(F-Skirt) BROOKFIELD (R-Skirt) NEW YORK SB
Aqua, Dark Aqua x1
Dark Green, Green, Light Blue, Light Green 1-2
Emerald Green 3-5
Dark Yellow Amber 200-250

[040] (Dome) [Letter or number]
(F-Skirt) BROOKFIELD (R-Skirt) NEW YORK SDP
Aqua, Dark Green Aqua x1
Green 1-2
Yellow Green 5-10
Ice Green 10-15

[050] (Dome) [Number] (F-Skirt) W.
BROOKFIELD (R-Skirt) NEW YORK SB
Light Aqua 1-2

[060] (F-Crown) (Arc)W. BROOKFIELD/45 CLIFF ST./N.Y. (F-Skirt) PAT'D NOV. 13TH 1883 (R-Crown) [Number] {MLOD} SB
Aqua 3-5

[070] (F-Crown) (Arc)W. BROOKFIELD/45 CLIFF ST./N.Y. (F-Skirt) PAT'D NOV. 13TH 1883 {MLOD} SB
Light Aqua, Light Blue 3-5

[080] (F-Crown) (Arc)W. BROOKFIELD/45 CLIFF ST./N.Y. (R-Crown) [Number] SB
Aqua 1-2
Lime Green 50-75

[085] (F-Crown) (Arc)W. BROOKFIELD/45 CLIFF ST./N.Y. SB
Yellow Green 300-350

[090] (F-Crown) (Arc)W. BROOKFIELD/NEW YORK (F-Skirt) PAT.FEB. 12TH 1884. (R-Crown) [Letter] {MLOD} SB
Light Blue 5-10

[100] (F-Crown) (Arc)W. BROOKFIELD/NEW YORK (F-Skirt) PAT.FEB. 12TH 1884. {MLOD} SB
Aqua, Blue Aqua, Green Aqua, Light Blue 5-10
Light Green 15-20
Dark Lime Green 125-150

[135] (F-Crown) (Arc)WM. BROOKFIELD/83 FULTON ST./N.Y. (F-Skirt) PAT'D NOV. 13 1883 {MLOD} SB
Aqua 3-5

[140] (F-Crown) (Arc)WM. BROOKFIELD/83 FULTON ST./N.Y. (F-Skirt) PAT'D NOV. 13TH 1883 {MLOD} SB
Aqua 3-5

[110] (F-Crown) (Arc)WM BROOKFIELD/83 FULTON ST./N.Y. (F-Skirt) PAT'D NOV. 13TH 1883 (R-Crown) [Number] {MLOD} SB
Light Aqua 3-5

[120] (F-Crown) (Arc)WM BROOKFIELD/NEW YORK (F-Skirt) PATD NOV. 13TH 1883 (R-Crown) [Number] {MLOD} SB
Aqua 3-5
Light Green 15-20

[130] (F-Crown) (Arc)WM BROOKFIELD/NEW YORK (F-Skirt) PAT'D NOV. 13TH 1883 {MLOD} SB
Aqua, Light Aqua, Light Blue 3-5
Light Green 15-20
Dark Lime Green 125-150

[150] (F-Crown) [Number]
(F-Skirt) BROOKFIILD {Note spelling} SB
Aqua 5-10

100

CD 162

[160] (F-Skirt) BROOKFIELD SB
Aqua x1

[170] (F-Skirt) BROOKFIELD SDP
Aqua x1

[180] (F-Skirt) BROOKFIELD (R-Skirt) NEW YORK SB
Light Aqua x1
Dark Green, Green, Light Blue 1-2
Dark Yellow Green, Ice Blue 3-5
Lime Green 10-15
Light Purple 100-125
Olive Amber 175-200

[190] (F-Skirt) BROOKFIELD (R-Skirt) NEW YORK/['B.G.M.CO.' blotted out] SB
Aqua, Light Blue Aqua 5-10

[200] (F-Skirt) BROOKFIELD (R-Skirt) №36 SB
Milky Green 200-250

[210] (F-Skirt) BROOKFIELD (R-Skirt) ['B.G.M.CO.' blotted out] SB
Aqua 5-10
Ice Blue 15-20

[215] (F-Skirt) BROOKFIELD {Narrow dome} SB
Aqua 10-15

[220] (F-Skirt) BROOKFIELD {Narrow dome} SDP
Aqua 15-20

[230] (F-Skirt) BROOKFIELD/[Number] SB
Aqua x1
Lime Green 10-15
Olive Green 20-30

[240] (F-Skirt) BROOKFIELD/[Number] (R-Skirt) №36 RB
Aqua, Blue Aqua 3-5
Green 10-15

[250] (F-Skirt) W. BROOKFIELD (R-Skirt) NEW YORK SB
Aqua x1
Light Steel Blue 5-10
Lime Green 10-15

CALIFORNIA

[010] (F-Skirt) CALIFORNIA SB
Sage Green 5-10
Gray, Green Aqua 10-15
Blue Aqua, Light Purple,
Light Yellow Green, Smoky Purple 15-20
Purple 20-30
Burgundy, Plum 30-40
Yellow 200-250
Purple/Aqua Two Tone,
Purple/Sage Two Tone 250-300

F.F.C.C.N.DE M.

[010] (Dome) ⚠ (F-Skirt) FF CC N DE M RDP
Clear 15-20

[020] (F-Skirt) FF CC N DE M RDP
Ice Green 15-20

[030] (F-Skirt) FF CC N DE M SDP
Ice Green, Light Green 15-20

FT.W.E.CO.

[010] (F-Crown) † (F-Skirt) FT.W.E.Co. (R-Skirt) PAT'D NOV. 13TH 1883 SB
Aqua 50-75

[020] (F-Skirt) FT.W.E.Co. (R-Skirt) PAT'D NOV. 13TH 1883 SB
Light Aqua 50-75

[030] (F-Skirt) FT W.E.Co. (R-Crown) † (R-Skirt) PAT'D NOV. 13TH 1883 {MLOD} SB
Light Aqua 50-75

GAYNER

[010] (F-Skirt) GAYNER (R-Skirt) NO.36-190 SDP
Aqua 3-5
Aqua w/ Milk Swirls 50-75

[020] (F-Skirt) GAYNER (R-Skirt) NO.36-190/[Number] SB
Aqua 3-5
Aqua w/ Milk Swirls 50-75

[030] (F-Skirt) GAYNER (R-Skirt) NO.36-190/[Number] SDP
Aqua 3-5
Sage Green 5-10

[035] (F-Skirt) ['Ⓛ' blotted out]/GAYNER (R-Skirt) NO.36-190/[Number] SDP
Aqua 10-15

GOOD

[010] (F-Skirt) R.GOOD JR. (R-Skirt) PETTICOAT SB
Aqua, Blue Aqua 10-15

[015] (F-Skirt) R.GOOD JR./DENVER, COLO. (R-Skirt) PETTICOAT SB
Aqua, Blue Aqua, Light Aqua 10-15

[020] (F-Skirt) R.GOOD JR./DENVER, COLO. (R-Skirt) [Glass dot]/PETTICOAT SB
Aqua, Blue Aqua 10-15
Gray, Light Purple 250-300
Light Sage Green, Off Clear, Purple 300-350
Dark Purple 350-400

101

CD 162

H.G.CO.

[010] (Dome) [Letter] (F-Skirt) H.G.CO. (R-Skirt) PETTICOAT SB
Aqua, Ice Green, Light Green Aqua 30-40
Dark Aqua, Light Aqua, Light Green 40-50
Light Lemon 75-100
Clear, Light Lime Green 200-250
Dark Orange Amber 1,000-1,250

[020] (Dome) [Letter] (F-Skirt) H.G.CO. (R-Skirt) PETTICOAT {Crown variant - top looks like a CD 134} SB
Aqua, Green Aqua 200-250

[030] (F-Skirt) H G CO/PATENT MAY 2 1893 {Note no periods} (R-Skirt) PETTICOAT SDP
Aqua, Blue Aqua, Green Aqua 5-10
Amber Two Tone, Whiskey Amber 200-250
Light Electric Blue 250-300
Peacock Blue, Purple 400-500

[040] (F-Skirt) H.G.CO. (R-Skirt) PETTICOAT SB
Aqua, Green Tint, Ice Green 30-40
Blue Aqua, Dark Aqua, Green Aqua,
Light Aqua, Light Green 40-50
Blue Tint, Ice Blue, Ice Steel Blue 50-75
Light Lemon 75-100
Light Blue 100-125
Clear, Lime Green 200-250
Hemingray Blue 250-300
Green 800-1,000
Dark Orange Amber,
Dark Red Amber,
Golden Amber, Orange Amber 1,000-1,250
Dark Yellow Amber 1,500-1,750
Depression Glass Green 2,000-2,500
Dark Peacock Blue,
Peacock Blue, Yellow 3,000-3,500
Light Purple 5,000-7,500
Purple 7,500-10,000

[050] (F-Skirt) H.G.CO. (R-Skirt) PETTICOAT {Crown variant - top looks like a CD 134} SB
Blue Aqua, Light Aqua, Light Blue Aqua .. 200-250
Light Green 300-350

[060] (F-Skirt) H.G.CO./PATENT MAY 2 1893 (R-Skirt) PETTICOAT RDP
Root Beer Amber 150-175

[070] (F-Skirt) H.G.CO./PATENT MAY 2 1893 (R-Skirt) PETTICOAT SDP
Aqua x1
Blue Aqua 3-5
Light Aqua, Light Blue Aqua 15-20
Hemingray Blue 50-75
Ice Aqua, Root Beer Amber 100-125
Dark Honey Amber, Whiskey Amber 125-150
Brown Amber, Olive Amber 150-175
Light Olive Amber 175-200
Dark Olive Amber, Dark Red Amber 200-250
Electric Blue, Light Electric Blue,
Red/Whiskey Amber Two Tone,
Sky Blue 250-300
Dark Electric Blue, Light Jade Green Milk,
Light Peacock Blue, Light Purple 300-350
Golden Amber, Purple, Red Amber 350-400
Amber Blackglass, Dark Orange Amber,
Honey Amber, Orange Amber,
Peacock Blue, Royal Purple 400-500
Burgundy, Clear, Lilac,
Purple Tint, Smoke 500-600
Milky Aqua, Pink, Yellow Amber,
Yellow Olive Amber 600-700
Jade Aqua Milk 800-1,000
Mustard Yellow 1,500-1,750
Green, Oxblood, Yellow 1,750-2,000
Dark Green, Milky Peacock Blue 2,000-2,500
Yellow Green 2,500-3,000

[080] (F-Skirt) H.G.CO./PATENT MAY 2 1893 (R-Skirt) PETTICOAT {Smooth base mold; transition style embossing} SDP
Aqua, Blue Aqua, Green Aqua,
Light Aqua, Light Green Aqua 30-40
Hemingray Blue, Ice Green 75-100
Ice Green, Sage Green 150-175
Aqua/Green Two Tone 200-250
Ice Blue Gray 300-350
Blue Tint, Ice Blue, Smoke Tint 350-400
Light Green, Orange Amber,
Pink Tint, Red Amber 400-500
Clear, Milky Aqua 500-600
Dark Olive Amber 800-1,000
Cornflower Blue 1,000-1,250
Green, Oxblood 1,750-2,000
Forest Green, Yellow Green 2,500-3,000

[090] (F-Skirt) H.G.CO./PATENT MAY 2 1893 (R-Skirt) PETTICOAT {'E' is engraved over an 'A'; last 'T' is engraved over a 'Y'} SDP
Amber Two Tone 200-250
Electric Blue 300-350
Purple 400-500

CD 162

[100] (F-Skirt) H.G.CO./PATENT MAY 2 1893 (R-Skirt) PETTICOAT {'OA' is engraved over 'AT'} SDP
Aqua x1
Honey Amber, Orange Amber 200-250
Electric Blue, Light Purple 300-350
Purple 350-400
Golden Amber, Peacock Blue,
Royal Purple 400-500
Lilac 500-600

[110] (F-Skirt) H.G.CO./PATENT MAY 2 1893 (R-Skirt) PETTICOAY {Note spelling} SDP
Aqua 10-15

[120] (F-Skirt) H.G.CO./PATENT MAY 2 1893 (R-Skirt) PETTIGOAT {Note spelling} SDP
Aqua 5-10

HAMILTON

[010] (F-Skirt) HAMILTON GLASS Co. {MLOD} RB
Light Aqua, Light Blue Aqua 20-30
Light Green 50-75

[020] (F-Skirt) HAMILTON GLASS Co. {MLOD} SB
Light Aqua 50-75

HAWLEY ⌗

[010] (F-Skirt) ⌗/HAWLEY PA./U.S.A. SB
Aqua, Blue Aqua 100-125
Green Aqua 125-150
Green 200-250

HEMINGRAY

[010] (Dome) 1 (F-Skirt) HEMINGRAY (R-Skirt) № 19 SB
Aqua, Light Aqua 10-15

[020] (Dome) 1 (F-Skirt) HEMINGRAY (R-Skirt) № 19 SDP
Aqua 5-10

[030] (Dome) 1 (F-Skirt) HEMINGRAY/MADE IN U.S.A. (R-Skirt) № 19 SB
Aqua, Green Aqua 10-15

[040] (Dome) 1 (F-Skirt) HEMINGRAY/MADE IN U.S.A. (R-Skirt) № 19 SDP
Aqua, Blue Aqua, Hemingray Blue 3-5
Light Olive Amber, Whiskey Amber 200-250
Golden Amber 250-300
Cobalt Blue, Dark Cobalt Blue 400-500
7-up Green 500-600
Sapphire Blue 600-700
Mustard Yellow, Yellow Olive Amber 800-1,000

[050] (F-Skirt) HEMINGRAY (R-Skirt) № 19 RDP
Light Apple Green 100-125

[060] (F-Skirt) HEMINGRAY (R-Skirt) № 19 SB
Blue Aqua, Dark Aqua, Green Aqua,
Hemingray Blue 5-10
Dark Yellow Green 2,000-2,500

[070] (F-Skirt) HEMINGRAY (R-Skirt) № 19 SDP
Green Aqua x1
Blue Aqua, Dark Aqua, Dark Green Aqua,
Light Aqua, Light Green Aqua 1-2
Hemingray Blue 5-10
Blue 10-15
Root Beer Amber 75-100
Brown Amber, Dark Red Amber 100-125
Aqua/Green Two Tone,
Dark Orange Amber, Light Green 150-175
Olive Amber, Sage Green 175-200
Aqua/Whiskey Amber Two Tone,
Dark Olive Green, Ice Aqua, Ice Green,
Red Amber, Sage Gray 200-250
Celery Green 250-300
Light Cobalt Blue 300-350
Cobalt Blue 350-400
Dark Cobalt Blue, Dark Green,
Dark Yellow Green, Green, Teal Green 400-500
7-up Green, Forest Green, Milky Aqua,
Yellow Green, Yellow Olive Green 500-600
Sapphire Blue 600-700

[075] (F-Skirt) HEMINGRAY (R-Skirt) [Blotted out embossing] № 19 SB
Aqua 5-10

[080] (F-Skirt) HEMINGRAY (R-Skirt) [Blotted out embossing] № 19 SDP
Aqua 1-2

[090] (F-Skirt) HEMINGRAY-19 (R-Skirt) MADE IN U.S.A. RDP
Aqua, Clear, Dark Aqua, Light Blue Aqua x1
Hemingray Blue, Ice Blue 1-2
Dark Blue, Light Green 3-5
Light 7-up Green 50-75
7-up Green, Dark Orange Amber,
Green, Root Beer Amber 100-125
Orange Amber, Red Amber 125-150
Golden Amber, Green/Aqua Two Tone,
Hemingray Blue/Green Two Tone,
Honey Amber 150-175
Aqua/Green/Amber Three Tone,
Clear/Aqua Two Tone,
Clear/Blue Two Tone,
Clear/Yellow Two Tone 200-250
Carnival, Sage/Gray Two Tone 250-300
Hemingray Blue/Amber Two Tone,
Light Cobalt Blue 300-350
Cobalt Blue, Yellow Amber 350-400
Dark Cobalt Blue 400-500
Dark Sapphire Blue 600-700
Yellow 1,750-2,000

103

CD 162

[100] (F-Skirt) HEMINGRAY-19
(R-Skirt) MADE IN U.S.A. SB
Aqua, Clear, Green Tint,
Hemingray Blue, Ice Blue 5-10

[110] (F-Skirt) HEMINGRAY-19
(R-Skirt) MADE IN U.S.A. SDP
Aqua x1
Hemingray Blue 1-2
Golden Amber, Whiskey Amber 175-200
Olive Amber 200-250
Cobalt Blue, Dark Cobalt Blue 350-400
Dark Cobalt Blue 400-500
Light Yellow Amber, Olive Green 500-600
Sapphire Blue 600-700
Jade Aqua Milk 800-1,000
Mustard Yellow 1,000-1,250

[115] (F-Skirt) HEMINGRAY-19
(R-Skirt) MADE IN U.S.A./[Blotted out embossing] SDP
Hemingray Blue 1-2

[120] (F-Skirt) HEMINGRAY-19
(R-Skirt) MADE IN U.S.A./[Numbers and dots] RDP
Green Tint 1-2

[130] (F-Skirt) HEMINGRAY-19
(R-Skirt) MADE IN U.S.A./['HEMINGRAY' blotted out] RDP
7-up Green 100-125

[140] (F-Skirt) HEMINGRAY-19/[Blotted out embossing] (R-Skirt) MADE IN U.S.A./[Blotted out embossing] SB
Hemingray Blue 5-10

[150] (F-Skirt) HEMINGRAY-19/[Number and dots] (R-Skirt) MADE IN U.S.A./[Number] RDP
Clear, Straw x1
Ice Blue, Ice Green 1-2
Carnival 250-300
Light Flashed Amber 600-700
Flashed Amber 800-1,000

[160] (F-Skirt) HEMINGRAY-19/[Numbers and dots] (R-Skirt) MADE IN U.S.A. RDP
Ice Blue 1-2
Carnival 250-300

[170] (F-Skirt) HEMINGRAY-19/[Numbers and dots] (R-Skirt) MADE IN U.S.A./[Number] SB
Ice Blue 5-10

[180] (F-Skirt) HEMINGRAY-19/[Numbers and dots] (R-Skirt) [Number]/MADE IN U.S.A. RDP
Clear x1

[190] (F-Skirt) HEMINGRAY-19/[Numbers and dots] (R-Skirt) [Number]/MADE IN U.S.A. SB
Ice Blue 5-10

[200] (F-Skirt) HEMINGRAY-19/[Number]
(R-Skirt) MADE IN U.S.A. SB
Ice Blue 5-10

[210] (F-Skirt) HEMINGRAY/MADE IN U.S.A.
(R-Skirt) № 19 SB
Aqua, Ice Blue 5-10
Hemingray Blue 10-15

[220] (F-Skirt) HEMINGRAY/MADE IN U.S.A.
(R-Skirt) № 19 SDP
Aqua 1-2
Hemingray Blue 3-5
Dark Honey Amber 150-175
Olive Amber, Whiskey Amber 200-250
Golden Amber, Honey Amber 250-300
Light Olive Green 350-400
Cobalt Blue, Dark Cobalt Blue,
Dark Mustard Green, Light Cobalt Blue 400-500
Sapphire Blue 600-700
Mustard Yellow 1,000-1,250

[230] (F-Skirt) HEMINGRAY/MAY 2 1893
(R-Skirt) PETTICOAT SDP
Aqua, Green Aqua 3-5
Root Beer Amber 300-350
Cobalt Blue 400-500
Sapphire Blue 700-800

[240] (F-Skirt) HEMINGRAY/PATENT MAY 2 1893 (R-Skirt) PETTICOAT SDP
Aqua, Green Aqua x1
Brown Amber, Root Beer Amber 100-125
Dark Red Amber 150-175
Aqua/Olive Amber Two Tone,
Olive Amber, Red Amber, Whiskey Amber 200-250
Orange Amber 250-300
Amber Blackglass 300-350
Dark Olive Green, Light Cobalt Blue 350-400
Cobalt Blue, Dark Cobalt Blue 400-500
Olive Green 500-600
Sapphire Blue 600-700
Yellow Olive Green 800-1,000

[250] (F-Skirt) HEMINGRAY/PATENT MAY 2 1893/[Blotted out embossing]
(R-Skirt) PETTICOAT SDP
Aqua x1

[260] (F-Skirt) HEMINGRAY/PATENTED
(R-Skirt) PETTICOAT SDP
Aqua, Green Aqua 5-10
Light Electric Blue, Powder Blue 250-300
Electric Blue 300-350

K.C.G.W.

[010] (F-Skirt) K.C.G.W. SB
Dark Green, Green, Light Aqua 20-30
Ice Blue 30-40

[020] (F-Skirt) K.C.G.W. (R-Skirt) [Number] SB
Light Green 20-30
Dark Green 30-40

CD 162

[030] (F-Skirt) ['K.C.G.W.' blotted out] SB
Aqua 10-15
Light Milky Aqua 150-175

[040] (F-Skirt) ['K.C.G.W.' blotted out]
(R-Skirt) [Number] SB
Aqua 10-15

LYNCHBURG Ⓛ

[010] (Dome) Ⓛ (F-Skirt) LYNCHBURG
(R-Skirt) NO.36/MADE IN [Number] U.S.A.
RDP
Aqua, Green, Sage Green 10-15
Smoky Sage Green 40-50

[020] (Dome) Ⓛ (F-Skirt) LYNCHBURG
(R-Skirt) NO.36/MADE IN [Number] U.S.A. SB
Aqua 5-10
Light Green 10-15
Green 15-20

[030] (Dome) Ⓛ (F-Skirt) LYNCHBURG
(R-Skirt) NO.36/MADE IN [Number] U.S.A.
SDP
Aqua, Blue Aqua 3-5
Blue, Green 5-10
Ice Blue 10-15
Yellow Green 75-100

[040] (F-Skirt) LYNCHBURG (R-Skirt) NO.
36-190/[Number] SDP
Light Aqua 50-75

[050] (F-Skirt) LYNCHBURG (R-Skirt/
Left) NO.36- (R-Skirt/Right) MADE IN/U.S.A.
SDP
Aqua 3-5

[060] (F-Skirt) LYИCHBURG {Note backwards
letter} (R-Skirt) NO.36 U.S.A./[Number] {Note
no 'MADE IN'} SDP
Aqua 5-10

[065] (F-Skirt) Ⓛ/LYNCHBURG
(R-Skirt) NO.36 MADE IN/['BROOKFIELD'
blotted out]/U.S.A. SDP
Aqua 5-10

[070] (F-Skirt) Ⓛ/LYNCHBURG
(R-Skirt) NO.36 U.S.A./[Number] {Note no
'MADE IN'} SB
Green 20-30

[080] (F-Skirt) Ⓛ/LYNCHBURG
(R-Skirt) NO.36 U.S.A./[Number] {Note no
'MADE IN'} SDP
Aqua 5-10
Green 10-15

[090] (F-Skirt) Ⓛ/LYNCHBURG
(R-Skirt) NO.36/MADE IN [Number] U.S.A.
SDP
Aqua, Light Aqua 3-5
Blue 5-10

[100] (F-Skirt) Ⓛ/LYNCHBURG (R-Skirt/
Left) NO.36- (R-Skirt/Right) MADE IN/U.S.A.
RDP
Aqua, Light Aqua, Sage Green 10-15
Green 20-30
Gingerale, Smoky Sage Green 40-50

[110] (F-Skirt) Ⓛ/LYNCHBURG (R-Skirt/
Left) NO.36- (R-Skirt/Right) MADE IN/U.S.A.
SB
Light Aqua 5-10

[115] (F-Skirt) Ⓛ/LYNCHBURG (R-Skirt/
Left) NO.36-/[Number] (R-Skirt/Right) MADE
IN/U.S.A. RDP
Blue Aqua 10-15

[120] (F-Skirt) Ⓛ/LYNCHBURG (R-Skirt/
Left) NO.36-/[Number] (R-Skirt/Right) MADE
IN/U.S.A. SB
Aqua, Light Aqua 5-10
Blue, Ice Blue 10-15
Dark Aqua, Green 15-20
Milky Aqua 50-75

[130] (F-Skirt) Ⓛ/LYNCHBURG (R-Skirt/
Left) NO.36-/[Number] (R-Skirt/Right) MADE
IN/U.S.A. SDP
Aqua, Blue Aqua 3-5
Green 5-10
Ice Blue 10-15
Milky Aqua 50-75
Dark Olive Green 400-500

[140] (F-Skirt) Ⓛ/LYИCHBURG {Note
backwards letter} (R-Skirt) NO.36-U.S.A./['190'
blotted out]/[Number] {'U.S.A.' is engraved over
the '190' from a Gayner mold} SDP
Aqua 5-10

[150] (F-Skirt) Ⓛ/LYИCHBURG {Note
backwards letter} (R-Skirt/Left) NO.36-/
[Number] (R-Skirt/Right) MADE IN .U.S.A
{Note periods} SDP
Light Green 5-10

[155] (F-Skirt) Ⓛ/LYNCHBURG/['GAYNER'
blotted out] (R-Skirt/Left) NO.36- (R-Skirt/
Right) MADE IN/U.S.A./['190' blotted out] SDP
Aqua 5-10

[160] (F-Skirt) Ⓛ/LYNCHBURG/['NEW
YORK' blotted out] (R-Skirt) NO.36 MADE IN/
['BROOKFIELD' blotted out]/U.S.A. SDP
Aqua 15-20
Aqua w/ Milk Swirls 50-75

105

CD 162

[170] (F-Skirt) Ⓛ/LYИCHBURⱭ {Note backwards and upside down letters} (R-Skirt/Left) NO.36-/[Number] (R-Skirt/Right) MADE IN/.U.S.A {Note periods} SDP
Light Aqua 10-15
Light Aqua w/ Milk Swirls 50-75

[180] (F-Skirt) Ⓛ/LYNCHBURⱭ {Note upside down letter} (R-Skirt/Left) NO.36- (R-Skirt/Right) MADE IN .U.S.A {Note periods} RDP
Sage Green 10-15

[190] (F-Skirt) Ⓛ/LYNCHBURⱭ {Note upside down letter} (R-Skirt/Left) NO.36-/[Number] (R-Skirt/Right) MADE IN/.U.S.A {Note periods} SB
Light Aqua 5-10

[200] (F-Skirt) Ⓛ/LYNCHBURⱭ {Note upside down letter} (R-Skirt/Left) NO.36-/[Number] (R-Skirt/Right) MADE IN/.U.S.A {Note periods} SDP
Green Aqua, Light Aqua, Light Green 10-15

MAYDWELL

[010] (F-Skirt) MAYDWELL-19 (R-Skirt) U.S.A. RDP
Clear, Dark Straw, Straw 1-2
Green Tint 3-5
Pink, Purple Tint 10-15

[020] (F-Skirt) MAYDWELL-19 (R-Skirt) U.S.A. SDP
Clear, Light Straw, Smoke 1-2
Gingerale 3-5
Pink, Purple Tint 10-15

[030] (F-Skirt) ['U.S.A.' blotted out]/MAYDWELL-19 (R-Skirt) ['MᶜLAUGHLIN 19' blotted out]/U.S.A. RDP
Dark Straw, Straw 1-2
Green Tint 3-5
Smoky Gray 5-10

MᶜLAUGHLIN

[010] (F-Skirt) MᶜLAUCHLIN {Note spelling} (R-Skirt) № 19 RDP
Light Green, Sage Green 5-10
Blue Aqua 10-15
Sky Blue, Steel Blue, Yellow Green 15-20
Delft Blue, Lime Green 20-30

[015] (F-Skirt) MᶜLAUCHLIN {Note spelling} (R-Skirt) № 19 SDP
Light Blue Aqua 10-15
Light Cornflower Blue, Yellow Green 15-20
Lime Green, Sky Blue 20-30

[020] (F-Skirt) MᶜLAUGHLIN (R-Skirt) № 19 FDP
Ice Green, Light Green 3-5
Light Blue 5-10
Blue 10-15
Apple Green, Lime Green 20-30
Emerald Green 30-40

[030] (F-Skirt) MᶜLAUGHLIN (R-Skirt) № 19 RDP
Aqua, Green Aqua, Ice Green,
Light Green, Sage Green 1-2
Light Blue 3-5
Light Yellow Green, Yellow Sage Green 5-10
Sage Green Tint 10-15
Blue Gray, Dark Forest Green, Green 20-30
Dark Emerald Green, Emerald Green 30-40
Cornflower Blue, Dark Delft Blue 40-50
Apple Green, Lemon 50-75
Green Straw, Off Clear, Yellow Straw 75-100

[040] (F-Skirt) MᶜLAUGHLIN (R-Skirt) № 19 SB
Aqua, Green Aqua, Light Blue Aqua,
Light Green, Sage Green 1-2
Steel Gray 3-5
Blue 10-15
Cornflower Blue, Delft Blue 40-50

[050] (F-Skirt) MᶜLAUGHLIN (R-Skirt) № 19 SDP
Aqua, Light Green 1-2
Sage Green, Smoke 3-5
Light Blue, Light Yellow Green, Steel Blue 5-10
Green, Sky Blue 20-30
Emerald Green, Lime Green 30-40
7-up Green, Apple Green 50-75

N ᴰᴱ M

[010] (Dome) ⚠ (F-Skirt) N ᴰᴱ M RDP
Clear 50-75

[020] (Dome) ⚠ (F-Skirt) N ᴰᴱ M SDP
Clear 50-75

[030] (Dome) ⚠ (F-Skirt) N ᴰᴱ M (R-Skirt) [Blotted out embossing] RDP
Clear 50-75

[040] (F-Skirt) N ᴰᴱ M SDP
Clear, Off Clear 50-75

N.E.G.M.CO.

[010] (F-Skirt) N.E.G.M.CO. SB
Aqua, Blue Aqua, Green Aqua 3-5
Green, Ice Blue 15-20
Yellow Green 50-75
Apple Green 75-100
Sapphire Blue 200-250
Olive Green 250-300

106

CD 162

NO EMBOSSING

[010] [No embossing] RDP
Light Aqua 15-20

NO EMBOSSING - CANADA

[010] [No embossing] {MLOD} {Hamilton style} RB
Light Aqua 30-40
Light Green 40-50
Clear 50-75

NO NAME

[010] (F-Skirt) NO.36-19 SB
Aqua 3-5

[020] (F-Skirt) № 19 {Hemingray product} SDP
Blue Aqua, Dark Aqua, Hemingray Blue 75-100
Light Cobalt Blue 350-400
Cobalt Blue 400-500

[030] (F-Skirt) PETTICOAT
(R-Skirt) PETTICOAT {Hemingray product} SDP
Aqua 400-500

[040] (F-Skirt) [Backwards number] [Vertical bar] (R-Skirt) [Backwards number] [Vertical bar] SB
Aqua, Light Green Aqua 10-15

[050] (F-Skirt) [Vertical bar] (R-Skirt) [Vertical bar] SB
Light Aqua, Light Green 10-15

NO NAME - CANADA

[010] (F-Skirt) 1678 RDP
Light Aqua, Light Green 75-100

[020] (F-Skirt) 1678 SDP
Ice Aqua, Light Aqua 40-50
Gray, Steel Blue 50-75
Clear 75-100
Royal Purple 100-125
Dark Purple, Light Green, Purple 125-150
Lavender, Light Purple 150-175

NO NAME - MEXICO

[010] (Dome) ⚠ SDP
Sage Green 10-15

[020] (F-Skirt) SM-2 RDP
Olive Amber Blackglass 1,000-1,250

O.V.G.CO.

[010] (F-Skirt) O.V.G.CO.
(R-Skirt) PETTICOAT SB
Aqua, Celery Green 30-40
Blue 40-50
Blue w/ Milk Swirls 100-125
Light Purple 2,500-3,000

PATENT - OTHER

[010] (F-Crown) [Number]
(F-Skirt) PAT'D.NOV. 13TH 1883. SB
Light Aqua 15-20
Light Green 20-30

S.S.& CO.

[010] (F-Skirt) S.S & Co. MFG'S/CHICAGO/
N° 19 (R-Skirt) DOUBLE/PETTICOAT SB
Aqua, Light Aqua, Light Blue 125-150
Ice Blue, Ice Green, Light Green 150-175
Clear 175-200
Lime Green 250-300
Apple Green 350-400
Yellow Green 600-700

STAR ✷

[010] (F-Skirt) ✷ SB
Aqua, Dark Aqua, Green Aqua,
Light Aqua, Light Blue 1-2
Blue, Green 3-5
Emerald Green 5-10
Dark Yellow Green, Yellow Green 10-15
7-up Green, Olive Green 20-30
Dark Olive Green 30-40
Sapphire Blue 350-400

[020] (F-Skirt) ✷ SDP
Aqua, Blue 5-10
Green 10-15

[030] (F-Skirt) ✷ (R-Skirt) ✷ SB
Aqua, Blue Aqua 1-2
Green 15-20
Yellow Green 30-40
Olive Green 50-75

T-H.E.CO.

[010] (F-Skirt) T-H.E.CO. (R-Skirt) PAT'D NOV. 13TH 1883 {MLOD} SB
Aqua, Light Blue 5-10

[020] (F-Skirt) T-H.E.CO. (R-Skirt) [Blotted out embossing]/PAT'D NOV. 13TH 1883 {MLOD} SB
Aqua 5-10

W.F.G.CO.

[010] (F-Skirt) W.F.G.CO./DENVER, COLO.
(R-Skirt) [Glass dot]/PETTICOAT SB
Aqua, Blue Aqua, Off Clear 5-10
Ice Blue 15-20
Light Sage Green 20-30
Light Steel Blue, Steel Blue 40-50
Clear, Ice Green 50-75
Purple 75-100
Dark Purple 100-125
Delft Blue, Gingerale 400-500
Purple/Straw Two Tone 600-700
Cornflower Blue 700-800

107

CD 162

W.G.M.CO.

[010] (F-Skirt) W.G.M.CO. SB
Light Purple 30-40
Purple 40-50
Pink, Royal Purple 50-75
Peach, Smoky Straw, Straw 75-100
Clear 200-250
Peach/Straw Two Tone,
Purple/Straw Two Tone 300-350

WESTINGHOUSE

[010] (F-Skirt) WESTINGHOUSE/NO.2 SB
Light Green, Off Clear 400-500
Lime Green 800-1,000
Blue Aqua 1,750-2,000
Peacock Blue 3,500-4,000

WHITALL TATUM ▽

[010] (F-Skirt) WHITALL TATUM CO. NO.4/
[Number] (R-Skirt) MADE IN U.S.A. ▽ SB
Clear, Light Straw 3-5

[020] (F-Skirt) WHITALL TATUM NO.4/
[Number] (R-Skirt) MADE IN U.S.A. ▽ /
[Number] {Dots may appear around the '▽'}
SB
Clear, Off Clear 3-5
Pink 5-10

CD 162.1

BROOKFIELD

[010] (Dome) [Number] (F-Skirt) BROOKFIELD
SB
Dark Green Aqua x1
Green 1-2
Emerald Green 3-5
Yellow Green 5-10
Dark Olive Green 20-30

[020] (Dome) [Number] (F-Skirt) BROOKFIELD
SDP
Aqua, Dark Green Aqua 1-2
Dark Olive Green 30-40

[030] (Dome) [Number] (F-Skirt) BROOKFIELD
(R-Skirt) NEW YORK SB
Aqua x1
Green 1-2
Emerald Green 3-5
Olive Amber 175-200

[035] (Dome) [Number] (F-Skirt) BROOKFIELD
(R-Skirt) NEW YORK SDP
Dark Green 5-10
Olive Green w/ Amber Swirls 75-100

[040] (F-Skirt) BROOKFIELD SB
Aqua, Dark Green Aqua x1
Blue 1-2

[050] (F-Skirt) BROOKFIELD SDP
Aqua, Dark Aqua 1-2

[060] (F-Skirt) BROOKFIELD (R-Skirt) NEW
YORK SB
Aqua x1
Blue, Dark Green 3-5

[070] (F-Skirt) BROOKFIELD (R-Skirt) NEW
YORK SDP
Aqua, Dark Aqua 1-2
Green 3-5
Dark Yellow Green 10-15
Dark Olive Green 30-40

[080] (F-Skirt) W. BROOKFIELD (R-Skirt) NEW
YORK SB
Light Aqua 3-5

STAR ✷

[010] (F-Skirt) ✷ SB
Aqua, Green Aqua 5-10

CD 162.3

BROOKFIELD

[005] (Dome) [Number] (F-Skirt) W.
BROOKFIELD (R-Skirt) NEW YORK SB
Light Aqua 5-10

[010] (F-Skirt) BROOKFIELD (R-Skirt) NEW
YORK SB
Aqua, Blue Aqua, Light Aqua 5-10
Light Green 10-15

[020] (F-Skirt) W. BROOKFIELD (R-Skirt) NEW
YORK SB
Light Aqua 5-10
Light Green 10-15
Gray, Ice Blue, Steel Blue 125-150
Off Clear 150-175
Light Purple 200-250
Purple 250-300

NO EMBOSSING

[010] [No embossing] {Pennycuick style} SB
Dark Aqua 400-500

CD 162.5

STAR ✶

[010] (F-Skirt) ✶ SB
Aqua, Blue Aqua, Green Aqua, Light Blue 3-5
Blue 10-15
Light Green, Light Steel Blue 15-20
Apple Green, Lime Green 50-75
Dark Blue, Dark Green 75-100
Light Purple/Blue Two Tone 300-350
Light Purple 350-400

[020] (F-Skirt) ✶ WDP
Aqua, Blue 300-350
Light Green 400-500

CD 162.4

NO EMBOSSING - CANADA

[010] [No embossing] {MLOD} RB
Aqua 5-10
Green Aqua 10-15
Blue Aqua 15-20
Apple Green, Green 40-50
Blue 50-75
Royal Purple 75-100
Milky Aqua, Purple 100-125
Light Purple 250-300
Yellow Olive Green 2,500-3,000

NO NAME - CANADA

[010] (F-Skirt) 1673 SB
Aqua 10-15
Lemon, Straw 40-50

[020] (F-Skirt) 1673 SDP
Aqua, Light Aqua 15-20

[023] (F-Skirt) 1673 {'3' is engraved over an '8'} SB
Aqua 10-15

[027] (F-Skirt) 1673 {'3' is engraved over an '8'} SDP
Aqua 15-20

[030] (F-Skirt) 1678 RDP
Straw 15-20
Aqua 75-100

[040] (F-Skirt) 1678 SB
Aqua, Blue Aqua, Dark Aqua,
Green Aqua, Light Aqua 5-10
Light Green 10-15
Straw 15-20
Dark Blue Aqua, Ice Blue,
Light Gray, Steel Blue 20-30
Lemon 30-40
Clear, Sage Green, Yellow 40-50
Blue, Green, Light Yellow Green,
Lime Green, Purple Tint 50-75
Powder Blue 75-100
Light Purple, Smoky Purple 150-175
Hemingray Blue, Milky Aqua, Purple 200-250
Milky Blue 500-600
Dark Green, Dark Olive Green,
Yellow Green 700-800
Emerald Green 1,000-1,250
Yellow Olive Green 1,250-1,500

[050] (F-Skirt) 1678 SDP
Aqua, Blue Aqua, Ice Aqua 10-15
Ice Green 20-30
Blue, Green, Lime Green 75-100
Gray 125-150
Milky Clear 175-200
Ice Gray 200-250
Light Purple,
Light Purple/Gray Two Tone 300-350
Purple 400-500
Milky Aqua 500-600

[060] (F-Skirt) 1678 {'8' is engraved over a '3'} SB
Dark Green Aqua, Light Aqua 5-10
Light Green 10-15
Light Yellow Green 50-75
Purple 200-250

[070] (F-Skirt) 1678 {'8' is engraved over a '3'} SDP
Aqua 5-10

CD 162.5

P.L.W.

[010] (Dome) P.L.W. (F-Skirt) B SB
Aqua 15-20
Green 30-40

P.R.R.

[010] (Dome) P.R.R. (F-Skirt) B SB
Aqua 5-10
Blue Aqua, Green 10-15
Dark Green 15-20
Yellow Green 40-50

109

CD 162.5

[020] (Dome) R.R.R. {Note spelling} (F-Skirt) B SB
Aqua 10-15

CD 162.7

DERF

[010] (F-Skirt) DERF (R-Skirt) TELGS NACLS MEXICO RDP
Aqua 10-15
Dark Green, Lime Green 20-30
Emerald Green 40-50

[020] (F-Skirt) DERF (R-Skirt) TELGS NACLS MEXICO SB
Aqua, Sage Green 15-20
Dark Green Aqua 20-30
Teal Green 30-40
Amber Blackglass 200-250

[030] (F-Skirt) DERF (R-Skirt) TELGS NACLS MEXICO SDP
Aqua, Dark Aqua, Dark Green,
Dark Green Aqua, Green Aqua 10-15
Dark Olive Green, Dark Yellow Green 50-75

[040] (F-Skirt) DERF (R-Skirt) TELGS NACLS MEXICO {MLOD} RDP
Green 20-30

[050] (F-Skirt) DERF (R-Skirt) TELGS NACLS MEXICO {MLOD} SDP
Aqua, Dark Aqua 15-20
Sage Green 20-30
Emerald Green 30-40

F.F.C.C.N.DE M.

[010] (F-Skirt) F.F.C.C.N DE M/F.D. RDP
Aqua, Clear 20-30
Lime Green 40-50
Dark Yellow Green 75-100

NO EMBOSSING - MEXICO

[010] [No embossing] {MLOD} RDP
Aqua 10-15
Lime Green, Yellow Green 30-40
Dark Olive Green 100-125
Light Gray 175-200
Olive Amber 200-250
Light Purple 350-400
Purple, Rose 400-500

[020] [No embossing] {MLOD} SB
Aqua, Light Green 10-15
Dark Olive Green 100-125

P.S.S.A.

[010] (F-Skirt) ['VB' in a circle]/PSSA #19 RDP
Sage Green Tint 20-30

CD 163

ARMSTRONG Ⓐ

[010] (F-Skirt) Armstrong's No.4 (R-Skirt) MADE IN U.S.A. Ⓐ [Numbers and dots] SB
Clear 1-2
Straw 3-5

HEMINGRAY

[010] (F-Skirt) HEMINGRAY-19 (R-Skirt) MADE IN U.S.A./[Numbers and dots] RDP
Clear x1
Green Tint, Straw 1-2

WHITALL TATUM ▽

[010] (F-Skirt) (Arc)WHITALL TATUM/No.4 (R-Skirt) (Arc)MADE IN U.S.A./[Numbers and dots]/Ⓐ/[Number] SB
Clear 3-5
Light Pink 5-10

CD 163.2

NO NAME

[010] (F-Skirt) [Vertical bar] (R-Skirt) [Vertical bar] SB
Light Aqua, Light Green 75-100

CD 164

CD 163.4

CD 163.4
3 1/8 x 3 1/2

WHITALL TATUM ▽

[010] (F-Skirt) WHITALL TATUM CO. No.4/
[Number] (R-Skirt) MADE IN U.S.A. ▽
{Previously shown as CD 163 in the 1995 edition} SB
Clear 3-5
Straw 5-10
Aqua, Ice Aqua 10-15

CD 164

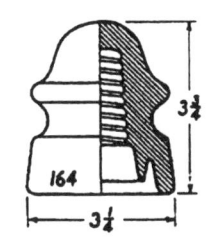

ARMSTRONG Ⓐ

[010] (F-Skirt) *Armstrong's* № 5
(R-Skirt) MADE IN U.S.A. Ⓐ [Numbers and dots] CB
Light Smoke 1-2

[020] (F-Skirt) *Armstrong's* № 5
(R-Skirt) MADE IN U.S.A. Ⓐ [Numbers and dots] SB
Clear, Off Clear 1-2
Light Green, Olive Green Tint 3-5

B

[005] (F-Skirt) B SB
Emerald Green 10-15

[010] (F-Skirt) B ['✶' blotted out] SB
Aqua, Dark Aqua 3-5
Green 5-10
Emerald Green 10-15
Yellow Green 15-20

[020] (F-Skirt) B ['✶' blotted out] SDP
Aqua 5-10
Green 10-15

B.C.DRIP

[010] (F-Skirt) B.C/DRIP SB
Light Aqua 700-800

[020] (F-Skirt) B.C/DRIP SDP
Aqua, Light Aqua 50-75

B.G.M.CO.

[010] (F-Skirt) B.G.M.CO. SB
Clear, Light Purple, Purple 500-600

BROOKFIELD

[010] (Dome) [Letter or number]
(F-Skirt) BROOKFIELD SB
Aqua x1
Light Aqua 1-2
Green, Green Aqua 3-5
Emerald Green 5-10
Yellow Green 15-20
Light Olive Green 20-30
Olive Green 30-40
Olive Amber 75-100
Dark Red Amber 400-500

[020] (Dome) [Letter or number]
(F-Skirt) BROOKFIELD SDP
Aqua x1
Green 5-10
Yellow Green 20-30
Olive Amber 100-125

[030] (Dome) [Number] (F-Skirt) BROOKFIELD
(R-Skirt) NEW YORK SB
Aqua x1
Blue Aqua 1-2
Green 3-5
Yellow Green 20-30
Olive Amber 75-100

[040] (Dome) [Number] (F-Skirt) BROOKFIELD
(R-Skirt) NEW YORK SDP
Green Aqua 1-2

[050] (F-Crown) (Arc)W. BROOKFIELD/NEW
YORK (F-Skirt) PAT'D NOV. 13TH 1883.
{MLOD} SB
Aqua, Blue Aqua, Light Aqua 15-20

[060] (F-Crown) (Arc)WM. BROOKFIELD/NEW
YORK (F-Skirt) PAT'D NOV. 13TH 1883 SB
Aqua 20-30

[070] (F-Skirt) BROOKFIELD (R-Skirt) NEW
YORK SB
Aqua x1
Light Blue, Light Green 3-5

[080] (F-Skirt) BROOKFIELD (R-Skirt) NEW
YORK SDP
Aqua x1
Green 5-10
Olive Green 30-40

[090] (F-Skirt) BROOKFIELD (R-Skirt) NEW
YORK/['B.G.M.CO.' blotted out] SB
Light Aqua, Light Green 5-10

CD 164

[100] (F-Skirt) BROOKFIELD (R-Skirt) №38 SB
Aqua 3-5
Green 15-20
Olive Green, Yellow Green 40-50
Olive Amber 100-125
Dark Yellow Amber 350-400

[110] (F-Skirt) BROOKFIELD/[Number] SB
Aqua, Blue Aqua, Green Aqua 1-2
Green 15-20

[120] (F-Skirt) W. BROOKFIELD (R-Skirt) NEW YORK SB
Aqua, Light Aqua 3-5
Ice Green 5-10
Green 10-15

DIAMOND ◇

[010] (F-Skirt) ◇ RB
Aqua, Light Green, Light Green Aqua,
Peach, Straw 15-20
Light Gray Green 20-30
Ice Yellow Green, Light Pink 30-40
Light Blue Gray, Light Steel Blue 40-50
Clear, Light Purple, Purple, Smoke 50-75

DOMINION ◉

[010] (F-Skirt) DOMINION 614 (R-Skirt) ◉ [Numbers and dots] RDP
Ice Green, Light Green, Straw 3-5

[020] (F-Skirt) DOMINION 614 (R-Skirt) ◉ [Numbers and dots] SDP
Clear, Light Green, Light Pink,
Light Straw, Straw 3-5

[040] (F-Skirt) DOMINION-614 (R-Skirt) ◇ SDP
Clear, Light Green,
Light Sage Green, Straw 3-5

[030] (F-Skirt) DOMINION-614 (R-Skirt) ◇ {Dots may appear over '◇'} RDP
Clear, Light Green, Light Sage Green,
Pink, Straw 3-5
Aqua 5-10

GAYNER

[010] (F-Skirt) GAYNER (R-Skirt) NO.38-20/ [Number] SDP
Aqua, Light Aqua 5-10

[020] (F-Skirt) GAYNER (R-Skirt) NO.38-20/ [Number] {Some specimens are tall, similar to a CD 166} SB
Aqua 5-10

H.G.CO.

[005] (F-Skirt) H G CO {Note no periods} (R-Skirt) PETTICOAT SB
Aqua 40-50

[010] (F-Skirt) H G CO/PATENT MAY 2 1893 {Note no periods} (R-Skirt) PETTICOAT SDP
Aqua 5-10
Hemingray Blue 50-75
Jade Green Milk 75-100

[020] (F-Skirt) H.G.CO. (R-Skirt) PETTICOAT SB
Aqua, Aqua Tint, Green Tint, Light Aqua 40-50
Blue Aqua, Ice Green, Light Blue,
Light Gray Green, Light Green,
Light Steel 50-75
Off Clear 75-100
Purple Tint, Steel Blue 100-125
Blue 125-150
Clear 175-200
Light Yellow Green 300-350
Lime Green 400-500
Green 500-600

[030] (F-Skirt) H.G.CO./PATENT MAY 2 1893 (R-Skirt) PETTICOAT SDP
Aqua x1
Blue Aqua, Green Aqua 1-2
Hemingray Blue, Light Jade Green Milk 50-75
Jade Green Milk, Milky Aqua 75-100
Ice Aqua 200-250
Lime Green 250-300
Green 300-350
Light Olive Green 400-500
Olive Green, Yellow Green 500-600
Emerald Green 800-1,000
Orange Amber 1,500-1,750
Peacock Blue 5,000-7,500

[040] (F-Skirt) H.G.CO./PATENT MAY 2 1893 (R-Skirt) PETTICOAT {Smooth base mold; transition style embossing} SDP
Aqua, Blue Aqua, Light Aqua 50-75
Blue 100-125
Light Green 250-300
Lime Green 300-350
Yellow Green 1,000-1,250
Purple 5,000-7,500

[050] (F-Skirt) H.G.CO./PATENT MAY 2 1893 (R-Skirt) PETTICOAY {Note spelling} SDP
Aqua 5-10

HAWLEY ⌘

[010] (F-Skirt) ⌘ (R-Skirt) HAWLEY PA. U.S.A. SB
Aqua 20-30

112

CD 164

[020] (F-Skirt) ⌷⊞⊏ /['STERLING' blotted out]
(R-Skirt) HAWLEY PA. U.S.A./['£' blotted out]
SB
Aqua 20-30
Blue Aqua 30-40

HEMINGRAY

[010] (F-Skirt) HEMINGRAY (R-Skirt) № 20
SDP
Aqua, Green Aqua, Light Aqua x1
Blue Aqua 1-2
Olive Green 400-500
Emerald Green 600-700

[020] (F-Skirt) HEMINGRAY-20
(R-Skirt) HEMINGRAY-20 RDP
Hemingray Blue 20-30

[030] (F-Skirt) HEMINGRAY-20
(R-Skirt) HEMINGRAY-20 SDP
Hemingray Blue 20-30

[040] (F-Skirt) HEMINGRAY-20
(R-Skirt) MADE IN U.S.A. RDP
Aqua x1
Blue Aqua, Blue Tint, Hemingray Blue,
Ice Blue, Ice Green 1-2
Light Hemingray Blue 5-10
Clear/Aqua Two Tone,
Clear/Blue Two Tone 150-175
Carnival 200-250
Orange Amber 400-500

[050] (F-Skirt) HEMINGRAY-20
(R-Skirt) MADE IN U.S.A. SB
Aqua, Ice Blue 15-20

[060] (F-Skirt) HEMINGRAY-20
(R-Skirt) MADE IN U.S.A. SDP
Aqua x1
Hemingray Blue 1-2

[065] (F-Skirt) H⊏MINGRAY-20 (R-Skirt) MAD⊏
IN U.S.A. {Note both 'E's are missing middle
bar} RDP
Aqua, Hemingray Blue, Ice Blue 5-10

[070] (F-Skirt) HEMINGRAY-20
(R-Skirt) MADE IN U.S.A./[Number] SB
Clear 15-20

[080] (F-Skirt) HEMINGRAY-20/[Numbers and
dots] (R-Skirt) MADE IN U.S.A. RDP
Ice Blue 1-2

[085] (F-Skirt) H⊏MINGRAY-20/[Numbers and
dots] (R-Skirt) MAD⊏ IN U.S.A. {Note both 'E's
are missing middle bar} RDP
Ice Blue 5-10

[090] (F-Skirt) HEMINGRAY-20/[Numbers and
dots] (R-Skirt) MADE IN U.S.A./[Number] RDP
Clear x1
Blue Tint, Ice Blue 1-2
Carnival 200-250

[095] (F-Skirt) H⊏MINGRAY-20/[Numbers and
dots] (R-Skirt) MAD⊏ IN U.S.A./[Number] {Note
both 'E's are missing middle bar} RDP
Ice Blue 5-10

[100] (F-Skirt) HEMINGRAY-20/['42' blotted
out] (R-Skirt) MADE IN U.S.A. SDP
Hemingray Blue 10-15

[110] (F-Skirt) HEMINGRAY/MADE IN U.S.A.
(R-Skirt) № 20 SDP
Aqua, Hemingray Blue 1-2

[120] (F-Skirt) HEMINGRAY/PATENTED MAY
2 1893 (R-Skirt) PETTICOAT SDP
Aqua x1
Aqua w/ Milk Swirls 75-100
Lime Green 300-350
Dark Yellow Green, Yellow Green 500-600
Emerald Green 700-800

[130] (F-Skirt) HEMINGRAY/[Blotted out
embossing] (R-Skirt) № 20 SDP
Aqua 1-2

[140] (F-Skirt) [Numbers and dots]/
HEMINGRAY-20 (R-Skirt) MADE IN U.S.A./
[Number] RDP
Ice Blue 1-2

[150] (F-Skirt) ['[Numbers]' blotted out]/
HEMINGRAY-20/[Numbers and dots]
(R-Skirt) MADE IN U.S.A./[Number] RDP
Light Carnival 175-200

K.C.G.W.

[010] (F-Skirt) K C G Co (R-Skirt) PERU SB
Aqua, Ice Aqua, Light Blue, Light Green 30-40
Lime Green 40-50

[015] (F-Skirt) K C G Go {Note spelling}
(R-Skirt) PERU SB
Green Aqua 30-40

[020] (F-Skirt) [Glass dot]/['K C G Co' blotted
out] (R-Skirt) [Glass dot]/['PERU' blotted out]
SB
Aqua, Light Green 20-30
Light Blue 30-40

LYNCHBURG Ⓛ

[010] (F-Skirt) LYNCHBURG
(R-Skirt) NO.38-20/[Number] SB
Aqua 10-15

113

CD 164

[020] (F-Skirt) LYNCHBURG
(R-Skirt) NO.38-20/[Number] SDP
Dark Green Aqua, Light Aqua 5-10
Yellow Green 30-40

[030] (F-Skirt) LYNCHBURG (R-Skirt/
Left) No.38-/[Number] (R-Skirt/Right) MADE
IN/U.S.A. SDP
Aqua 5-10

[040] (F-Skirt) LYNCHBURG/Ⓛ [Number]
(R-Skirt/Left) No.38-/[Number] (R-Skirt/
Right) MADE IN/U.S.A. SB
Aqua 10-15

[050] (F-Skirt) LYNCHBURG/Ⓛ [Number]
(R-Skirt/Left) No.38-/[Number] (R-Skirt/
Right) MADE IN/U.S.A. SDP
Aqua, Blue Aqua 5-10

[060] (F-Skirt) Ⓛ/LYNCHBURG
(R-Skirt) NO.38-20/[Number] RDP
Aqua, Blue Aqua 20-30

[070] (F-Skirt) Ⓛ/LYNCHBURG
(R-Skirt) NO.38-20/[Number] SB
Aqua, Light Aqua 5-10
Green 20-30
Yellow Green 30-40

[080] (F-Skirt) Ⓛ/LYNCHBURG
(R-Skirt) NO.38-20/[Number] SDP
Aqua, Light Aqua 3-5
Blue Aqua 5-10
Sage Green 10-15
Green 15-20
Yellow Green 20-30
Dark Yellow Green 30-40

[085] (F-Skirt) Ⓛ/LYNCHBURG (R-Skirt/
Left) No.38- (R-Skirt/Right) MADE IN/U.S.A.
RDP
Aqua 10-15

[090] (F-Skirt) Ⓛ/LYNCHBURG (R-Skirt/
Left) No.38-/[Number] (R-Skirt/Right) MADE
IN/U.S.A. RDP
Aqua, Light Blue Aqua 10-15

[100] (F-Skirt) Ⓛ/LYNCHBURG (R-Skirt/
Left) No.38-/[Number] (R-Skirt/Right) MADE
IN/U.S.A. SB
Aqua, Blue Aqua, Green Aqua, Light Aqua 5-10
Green 10-15

[110] (F-Skirt) Ⓛ/LYNCHBURG (R-Skirt/
Left) No.38-/[Number] (R-Skirt/Right) MADE
IN/U.S.A. SDP
Aqua, Light Aqua 3-5
Dark Green Aqua 5-10
Light Green, Sage Green 10-15
Green 15-20
Yellow Green 20-30
Dark Yellow Green 30-40

[120] (F-Skirt) Ⓛ/LYNCHBURG (R-Skirt/
Left) NO.38-/[Number] [Number] (R-Skirt/
Right) MADE IN U.S.A.. {Note periods} SB
Green 10-15
Yellow Green 30-40

[130] (F-Skirt) Ⓛ/LYNCHBURG (R-Skirt/
Left) NO.38-/[Number] [Number] (R-Skirt/
Right) MADE IN U.S.A.. {Note periods} SDP
Aqua 3-5
Green 15-20
Yellow Green 20-30
Dark Yellow Green 30-40

MAYDWELL

[010] (F-Skirt) MAYDWELL-20 (R-Skirt) U.S.A.
SDP
Clear, Off Clear, Smoke, Straw 1-2
Light Sage Green, Light Yellow 3-5
White Milk 10-15
White Milk w/ Blue Tint,
White Milk w/ Green Tint 20-30
White Milk w/ Pink Tint 30-40

[020] (F-Skirt) MAYDWELL-20 (R-Skirt) U.S.A./
[Number] RDP
Clear, Green Tint, Off Clear, Smoke, Straw 1-2
Light Yellow, Peach 3-5
Pink 5-10
White Milk 10-15
White Milk w/ Blue Tint,
White Milk w/ Green Tint 20-30
White Milk w/ Pink Tint,
White Milk w/ Yellow Swirls 30-40
Apple Green 300-350

[030] (F-Skirt) MAYDWELL-20 (R-Skirt) U.S.A./
['McLAUGHLIN' blotted out] RDP
White Milk 10-15

[040] (F-Skirt) MAYDWELL-20/['McLAUGHLIN'
blotted out] (R-Skirt) U.S.A. RDP
Clear 1-2
White Milk 10-15

McLAUGHLIN

[010] (F-Skirt) McLAUCHLIN {Note spelling}
(R-Skirt) NO - 20 RDP
Aqua 3-5
Light Green, Sage Green 5-10
Green 15-20
Light Apple Green 20-30

[020] (F-Skirt) McLAUCHLIN {Note spelling}
(R-Skirt) NO - 20 SDP
Aqua, Light Blue 5-10
Lime Green 40-50

CD 164

[030] (F-Skirt) McLAUCHLIN {Note spelling} (R-Skirt) No.-20 RDP
Aqua, Sage Green 5-10
Green 15-20
7-up Green 40-50

[040] (F-Skirt) McLAUCHLIN {Note spelling} (R-Skirt) No.-20 SDP
Light Blue 10-15
Cornflower Blue 20-30
Lime Green 40-50

[050] (F-Skirt) McLAUGHLIN (R-Skirt) № 20 RDP
Aqua, Green Aqua 3-5
Light Blue, Light Green 5-10
Lime Green, Steel Blue 10-15
Purple Tint, Yellow 40-50

[060] (F-Skirt) McLAUGHLIN (R-Skirt) № 20 SB
Aqua, Dark Aqua, Light Green 3-5
Green, Sage Green, Steel Blue 5-10
Light Blue 10-15
Smoke 15-20
Off Clear, Straw 20-30
Delft Blue 40-50

[070] (F-Skirt) McLAUGHLIN (R-Skirt) № 20 SDP
Aqua, Light Green 3-5
Light Blue, Sage Green 5-10
Steel Blue 10-15
Lime Green 20-30
7-up Green, Delft Blue 40-50

[080] (F-Skirt) McLAUGHLIN (R-Skirt) №-20 FDP
Green Aqua 3-5
Light Blue 5-10
Light Cornflower Blue 40-50

[090] (F-Skirt) McLAUGHLIИ {Note backwards letter} (R-Skirt) NO - 20 FDP
Aqua, Light Green 3-5

[100] (F-Skirt) McLAUGHLIИ {Note backwards letter} (R-Skirt) NO - 20 RDP
Aqua, Light Green, Sage Green,
Steel Green 3-5
Green, Light Blue 5-10
Apple Green,
Light Cornflower Blue, Lime Green 20-30
7-up Green 40-50

[105] (F-Skirt) McLAUGHLIИ {Note backwards letter} (R-Skirt) NO - 20 SDP
Light Cornflower Blue, Lime Green 20-30

[110] (F-Skirt) McLAUGHLIИ {Note backwards letter} (R-Skirt) № 20 RDP
Light Aqua, Light Green, Light Sage Green 3-5
Light Blue 5-10
Green 10-15
Apple Green 20-30
7-up Green 40-50

[120] (F-Skirt) McLAUGHLIИ {Note backwards letter} (R-Skirt) № 20 SB
Aqua, Dark Aqua, Light Sage Green 3-5
Light Blue 5-10
Delft Blue 20-30

[130] (F-Skirt) McLAUGHLIИ {Note backwards letter} (R-Skirt) № 20 SDP
Light Green, Sage Aqua 3-5
Ice Blue, Steel Blue 5-10
Lime Green 20-30
7-up Green 40-50

[140] (F-Skirt) McLAUGHLIИ {Note backwards letter} (R-Skirt) №-20 FDP
Aqua, Light Green, Light Green Aqua 3-5

[150] (F-Skirt) McLAUGHLIN-20 FDP
Ice Green, Light Green 3-5
Blue, Emerald Green 10-15
Light Cornflower Blue, Yellow Green ... 15-20
Apple Green 20-30
Delft Blue, Off Clear 30-40
7-up Green, Lime Green 40-50
Clear 50-75

[160] (F-Skirt) McLAUGHLIN-20 RDP
Aqua, Light Green, Sage Green 3-5
Green 5-10
Emerald Green, Light Yellow Green 10-15
Cornflower Blue, Lemon,
Light 7-up Green, Yellow Green 20-30
Off Clear 30-40
7-up Green, Clear 40-50
Light Citrine 150-175
Citrine 350-400
Dark Citrine 700-800

[170] (F-Skirt) McLAUGHLIN-20 SDP
Aqua 3-5
Emerald Green, Green, Yellow Green 10-15
Apple Green 20-30
Citrine 350-400

NO EMBOSSING

[010] [No embossing] {California product} SB
Aqua, Sage Green 15-20

NO NAME

[005] (F-Skirt) NO 38-20 {Gayner/Lynchburg product} SB
Aqua 3-5

115

CD 164

[010] (F-Skirt) NO.38-20/[Number] {Gayner/Lynchburg product} SB
Aqua, Blue 3-5

[020] (F-Skirt) PETTICOAT SDP
Purple 500-600
Royal Purple 800-1,000
Purple Blackglass 1,000-1,250

[030] (F-Skirt) [Vertical bar] 1 {'1' is upside down and backwards} (R-Skirt) [Vertical bar] SB
Green Aqua 20-30

STAR ✶

[010] (F-Skirt) ✶ SB
Aqua, Green Aqua 1-2
Blue, Blue Aqua 3-5
Green, Lime Green 10-15
Yellow Green 15-20
7-up Green, Dark Yellow Green,
Emerald Green 20-30
Aqua/Yellow Green Two Tone,
Olive Green 30-40

[020] (F-Skirt) ✶ SDP
Aqua 10-15

[030] (F-Skirt) ✶ (R-Skirt) [Blotted circle] SB
Dark Aqua 3-5

STERLING £

[010] (F-Skirt) STERLING (R-Skirt) £ SB
Aqua, Green Aqua 15-20

WHITALL TATUM 𝕎

[010] (F-Skirt) WHITALL TATUM CO. No.5 (R-Skirt) MADE IN U.S.A. [Numbers and dots] SB
Clear, Light Pink 3-5

[020] (F-Skirt) WHITALL TATUM No.5 (R-Skirt) MADE IN U.S.A. 𝕎 /[Number] SB
Clear 3-5

[030] (F-Skirt) WHITALL TATUM No.5/[Number] (R-Skirt) MADE IN U.S.A. [Numbers and dots] SB
Clear, Light Pink, Straw 3-5

[040] (F-Skirt) WHITALL TATUM No.5/[Number] (R-Skirt) MADE IN U.S.A. 𝕎 SB
Clear, Smoke, Straw 3-5

[050] (F-Skirt) WHITALL TATUM No.5/[Number] (R-Skirt) MADE IN U.S.A. 𝕎 /[Numbers and dots] SB
Clear, Olive Green Tint, Smoke, Straw 3-5

CD 164.4

NO EMBOSSING

[010] [No embossing] {Fry glass} SB
Black Opalescent 700-800
Straw 2,000-2,500
Emerald Green Blackglass,
White Opalescent 2,500-3,000
Cobalt Blue Opalescent,
Opaque Cobalt Blue 4,000-4,500

CD 165

HEMINGRAY

[010] (F-Skirt) HEMINGRAY-20 (R-Skirt) MADE IN U.S.A./[Numbers and dots] RDP
Clear, Light Straw 3-5

[020] (F-Skirt) HEMINGRAY-20 (R-Skirt) MADE IN U.S.A./[Numbers and dots] SB
Clear 15-20

CD 165.1

WHITALL TATUM 𝕎

[010] (F-Skirt) № 5/WHITALL TATUM CO. (R-Skirt) MADE IN U.S.A. SB
Light Aqua 3-5

[020] (F-Skirt) № 5/WHITALL TATUM CO. (R-Skirt) MADE IN U.S.A./[Number] SB
Aqua, Light Aqua 3-5

[030] (F-Skirt) WHITALL TATUM CO. № 5/[Number] (R-Skirt) MADE IN U.S.A. 𝕎 SB
Aqua, Ice Aqua 3-5
Ice Blue, Peach, Straw 5-10
Clear 10-15

CD 168

CD 166

CALIFORNIA

[010] (F-Skirt) A011/CALIFORNIA SB
Sage Green, Smoke	10-15
Light Purple, Purple	15-20
Burgundy	20-30
Pink	30-40
Peach	100-125
Yellow	125-150
Light Yellow Olive Green	150-175
Purple/Sage Two Tone, Purple/Yellow Two Tone	250-300

[020] (F-Skirt) CALIFORNIA SB
Aqua, Green 15-20

E.L.CO.

[010] (F-Skirt) E.L.Co. SB {Previously shown as CD 166.2 in the 1995 edition}
Aqua 30-40

CD 166.2

NO EMBOSSING

[010] [No embossing] CD
Dark Aqua 15-20

[020] [No embossing] SB
Aqua, Dark Aqua, Green Aqua,
Light Aqua 10-15
Blue 20-30
Light Gray Aqua 100-125
Apple Green, Green 200-250
Dark Yellow Green, Emerald Green ... 400-500
Olive Green 700-800
Olive Amber 1,000-1,250

CD 167

ARMSTRONG Ⓐ

[010] (F-Skirt) (Arc)ARMSTRONG/[Numbers and dots] Ⓐ (R-Skirt) 51-C1/MADE IN U.S.A. SB
Clear, Off Clear 3-5
Root Beer Amber 5-10

[020] (F-Skirt) (Arc)ARMSTRONG/[Numbers and dots] Ⓐ/[Number] (R-Skirt) 51-C1/MADE IN U.S.A. CB
Root Beer Amber 5-10
Red Amber 10-15

[030] (F-Skirt) (Arc)ARMSTRONG/[Numbers and dots] Ⓐ/[Number] (R-Skirt) 51-C1/MADE IN U.S.A. SB
Clear, Green Tint, Light Straw 3-5
Root Beer Amber 5-10
Red Amber 10-15

OWENS ILLINOIS ⓘ

[010] (F-Skirt) [Numbers] ⓘ [Numbers] (R-Skirt) 51-C1 SB
Clear 5-10

[020] (F-Skirt) [Number] ⓘ [Number and dots] (R-Skirt) 51-C1 CB
Clear, Light Straw 5-10

CD 168

HEMINGRAY

[010] (F-Skirt) HEMINGRAY/MADE IN U.S.A. (R-Skirt) D-510 CB
Clear 1-2

[020] (F-Skirt) HEMINGRAY/MADE IN U.S.A. (R-Skirt) D-510 SB
Ice Blue 3-5
Aqua 5-10
Light Carnival 20-30
Carnival 40-50

[030] (F-Skirt) LOWEX/HEMINGRAY/MADE IN U.S.A. (R-Skirt) D-510 SB
Off Clear 1-2

[040] (F-Skirt) [Numbers and dots]/HEMINGRAY/MADE IN U S A {Note no periods} (R-Skirt) [Number]/D-510 SB
Carnival 40-50

[050] (F-Skirt) [Numbers and dots]/HEMINGRAY/MADE IN U.S.A. (R-Skirt) D-510 CB
Clear 1-2

[060] (F-Skirt) [Numbers and dots]/HEMINGRAY/MADE IN U.S.A. (R-Skirt) LOWEX/D-510 SB
Dark Olive Amber 10-15
Carnival 40-50

CD 168

[070] (F-Skirt) [Numbers and dots]/
HEMINGRAY/MADE IN U.S.A.
(R-Skirt) LOWEX/[Number]/D-510 SB
Clear 3-5

[080] (F-Skirt) [Numbers and dots]/
HEMINGRAY/MADE IN U.S.A. (R-Skirt) REG.
IN U.S. PAT. OFF./LOWEX/D-510 SB
Clear 1-2
Ice Green 3-5

[090] (F-Skirt) [Numbers and dots]/
HEMINGRAY/MADE IN U.S.A. (R-Skirt) REG.
U.S. PAT. OFF./LOWEX/D-510 SB
Clear, Green Tint, Off Clear, Straw 1-2
Ice Green 3-5
Clear/Yellow Two Tone,
Dark Olive Amber, Red Amber 10-15

[100] (F-Skirt) [Numbers and dots]/
HEMINGRAY/MADE IN U.S.A. (R-Skirt) REG.
U.S. PAT. OFF./LOWEX/['[Number]' blotted
out]/D-510 SB
Dark Olive Amber 10-15

[110] (F-Skirt) [Numbers and dots]/
HEMINGRAY/MADE IN U.S.A.
(R-Skirt) [Number]/D-510 SB
Clear 1-2
Ice Blue 3-5
Dark Olive Amber 10-15
Dark Olive Amber Two Tone 15-20
Light Carnival, Orange Amber,
Red Amber 20-30
Carnival 40-50
Opaque Blackglass 175-200
White Milk 350-400

[120] (F-Skirt) [Numbers and dots]/
HEMINGRAY/MADE IN U.S.A. (R-Skirt) ['REG.
U.S. PAT. OFF.' blotted out]/['LOWEX' blotted
out]/D-510 CB
Clear, Straw 1-2

[130] (F-Skirt) [Numbers]/HEMINGRAY/MADE
IN U.S.A. (R-Skirt) D-510 SB
Blue Tint 1-2
Ice Blue 3-5
Carnival 40-50

[140] (F-Skirt) [Number]/HEMINGRAY/MADE
IN U.S.A. (R-Skirt) ['LOWEX' blotted out]/D-510
SB
Clear, Yellow Tint 1-2
Dark Olive Amber, Dark Olive Green 10-15

[150] (F-Skirt) ['LOWEX' blotted out]/
HEMINGRAY/MADE IN U.S.A. (R-Skirt) D-510
CB
Clear 1-2

WHITALL TATUM ▽

[010] (F-Skirt) WHITALL TATUM CO. № 11/
[Number] (R-Skirt) MADE IN U.S.A. ▽ SB
Ice Blue, Light Aqua, Light Green 20-30

CD 168.5

NO NAME

[020] (F-Skirt) [Glass dot] SB
Dark Aqua 2,000-2,500

CD 168.6

NO NAME

CD 168.6
$3\ 1/4\ \times\ 3\ 7/8$

[010] (F-Skirt) [4 small dots in a vertical line]
(R-Skirt) [4 small dots in a vertical line]
{Previously shown as CD 168.5 in the 1995
edition} SB
Dark Aqua 2,000-2,500

CD 169

HEMINGRAY

[010] (F-Skirt) HEMINGRAY (R-Skirt) TYPE.1
SDP
Aqua 3,000-3,500

WHITALL TATUM ▽

[010] (F-Skirt) NO.4/WHITALL TATUM CO.
(R-Skirt) MADE IN U.S.A./[Number] SB
Light Aqua 3-5
Ice Blue 5-10

CD 176

CD 169.5

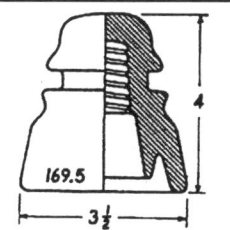

BROOKFIELD

[010] (F-Skirt) BROOKFIELD SB
Aqua, Green Aqua 10-15
Green 20-30
Dark Green 40-50

CD 170

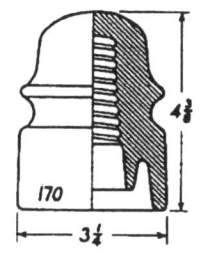

NO EMBOSSING

[010] [No embossing] SB
Aqua, Dark Aqua, Green Aqua, Light Aqua . 10-15
Blue Aqua 20-30
Light Blue 50-75
Dark Yellow Green 800-1,000

CD 170.1

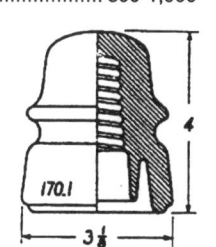

NO EMBOSSING

[010] [No embossing] CD
Dark Aqua, Light Aqua 20-30
Light Blue 30-40
Milky Light Blue 500-600

[020] [No embossing] SB
Aqua, Dark Blue Aqua,
Green Aqua, Light Aqua 10-15
Light Blue Milk, Light Gray 500-600
Emerald Green 800-1,000
Light Lavender 2,500-3,000
Dark Golden Amber 5,000-7,500

CD 173

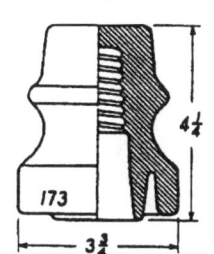

BROOKFIELD

[010] (Dome) BROOKFIELD SB
Aqua 3,500-4,000

CD 174

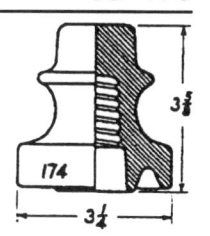

BROOKFIELD

[010] (F-Skirt) BROOKFIELD SB
Aqua 3,500-4,000

CD 175

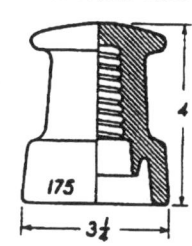

HEMINGRAY

[010] (F-Skirt) HEMINGRAY-25
(R-Skirt) MADE IN U.S.A. SB
Aqua, Blue Aqua, Ice Blue 20-30

[020] (F-Skirt) HEMINGRAY-25 /[Numbers and dots] (R-Skirt) MADE IN U.S.A. /[Number] SB
Clear 20-30

CD 175.5

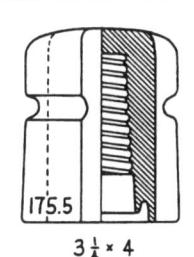

NO EMBOSSING

[010] [No embossing] {Specimen only} SB
Aqua 1,000-1,250

CD 176

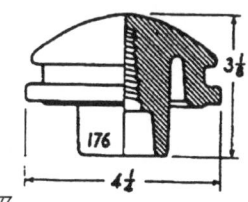

WHITALL TATUM ▽

[010] (F-Below wire groove) WHITALL TATUM CO. NO.12 MADE IN U.S.A. (R-Below wire groove) PATENT NO.1708038 SB
Clear, Ice Aqua, Light Straw 10,000-15,000

119

CD 176

[020] (F-Below wire groove) WHITALL TATUM CO. NO.12 MADE IN U.S.A. (R-Below wire groove) PATENT NO.1708038 {Copper top} SB
Ice Aqua 10,000-15,000

CD 178

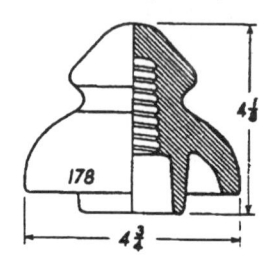

CALIFORNIA

[010] (F-Skirt) CALIFORNIA (R-Skirt) 303/ SANTA ANA SB
Sage Green 75-100
Smoke 100-125
Purple 125-150

HEMINGRAY

[010] (F-Skirt) HEMINGRAY (R-Skirt) PATENT MAY 2 1893 SDP
Aqua, Blue Aqua 150-175

NO EMBOSSING

[010] [No embossing] SB
Aqua 15-20

SANTA ANA

[010] (F-Skirt) 303/SANTA ANA SB
Purple, Smoke 125-150

[015] (F-Skirt) SANTA ANA RB
Aqua, Dark Aqua 20-30

[020] (F-Skirt) SANTA ANA SB
Blue, Green, Lime Green, Yellow Green 50-75
Light Purple, Purple, Smoke 75-100
Olive Green 150-175

[030] (F-Skirt) ['SANTA ANA' blotted out] SB
Dark Aqua 20-30

CD 178.5

LOCKE

CD 178.5
$5\ 1/4 \times 3\ 1/2$

[010] (F-Skirt) MAR. 31, 1914 F M LOCKE/ VICTOR/N.Y. (R-Skirt) PAT. FEB 2, 1915 {Embossing is upside down} SB
Light Opalescent 15,000-20,000

PATENT - OTHER

[010] (F-Skirt) PAT. FEB 2, 1915 {Embossing is upside down} SB
Opalescent 15,000-20,000

CD 180

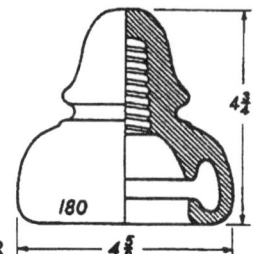

LIQUID INSULATOR

[010] (F-Skirt) LIQUID INSULATOR SB
Light Aqua 5,000-7,500

CD 180.5

FLUID INSULATOR

CD 180.5
$5 \times 5\ 5/8$

[010] (F-Skirt) FLUID INSULATOR SB {Previously shown as CD 180.1 in the 1995 edition}
Ice Aqua, Ice Gray Green 7,500-10,000
Green 10,000-15,000

CD 181

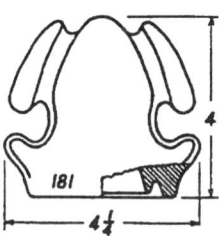

NO EMBOSSING

[010] [No embossing] SB
Light Aqua 4,500-5,000

CD 185

CD 181.5

NO EMBOSSING

CD 181.5
4 $\frac{7}{8}$ x 4 $\frac{3}{4}$

[010] [No embossing] SB
Dark Aqua .... 20,000+

CD 182

DRY SPOT INSULATOR

[010] (F-Skirt) DRY SPOT INSULATOR № 10 (R-Skirt) MADE IN U.S.A. SB
Light Straw 3,500-4,000

[020] (F-Skirt) DRY SPOT INSULATOR № 10 (R-Skirt) ▽/MADE IN U.S.A. SB
Off Clear, Straw 3,500-4,000

CD 183

HEMINGRAY

[010] (F-Skirt) HEMINGRAY-71 (R-Skirt) MADE IN U.S.A. CB
Clear 5-10

[020] (F-Skirt) HEMINGRAY-71 (R-Skirt) MADE IN U.S.A./[Numbers and dots] CB
Clear 5-10

[030] (F-Skirt) HEMINGRAY-71 (R-Skirt) MADE IN U.S.A./[Numbers and dots] SB
Clear 5-10

[040] (F-Skirt) HEMINGRAY-71/LOWEX (R-Skirt) MADE IN U.S.A./[Numbers and dots] SB
Clear 5-10
Ice Green 10-15

[050] (F-Skirt) HEMINGRAY-71/[Numbers and dots] (R-Skirt) MADE IN U.S.A./[Number] SB
Clear 5-10
Ice Blue, Light Lemon 10-15

[060] (F-Skirt) HEMINGRAY-71/[Numbers] (R-Skirt) MADE IN U.S.A. SB
Ice Blue 10-15

[070] (F-Skirt) HEMINGRAY-71/['LOWEX' blotted out] (R-Skirt) MADE IN U.S.A./[Numbers and dots] CB
Clear 5-10

CD 184

CLIMAX

CD 184
4 $\frac{3}{4}$ x 4 $\frac{1}{4}$

[010] (F-Center) CLIMAX {Specimen only} SB
Blue Aqua .... 20,000+

CD 185

B

[010] (F-Above wire groove) B SB
Aqua 100-125

[020] (F-Above wire groove) "B-1" SB
Dark Aqua 200-250
Aqua 250-300

HEMINGRAY

[010] (F-Skirt) HEMINGRAY (R-Skirt) № 95 SB
Aqua 150-175
Hemingray Blue 175-200

[020] (F-Skirt) HEMINGRAY (R-Skirt) № 95 SDP
Aqua 75-100

[030] (F-Skirt) HEMINGRAY-95 (R-Skirt) MADE IN U.S.A. SDP
Aqua 75-100
Hemingray Blue 100-125

CD 185

[040] (F-Skirt) HEMINGRAY-95
(R-Skirt) MADE IN U.S.A./['HEMINGRAY-95'
blotted out] SDP
Aqua 75-100

JEFFREY MFG.CO.

[003] (F-Above wire groove) SPECIAL MINE
INSULATOL/JEFFREY MFG.CO. {Note
spelling} (F-Skirt) COLUMBUS OHIO/PATENT
SEPT 25 1894 {Drip points are on the top of the
insulator} SDP
Hemingray Blue 200-250

[007] (F-Above wire groove) SPECIAL MINE
INSULATOR/JEFFREY MFG.CO.
(F-Skirt) COLUMBUS OHIO/PATENT
APPLIED FOR SB
Aqua 175-200

[010] (F-Above wire groove) SPECIAL MINE
INSULATOR/JEFFREY MFG.CO.
(F-Skirt) COLUMBUS OHIO/PATENT SEPT
25 1894 SB
Aqua 150-175
Ice Aqua 250-300

[020] (F-Above wire groove) SPECIAL MINE
INSULATOR/JEFFREY MFG.CO.
(F-Skirt) COLUMBUS OHIO/PATENT SEPT
25 1894 {Drip points may be on the top or on
the bottom} SDP
Aqua 150-175
Hemingray Blue 200-250

[030] (F-Above wire groove) SPECIAL MINE
INSULATOR/JEFFREY MFG.CO.
(F-Skirt) COLUMBUS OHIO/PATENT SEPT
25 1894/['APPLIED FOR' blotted out] SB
Ice Aqua 250-300

[040] (F-Above wire groove) SPECIAL MINE
INSULR/[Blotted out embossing]/JEFFREY
MFG.CO. {Note spelling} (F-Skirt) COLUMBUS
OHIO/PATENT SEPT 25 1894 {Drip points are
on the bottom of the insulator} SDP
Aqua 200-250

MINE INSULATOR

[010] (F-Skirt) MINE/INSULATOR SB
Aqua 100-125

NO EMBOSSING

[010] [No embossing] SB
Aqua 75-100

STAR ✹

[010] (F-Skirt) ✹ SB
Aqua, Blue Aqua 350-400
Green, Lime Green 700-800
Yellow Green 800-1,000

CD 186

HEMINGRAY

[010] (F-Skirt) HEMINGRAY-19/[Number and
dots] (R-Skirt) MADE IN U.S.A./[Number] SB
Clear, Green Tint, Light Lemon 1,750-2,000

CD 186.1

CD 186.1
$3\ ^{1}/_{4}$ x $3\ ^{7}/_{8}$

HEMINGRAY

[010] (F-Skirt) HEMINGRAY-19/[Numbers and
dots] (R-Skirt) MADE IN U.S.A./[Number] SB
Clear 1,500-1,750

CD 186.2

HEMINGRAY

[010] (F-Skirt) HEMINGRAY-19/[Number and
dots] (R-Skirt) MADE IN U.S.A./[Number] {A six
sided nut with machine threading is embedded
in the dome} SB
Clear, Green Tint 500-600

CD 187

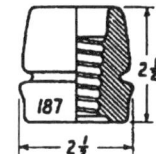

CALIFORNIA

[010] (F-Skirt) CALIFORNIA SB
Smoke 75-100
Light Purple, Purple 100-125
Light Sage Green 125-150

GREELEY

[010] (F-Above wire groove) GREELEY/N.Y.
(R-Above wire groove) PATENTED/NOV. 23rd 1886 SB
Light Aqua 300-350

NO NAME

[010] (F-Skirt) [Vertical bar] SB
Aqua, Blue, Blue Aqua,
Green Aqua, Light Aqua 10-15
Light Green 15-20
Yellow Green 30-40

PAT'D

[010] (F-Above wire groove) PAT'D SB
Green 15-20

PATENT - OTHER

[010] (F-Above wire groove) PAT'D/NOV. 23D 1886 SB
Aqua, Ice Green, Light Aqua, Light Blue 5-10
Dark Blue Aqua 15-20
Yellow Green 30-40
Emerald Green 75-100
Olive Green 125-150
Clear 250-300
Amber Blackglass, Light Purple,
Olive Blackglass 400-500

[020] (F-Above wire groove) PAT'D/NOV. 23D 1886 {Two segmented threads} SB
Light Aqua 150-175

CD 188

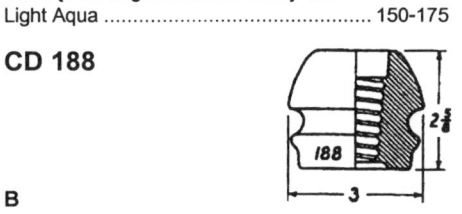

B

[010] (F-Above wire groove) B SB
Dark Aqua 10-15
Emerald Green 50-75
Dark Yellow Green, Olive Green 75-100

CD 190/191

NO EMBOSSING

[010] [No embossing] SB
Aqua 10-15

PATENT - OTHER

[010] (F-Above wire groove) PAT'D/NOV. 23D 1886 SB
Aqua, Dark Aqua, Light Aqua, Light Blue 10-15
Green, Ice Green 20-30
Aqua/Yellow Green Two Tone,
Emerald Green 50-75

CD 190/191

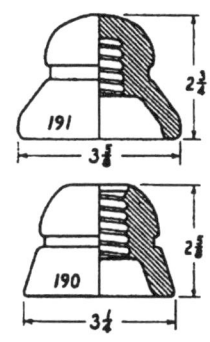

A.T.& T.CO.

[010] {191 Top} (Dome) [Number]
(F-Skirt) A.T.& T.Co. {190 Bottom}
(F-Skirt) [Number] A.T.& T.Co. {Number on the skirt may appear before or after 'A.T.& T.Co.'} SB
Aqua, Blue Aqua, Dark Aqua,
Light Aqua, Light Blue 10-15
Green 40-50
Emerald Green 200-250

[020] {191 Top} (F-Skirt) A.T.& T.Co. {190 Bottom} (F-Skirt) A.T.& T.Co. SB
Ice Green, Light Aqua, Light Blue,
Light Green 10-15

[030] {191 Top} (F-Skirt) A.T.& T.Co. {190 Bottom} (F-Skirt) [Number] A.T.& T.Co. SB
Aqua, Dark Green Aqua 10-15
Green 40-50

[040] {191 Top} (F-Skirt) A.T.& T.Co. {MLOD} {190 Bottom} (F-Skirt) [Number] A.T.& T.Co. SB
Green Aqua, Ice Green, Light Aqua 10-15
Green 40-50

[050] {191 Top} {Unverified} {190 Bottom} (F-Skirt) [Number] ['A.T.& T.Co' blotted out] SB
Dark Yellow Green 250-300

CD 190/191

AM.TEL.& TEL.CO.

[010] {191 Top} (F-Skirt) AM.TEL.& TEL.Co. (R-Skirt) TWO PIECE/TRANSPOSITION {190 Bottom} (F-Skirt) AM.TEL.& TEL.Co. (R-Skirt) TWO PIECE/TRANSPOSITION SB
Aqua, Blue 10-15
Hemingray Blue 15-20
Ice Aqua 20-30
Green 50-75
Milky Aqua 350-400
Jade Green Milk 600-700

[020] {191 Top} (F-Skirt) AM.TEL.& TEL.Co. (R-Skirt) TWO PIECE/TRANSPOSITON {Note spelling} {190 Bottom} (F-Skirt) AM.TEL.& TEL.Co. (R-Skirt) TWO PIECE/TRANSPOSITON {Note spelling} SB
Aqua 20-30

[030] {191 Top} (F-Skirt) AM.TEL.& TEL.Co. {190 Bottom} (F-Skirt) AM.TEL.& TEL.Co. SB
Light Aqua, Light Green Aqua 20-30

[040] {191 Top} (F-Skirt) AM.TEL.& TEL.Co. {Note 'E' is missing middle bar} {190 Bottom} (F-Skirt) AM.TEL.& TEL.Co. SB
Light Aqua 20-30

B

[010] {191 Top} (Dome) [Number] (F-Skirt) B {190 Bottom} (F-Skirt) B SB
Emerald Green 150-175

[020] {191 Top} (F-Skirt) B (R-Skirt) [Number] {190 Bottom} (F-Skirt) B (R-Skirt) [Number] SB
Green 40-50

[030] {191 Top} (F-Skirt) B {190 Bottom} (F-Skirt) B SB
Aqua, Blue Aqua, Dark Aqua 10-15
Green 50-75
Light Aqua 100-125
Emerald Green 175-200
Yellow Green 200-250
Olive Green 300-350

CALIFORNIA

[010] {191 Top} (F-Skirt) CALIFORNIA {190 Bottom} (F-Skirt) A021/CALIFORNIA SB
Light Purple, Smoke 2,500-3,000

DIAMOND ◇

[010] {191 Top} [No embossing] {190 Bottom} (F-Above wire groove) ◇ (F-Skirt) ◇ (R-Above wire groove) ◇ (R-Skirt) ◇ SB
Aqua, Ice Blue, Light Aqua 20-30
Ice Green, Light Yellow,
Light Yellow Green, Lime Green 30-40
Royal Purple 100-125
Aqua Tint 125-150
Light Purple 175-200
Gray, Gray Tint, Off Clear, Smoke 250-300

[020] {191 Top} [No embossing] {190 Bottom} (F-Skirt) ◇ (R-Skirt) ◇ SB
Light Blue 150-175

HEMINGRAY

[010] {191 Top} (F-Skirt) HEMINGAY-50 {Note spelling} (R-Skirt) MADE IN U.S.A. {190 Bottom} (F-Skirt) HEMINGAY-50 {Note spelling} (R-Skirt) MADE IN U.S.A. SB
Aqua, Blue Aqua, Hemingray Blue 30-40

[020] {191 Top} (F-Skirt) HEMINGRAY (R-Skirt) № 50 {190 Bottom} (F-Skirt) HEMINGRAY (R-Skirt) № 50 SB
Aqua, Green Aqua 10-15

[030] {191 Top} (F-Skirt) HEMINGRAY-50 (R-Skirt) MADE IN U.S.A. {190 Bottom} (F-Skirt) HEMINGRAY-50 (R-Skirt) MADE IN U.S.A. SB
Clear 5-10
Aqua, Blue, Blue Aqua, Green Aqua,
Green Tint, Ice Blue, Ice Green 10-15
Hemingray Blue 15-20

[040] {191 Top} (F-Skirt) HEMINGRAY-50/ [Numbers and dots] (R-Skirt) MADE IN U.S.A./ [Number] {190 Bottom} (F-Skirt) HEMINGRAY-50/[Numbers and dots] (R-Skirt) MADE IN U.S.A./[Number] CB
Clear 5-10

[050] {191 Top} (F-Skirt) HEMINGRAY-50/ [Numbers and dots] (R-Skirt) MADE IN U.S.A./ [Number] {190 Bottom} (F-Skirt) HEMINGRAY-50/[Numbers and dots] (R-Skirt) MADE IN U.S.A./[Number] SB
Clear 5-10
Aqua Tint, Blue Tint 10-15

[060] {191 Top} (F-Skirt) HEMINGRAY-50/ [Numbers] (R-Skirt) MADE IN U.S.A. {190 Bottom} (F-Skirt) HEMINGRAY-50/[Numbers] (R-Skirt) MADE IN U.S.A. SB
Ice Blue 10-15

CD 192.1/193.1

[070] {191 Top} {Unverified} {190 Bottom} (F-Skirt) HEMINGRAY-50 (R-Skirt) MADE IN U.S.A./['№ 50' blotted out] SB
Aqua 10-15
Hemingray Blue 15-20

K

[010] {191 Top} (F-Skirt) K {190 Bottom} (F-Skirt) K SB
Aqua 400-500
Ice Blue, Light Green 500-600
Green 600-700

NO EMBOSSING

[010] {191 Top} [No embossing] {190 Bottom} [No embossing] {Denver product} SB
Aqua, Light Blue, Purple,
Royal Purple, Steel Blue,
Steel Blue Violet 1,000-1,250

P.R.R.

[010] {191 Top} (Dome) P.R.R. {190 Bottom} (F-Skirt) B SB
Dark Aqua 75-100
Green 125-150
Emerald Green 250-300

PRISM ▭

[010] {191 Top} (F-Skirt) ▭ {190 Bottom} (F-Skirt) ▭ SB
Aqua, Blue Aqua 75-100
Ice Blue 100-125
Green 175-200
Yellow Green 350-400
Dark Yellow Green, Olive Green 600-700

TWO PIECE TRANSPOSITION

[010] {191 Top} (F-Skirt) TWO PIECE/ TRANSPOSITION (R-Skirt) [Blotted out embossing] {190 Bottom} (F-Skirt) TWO PIECE/TRANSPOSITION (R-Skirt) [Blotted out embossing] SB
Aqua 15-20

[020] {191 Top} (F-Skirt) TWO PIECE/ TRANSPOSITION {190 Bottom} (F-Skirt) TWO PIECE/TRANSPOSITION SB
Aqua, Blue, Blue Aqua,
Green Aqua, Light Aqua 10-15
Hemingray Blue 15-20
Ice Blue 30-40
Milky Green Aqua 350-400

[030] {191 Top} (F-Skirt) TWO PIECE/ TRANSPOSITON {Note spelling} {190 Bottom} (F-Skirt) TWO PIECE/TRANSPOSITON {Note spelling} SB
Aqua 10-15
Milky Aqua 400-500

[040] {191 Top} {Unverified} {190 Bottom} (F-Skirt) TWO PIECE/TRANSPOSITION (R-Skirt) [Blotted out embossing] SB
Gray Green 350-400

CD 192/193

CD 193
3 $^3/_8$ x 2 $^3/_4$

CD 192
3 $^1/_4$ x 2 $^1/_2$

AM.TEL.& TEL.CO.

[010] {193 Top} (F-Skirt) AM.TEL.& TEL.Co. {192 Bottom} (F-Skirt) AM.TEL.& TEL.Co. SB
Aqua 150-175

NO EMBOSSING

[010] {193 Top} [No embossing] {192 Bottom} [No embossing] SB
Aqua 150-175
Green 200-250

CD 192.1/193.1

CD 193.1
3 x 2 $^3/_4$

CD 192.1
2 $^3/_4$ x 2 $^1/_2$

AM.TEL.& TEL.CO.

[010] {193.1 Top} (F-Skirt) AM.TEL.& TEL.Co. {192.1 Bottom} (F-Skirt) AM.TEL.& TEL.Co. SB
Aqua 2,000-2,500

NO EMBOSSING

[010] {193.1 Top} [No embossing] {192.1 Bottom} {Unverified} SB
Light Aqua 200-250

CD 192.1/193.1

[020] {193.1 Top} [No embossing] {MLOD} {192.1 Bottom} {Unverified} SB
Light Aqua 200-250
Light Green 250-300

CD 194/195

HEMINGRAY

[010] {195 Top} (F-Skirt) HEMINGRAY-54-A (R-Skirt) MADE IN U.S.A./[Numbers] {194 Bottom} (F-Skirt) HEMINGRAY-54-B (R-Skirt) MADE IN U.S.A./[Numbers] SB
Light Purple, Purple 125-150
Off Clear 300-350

CD 194.5/195.5

PYREX

2 3/8 x 2 3/4

[010] {195.5 Top} (F-Skirt) PYREX/T.M. REG. U.S. PAT. OFF. MADE IN U.S.A. (R-Skirt) PAT.APP.FOR {194.5 Bottom} (F-Below wire groove) CODE NO 70005 (F-Skirt) PYREX/T.M. REG. U.S. PAT. OFF. MADE IN U.S.A. SB
Light Carnival {Top}, Clear {Bottom} . 7,500-10,000

CD 196

B

[010] (F-Skirt) B SDP
Aqua, Dark Aqua 100-125
Green 200-250
Emerald Green 600-700

H.G.CO.

[005] (Dome) H (F-Skirt) H.G.CO. (R-Skirt) PAT.MAY 2 1893 SDP
Ice Aqua 75-100

[010] (F-Skirt) H.G.CO. (R-Skirt) PAT.MAY 2 1893 SDP
Aqua, Light Aqua 50-75
Hemingray Blue, Ice Aqua,
Ice Blue, Ice Green 75-100
Gray Green, Milky Aqua 800-1,000
Gray Purple, Purple/Sage Two Tone .. 2,500-3,000
Light Purple 5,000-7,500

[015] (F-Skirt) H.G.CO. (R-Skirt) PAT^D MAY 2 1893 {Variation has RDP on the umbrella and SDP on the skirt} SDP
Aqua 75-100

[020] (F-Skirt) H.G.CO. (R-Skirt) PAT^D MAY 2 1893 SDP
Dark Aqua, Light Aqua 50-75
Ice Aqua, Ice Blue 75-100

[030] (F-Skirt) H.G.CO. (R-Umbrella) PATENT APPLIED FOR (R-Skirt) PAT^D MAY 2 1893 SDP
Aqua, Light Aqua 200-250

[040] (F-Skirt) № 51/H.G.CO. (R-Skirt) PAT.MAY 2 1893 SDP
Aqua 75-100
Dark Blue Aqua,
Hemingray Blue, Ice Aqua 100-125

HEMINGRAY

[010] (F-Umbrella) HEMINGRAY (F-Skirt) HEMINGRAY (R-Skirt) № 51 SDP
Aqua, Blue Aqua 75-100

[020] (F-Umbrella) HEMINGRAY (F-Skirt) № 51 (R-Skirt) HEMINGRAY SDP
Blue Aqua 75-100
Hemingray Blue 100-125

NO EMBOSSING

[010] [No embossing] SDP
Aqua, Blue Aqua 100-125

O.V.G.CO.

[010] (F-Skirt) O.V.G.CO. SB
Aqua 3,000-3,500
Light Blue Aqua 3,500-4,000

CD 200

CD 196.2

H.G.CO. CD 196.2 $3\,^7/_8 \times 4\,^1/_2$

[010] (F-Skirt) H.G.CO. (R-Umbrella) PATENT APPLIED FOR (R-Skirt) PATD MAY 2 1893 SDP
Aqua 500-600

CD 196.5

H.G.CO. CD 196.5 $4\,^1/_8 \times 4\,^1/_2$

[010] (F-Skirt) H.G.CO. (R-Skirt) PAT MAY 2 1893 SDP
Aqua 1,250-1,500

CD 197

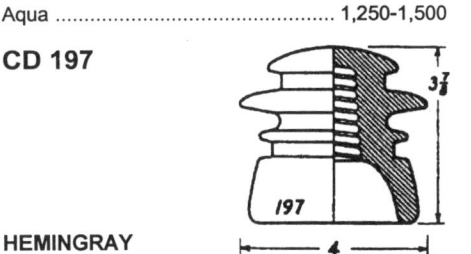

HEMINGRAY |— 4 —|

[010] (F-Skirt) HEMINGRAY-53 (R-Skirt) MADE IN U.S.A./[Numbers and dots] CB
Clear, Straw 1-2
Green Tint 3-5

[020] (F-Skirt) HEMINGRAY-53 (R-Skirt) MADE IN U.S.A./[Numbers and dots] SB
Clear 1-2
Pink Tint 3-5

[030] (F-Skirt) HEMINGRAY-53S/[Number and dots] (R-Skirt) MADE IN U.S.A./[Number] SB
Clear, Off Clear 1-2
Blue Tint, Ice Blue 3-5

WHITALL TATUM 〈W〉

[010] (F-Skirt) (Arc)WHITALL TATUM/N⁰.15 (R-Skirt) (Arc)MADE IN U.S.A./[Number and dots] ['W/T' in a circle] SB
Clear, Smoke 1-2

[020] (F-Skirt) (Arc)WHITALL TATUM/N⁰.15 (R-Skirt) (Arc)MADE IN U.S.A./[Number and dots]/Ⓐ/[Number] SB
Clear, Smoke 1-2

[040] (F-Skirt) WHITALL TATUM CO. NO.15/ [Number] (R-Skirt) MADE IN U.S.A. 〈W〉 SB
Clear, Straw 1-2

[030] (F-Skirt) WHITALL TATUM CO. NO.15/ [Number] (R-Skirt) MADE IN U.S.A./〈W〉/ [Number] SB
Light Straw 1-2

CD 199

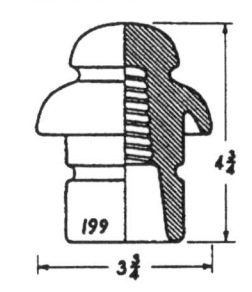

PRISM ▣

[010] (F-Umbrella) ▣ SB
Blue 350-400
Light Green 600-700
Lime Green 800-1,000
Yellow Green 2,000-2,500

CD 200

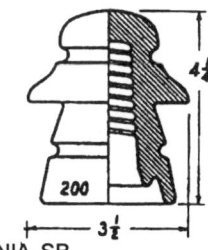

CALIFORNIA

[010] (F-Skirt) CALIFORNIA SB
Purple 1,250-1,500
Dark Burgundy, Peach, Yellow 1,500-1,750
Purple/Peach Two Tone 1,750-2,000

NO NAME

[010] (F-Skirt) N⁰ 2 (R-Skirt) TRANSPOSITION SB
Aqua, Dark Aqua, Light Green Aqua 20-30
Green 50-75
Olive Green 800-1,000

127

CD 200

[020] (F-Skirt) № 2 (R-Skirt) TRANSPOSITION SDP
Aqua, Dark Green Aqua 800-1,000
Green 1,000-1,250

STAR ✱

[010] (F-Skirt) ✱ SB
Blue Aqua, Light Aqua 125-150
Light Blue 150-175
Light Green 200-250
Green, Lime Green 400-500
Yellow Green 600-700

CD 201

CALIFORNIA

[010] (F-Skirt) CALIFORNIA SB
Blue Aqua, Light Green 2,000-2,500

H.G.CO.

[010] (F-Skirt) H.G.CO. (R-Skirt) PATENT MAY 2 1893 SDP
Aqua 2,000-2,500

[015] (F-Umbrella) № 2 TRANSPOSITION/[3 blot out dots] (F-Skirt) H.G.Co (R-Skirt) PATENT MAY 2 1893 SDP
Aqua, Light Aqua 2,000-2,500

[020] (F-Umbrella) [3 blot out dots] (F-Skirt) H.G.Co SB
Blue Aqua, Hemingray Blue 2,000-2,500

HEMINGRAY

[010] (F-Umbrella) № 2 TRANSPOSITION (F-Skirt) HEMINGRAY (R-Skirt) PATENT MAY 2 1893 SDP
Aqua, Blue Aqua 20-30
Milky Aqua 350-400

[020] (F-Umbrella) № 2 TRANSPOSITION (F-Skirt) HEMINGRAY (R-Skirt) PATENT/MAY 2 1893 SDP
Aqua, Dark Blue Aqua, Green Aqua 20-30

[030] (F-Umbrella) № 2 TRANSPOSITION (F-Skirt) HEMINGRAY/['H.G.CO.' blotted out] (R-Skirt) PATENT MAY 2 1893 SDP
Aqua 20-30

[040] (F-Umbrella) № 2 TRANSPOSITION (F-Skirt) HEMINGRAY/['H.G.CO.' blotted out] (R-Skirt) PATENT/MAY 2 1893 SDP
Aqua, Dark Blue Aqua 20-30

CD 202

HEMINGRAY

[005] (F-Skirt) HEMINGRAY-53 (R-Skirt) MADE IN U.S.A. RDP
Aqua, Blue Aqua, Hemingray Blue, Ice Aqua 10-15

[010] (F-Skirt) HEMINGRAY-53 (R-Skirt) MADE IN U.S.A. SDP
Aqua, Blue Aqua 5-10
Hemingray Blue 10-15

[020] (F-Skirt) HEMINGRAY-53/MADE IN U.S.A. (R-Skirt) PATENTED/MAY 2 1893 SDP
Hemingray Blue 10-15

[030] (F-Skirt) HEMINGRAY-53/MADE IN U.S.A./['№ 14' blotted out] (R-Skirt) PATENTED/MAY 2 1893 SDP
Hemingray Blue 10-15

[040] (F-Skirt) HEMINGRAY-53/[Numbers and dots] (R-Skirt) MADE IN U.S.A./[Number] RDP
Aqua, Clear 5-10
Ice Blue 10-15

[050] (F-Skirt) HEMINGRAY/№ 14 (R-Skirt) PATENTED/MAY 2 1893 SDP
Aqua, Blue Aqua 5-10
Green 20-30

[060] (F-Skirt) HEMINGRAY/№ 53 (R-Skirt) PATENTED/MAY 2 1893 SDP
Aqua 5-10

[070] (F-Skirt) HEMINGRAY/№ 53/['14' blotted out] (R-Skirt) PATENTED/MAY 2 1893 SDP
Aqua, Blue Aqua 5-10

K

[010] (F-Skirt) K SB
Aqua, Blue Aqua, Green Aqua, Light Blue .. 30-40
Green 250-300
Yellow Green 500-600
Dark Yellow Green 700-800

CD 204

LOCKE

[010] (Dome) [Number] (F-Skirt) FRED M. LOCKE VICTOR N.Y. (R-Skirt) № 14/ PAT.MAY 22 1894 SB
Aqua 5-10
Green 15-20

[020] (Dome) [Number] (F-Skirt) LOCKE VICTOR N.Y. (R-Skirt) № 14/PAT.MAY 22 1894 SB
Dark Aqua 5-10
Green 15-20
Emerald Green 75-100
Yellow Green 100-125
Olive Green 200-250

[030] (F-Skirt) FRED M LOCKE VICTOR N.Y. (R-Skirt) № 14/PAT MAY 22 1894 SB
Blue Aqua, Green Aqua, Light Aqua 5-10

[040] (F-Skirt) FRED M. LOCKE VICTOR N.Y. (R-Skirt) PAT.MAY 22 1894 SB
Blue, Light Aqua 5-10
Light Green 15-20

[050] (F-Skirt) LOCKE VICTOR N.Y. (R-Skirt) № 14/PAT.MAY 22 1894 SB
Aqua 5-10
Green 15-20

NO NAME

[010] (F-Skirt) No.51 SB
Aqua 50-75
Dark Green 125-150

[020] (F-Skirt) No.51 (R-Skirt) [Blotted out embossing] SB
Aqua 50-75

[030] (F-Skirt) № 14/[Blotted out embossing] (R-Skirt) [Blotted out embossing] SB
Blue Aqua, Dark Aqua 5-10
Green 15-20
Emerald Green 40-50
Yellow Green 50-75
Olive Green 200-250
Olive Amber 350-400

CD 203

ARMSTRONG Ⓐ

[010] (F-Skirt) Ⓐrmstrong T.W. (R-Skirt) MADE IN U.S.A. [Numbers and dots] SB
Clear, Off Clear, Smoky Straw 1-2

[020] (F-Skirt) *Armstrong's* T.W. (R-Skirt) MADE IN U.S.A. Ⓐ [Number and dots] SB
Clear, Light Smoke, Light Straw 1-2
Green Tint, Light Pink 3-5

HEMINGRAY

[010] (F-Skirt) HEMINGRAY-56 (R-Skirt) MADE IN U.S.A./[Numbers and dots] CB
Blue Tint, Clear, Green Tint, Light Straw x1
Pink Tint 1-2

KERR

[010] (F-Skirt) KERR T.W. (R-Skirt) MADE IN U.S.A. [Number and dots] SB
Clear, Green Tint 1-2

CD 203.2

ARMSTRONG Ⓐ

[010] (F-Skirt) *Armstrong's* T.W. (R-Skirt) MADE IN U.S.A. Ⓐ [Numbers and dots] SB
Clear 2,500-3,000

CD 204

LOCKE

[010] (F-Skirt) F.M. LOCKE & CO. VICTOR N.Y./PAT'D MAY 22, 1894 SB
Blue, Light Aqua, Light Green 3,000-3,500

NO EMBOSSING

[010] [No embossing] SB
Blue Aqua 2,500-3,000

129

CD 205

CD 205

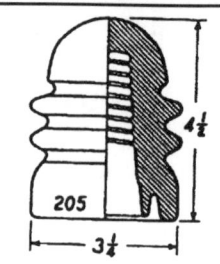

BROOKFIELD

[010] (F-Skirt) BROOKFIELD SB
Dark Aqua 5-10
Dark Green 15-20
Yellow Green 50-75

[020] (F-Skirt) BROOKFIELD (R-Skirt) № 3 TRANS. SB
Aqua, Dark Blue Aqua, Green Aqua 5-10
Dark Green, Green, Teal Green 15-20
Emerald Green 20-30
Dark Yellow Green 75-100
Dark Olive Green 100-125

[030] (F-Skirt) BROOKFIELD (R-Skirt) № 53 SB
Aqua, Dark Aqua 5-10
Green 15-20

[040] (F-Skirt) BROOKFIELD (R-Skirt) № 53/ ['№ 3 TRANS.' blotted out] SB
Dark Green Aqua 5-10

[050] (F-Skirt) BROOKFIELD (R-Skirt) ['№ 3 TRANS.' blotted out] SB
Dark Aqua 5-10
Dark Green 15-20
Emerald Green 20-30

GAYNER

[010] (F-Skirt) GAYNER (R-Skirt) NO.530/ [Number] SB
Aqua, Light Aqua 10-15
Blue 15-20

[020] (F-Skirt) GAYNER (R-Skirt) NO.530/ [Number] SDP
Light Aqua 10-15
Celery Green, Sage Green 15-20

HEMINGRAY

[010] (F-Skirt) HEMINGRAY (R-Skirt) № 55 SB
Aqua, Blue Aqua, Dark Aqua 5-10
Green 50-75

[020] (F-Skirt) HEMINGRAY-55 (R-Skirt) MADE IN U.S.A. RDP
Aqua, Hemingray Blue 5-10
Clear, Light Blue 10-15
Ice Aqua, Ice Blue 15-20

[030] (F-Skirt) HEMINGRAY-55 (R-Skirt) MADE IN U.S.A. SB
Aqua, Blue Aqua 3-5
Hemingray Blue 5-10
Ice Blue 15-20

[040] (F-Skirt) HEMINGRAY/MADE IN U.S.A. (R-Skirt) № 55 SB
Dark Blue, Hemingray Blue 15-20
Green 50-75

[050] (F-Skirt) HEMINGRAY/MADE IN U.S.A. (R-Skirt) № 55 SDP
Aqua 10-15

LYNCHBURG Ⓛ

[010] (F-Skirt) LYNCHBURG (R-Skirt) NO.53/ MADE IN U.S.A./[Number] SB
Light Aqua 10-15
Light Green 15-20

[020] (F-Skirt) LYNCHBURG (R-Skirt) NO.530/ MADE IN U.S.A./[Number] SB
Aqua, Green Aqua 10-15
Light Green 15-20

[030] (F-Skirt) LYNCHBURG (R-Skirt) NO.530/ MADE IN U.S.A./[Number] SDP
Aqua, Blue Aqua 10-15

[040] (F-Skirt) LYNCHBURG/['GAYNER' blotted out] (R-Skirt) NO.53/MADE IN U.S.A./ [Number] SB
Light Blue 10-15

[045] (F-Skirt) LYNCHBURG/['GAYNER' blotted out] (R-Skirt) NO.530/MADE IN U.S.A./ [Number] SB
Green Aqua 10-15

[050] (F-Skirt) LYNCHBURG/['GAYNER' blotted out] (R-Skirt) NO.530/MADE IN U.S.A./ [Number] SDP
Green Aqua 10-15

CD 205.5

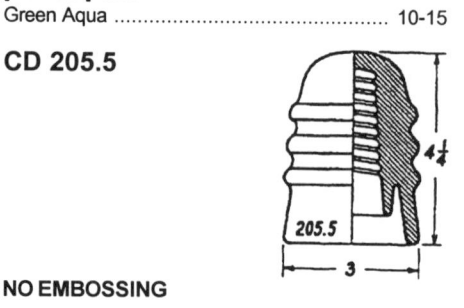

NO EMBOSSING

[010] [No embossing] {This is a CD 116 with an inner skirt} SB
Aqua 800-1,000

CD 206

NO EMBOSSING

[010] [No embossing] FDP
Light Aqua, Light Blue 400-500
Light Green 700-800

[020] [No embossing] SDP
Light Straw, Off Clear 400-500

[030] [No embossing] {Rounded turrets} SDP
Straw 400-500

CD 206.5

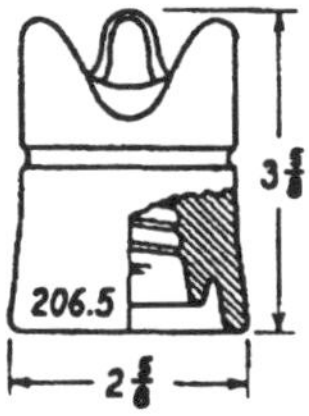

HARLOE ⌗

[010] (F-Skirt) HARLOE'S PATENT/MAR 21,
99/MAR 12, 01/DEC 9, 02/P2S SB
Aqua 3,500-4,000

[020] (F-Skirt) HARLOE'S PATENT/MAR 21,
'99/MAR 12, '01/DEC 9, '02 P2 S
(R-Skirt) [Blotted out embossing]/⌗/
HAWLEY, PA./U.S.A. SB
Aqua 3,500-4,000

[030] (F-Skirt) HARLOE'S PATENT/MAR 21,
'99/MAR 12, '01/DEC 9, '02 P2S
(R-Skirt) ⌗/HAWLEY, PA./U.S.A. [Number]
SB
Aqua, Light Aqua 3,500-4,000

CD 207

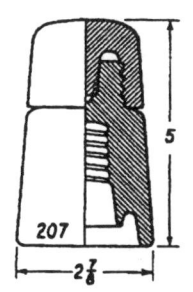

B

[010] (F-Crown) B SB
Aqua .... 20,000+

CD 208

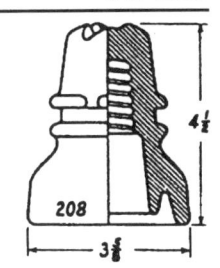

BROOKFIELD

[010] (F-Skirt) BROOKFIELD SB
Aqua 20-30
Green 50-75

CALIFORNIA

[010] (F-Skirt) CALIFORNIA SB
Light Purple 50-75
Smoke 100-125
Yellow 200-250

HEMINGRAY

[010] (F-Skirt) HEMINGRAY (R-Skirt) № 44
SDP
Aqua, Blue Aqua, Green Aqua 10-15
Green 40-50

[015] (F-Skirt) HEMINGRAY (R-Skirt) № 44/
[Blotted out embossing] SDP
Aqua, Blue Aqua 10-15

[020] (F-Skirt) HEMINGRAY-44
(R-Skirt) MADE IN U.S.A. RDP
Aqua, Blue Aqua 15-20
Hemingray Blue 20-30

[030] (F-Skirt) HEMINGRAY-44
(R-Skirt) MADE IN U.S.A. SDP
Aqua 10-15
Hemingray Blue 20-30

[040] (F-Skirt) HEMINGRAY/MADE IN U.S.A.
(R-Skirt) № 44 SDP
Aqua, Green Aqua 10-15
Hemingray Blue 20-30

[050] (F-Skirt) HEMINGRAY/MADE IN U.S.A.
(R-Skirt) № 44/[Blotted out embossing] SDP
Aqua 10-15
Hemingray Blue 20-30

[060] (F-Skirt) HEMINGRAY/MADE IN U.S.A./
[Blotted out embossing] (R-Skirt) № 44 SDP
Hemingray Blue 20-30

[070] (F-Skirt) HEMINGRAY/MADE N U.S.A.
{Note spelling} (R-Skirt) № 44 SDP
Hemingray Blue 20-30

CD 210

CD 210

NO EMBOSSING

[010] [No embossing] {Unembossed California} SB
Sage Green 20-30

POSTAL

[010] (Top of ear) POSTAL SB
Dark Aqua 3-5
Emerald Green 15-20
Olive Green 75-100

[020] (Top of ear) POSTAL SDP
Aqua, Dark Aqua 5-10

[030] (Top of ear) POSTAL (F-Skirt) [Vertical bar] SDP
Aqua, Dark Aqua 5-10

CD 211

BROOKFIELD

[010] (F-Skirt) BROOKFIELD (R-Skirt) NO LEAK D./PATENTED {A glass insert is fitted inside the skirt} SB
Dark Aqua (without glass insert),
Green Aqua (without glass insert) 200-250
Dark Aqua (with glass insert),
Green Aqua (with glass insert) 400-500

CD 213

HEMINGRAY

[010] (F-Skirt) HEMINGRAY (R-Skirt) № 43 RDP
Aqua 10-15

[020] (F-Skirt) HEMINGRAY (R-Skirt) № 43 SDP
Aqua 10-15

[030] (F-Skirt) HEMINGRAY (R-Skirt) [Vertical bar] № 43 RDP
Aqua 10-15
Hemingray Blue 15-20

[040] (F-Skirt) HEMINGRAY (R-Skirt) [Vertical bar] № 43 SDP
Aqua 10-15

[045] (F-Skirt) HEMINGRAY {Large 3/8" tall embossing} (R-Skirt) № 43 SDP
Aqua 10-15

[050] (F-Skirt) HEMINGRAY/MADE IN U.S.A. (R-Skirt) № 43 RDP
Aqua, Blue Aqua 10-15
Hemingray Blue 15-20
Emerald Green 200-250

[060] (F-Skirt) HEMINGRAY/MADE IN U.S.A. (R-Skirt) № 43 SDP
Aqua 10-15
Hemingray Blue 15-20

[070] (F-Skirt) HEMNGRAY/MADE IN U.S.A. {Note spelling} (R-Skirt) № 43 SDP
Hemingray Blue 20-30

CD 214

ARMSTRONG Ⓐ

[010] (F-Skirt) *Armstrong's* № 10 (R-Skirt) MADE IN U.S.A. Ⓐ [Number and dots] SB
Clear, Light Straw 1-2
Olive Green Tint 3-5

HEMINGRAY

[010] (F-Skirt) HEMINGRAY-43 (R-Skirt) MADE IN U.S.A. RDP
Aqua, Blue Aqua, Dark Green Aqua 1-2
Blue, Ice Aqua, Ice Blue 3-5
7-up Green 50-75
Carnival 350-400

[015] (F-Skirt) HEMINGRAY-43 (R-Skirt) MADE IN U.S.A. {Short, square wire groove variant} RDP
Aqua 10-15

CD 217

[020] (F-Skirt) HEMINGRAY-43
(R-Skirt) MADE IN U.S.A./[Numbers and dots]
CB
Clear, Light Straw, Yellow Tint x1

[030] (F-Skirt) HEMINGRAY-43
(R-Skirt) MADE IN U.S.A./[Numbers and dots]
RDP
Clear, Straw x1

[040] (F-Skirt) HEMINGRAY-43/[Numbers and dots] (R-Skirt) MADE IN U.S.A./[Number] RDP
Off Clear x1
Pink Tint 1-2

TELEGRAFOS NACIONALES

[010] (F-Skirt) TELEGRAFOS
(R-Skirt) NACIONALES RDP
Dark Orange Amber, Dark Red Amber 75-100
Golden Amber 100-125
Yellow Amber 125-150
Yellow 200-250

[020] (F-Skirt) TELEGRAFOS
(R-Skirt) NACIONALES {Long drip point variant} RDP
Dark Orange Amber 100-125
Golden Amber 125-150

[030] (F-Skirt) TELEGRAFOS
(R-Skirt) NACIONALES {MLOD} RDP
Golden Amber 100-125

WHITALL TATUM ▽

[005] (F-Skirt) WHITALL TATUM Co. Nº 10
(R-Skirt) MADE IN U.S.A. ▽ /[Number] SB
Clear 3-5

[010] (F-Skirt) WHITALL TATUM Co. Nº 10/[Number] (R-Skirt) MADE IN U.S.A. ▽ SB
Light Aqua 3-5

[020] (F-Skirt) WHITALL TATUM Co. Nº 10/[Number] (R-Skirt) MADE IN U.S.A. ▽ /[Number] SB
Clear, Light Aqua, Light Green,
Smoke, Straw 3-5

[025] (F-Skirt) WHITALL TATUM IΩ 10/[Number] {Note 'I' instead of 'N'}
(R-Skirt) MADE IN U.S.A. ▽ /[Number] SB
Clear 3-5

[030] (F-Skirt) WHITALL TATUM Nº 10/[Number] (R-Skirt) MADE IN U.S.A. ▽ /[Number] SB
Clear, Smoke 3-5

CD 216

ARMSTRONG Ⓐ

[010] (F-Skirt) *Armstrong's* 51-2U
(R-Skirt) MADE IN U.S.A. Ⓐ [Numbers and dots] SB
Clear 3-5
Root Beer Amber 5-10

HEMINGRAY

[010] (F-Skirt) HEMINGRAY-661
(R-Skirt) MADE IN U.S.A./[Number] [Number] SB
Clear 1-2
Dark Flashed Amber, Light Flashed Amber . 30-40
Flashed Red Amber 50-75

LOWEX

[010] (F-Skirt) LOWEX/REG. U.S. PAT. OFF.
(R-Skirt) [Number] MADE IN U.S.A. [Number] SB
Light Straw 40-50
Flashed Amber 75-100

WHITALL TATUM ▽

[010] (F-Skirt) (Arc)WHITALL TATUM/512U
(R-Skirt) (Arc)MADE IN U.S.A./[Numbers]/Ⓐ SB
Clear, Smoke 3-5
Root Beer Amber 5-10
Red Amber 10-15

CD 217

ARMSTRONG Ⓐ

[010] (F-Skirt) *Armstrong's* 51 C3
(R-Skirt) MADE IN U.S.A. Ⓐ [Numbers and dots] SB
Root Beer Amber 5-10

133

CD 218

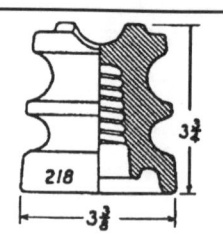

CD 218

HEMINGRAY

[010] (F-Skirt) HEMINGRAY-660
(R-Skirt) MADE IN U.S.A./[Numbers and dots] CB
Clear, Green Tint, Ice Green, Straw 3-5

CD 219

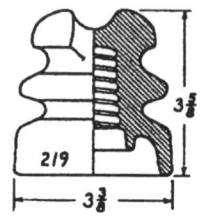

HEMINGRAY

[010] (F-Skirt) HEMINGRAY-66/[Numbers]
(R-Skirt) MADE IN U.S.A./[Number] SB
Clear, Light Lemon 30-40
Red Amber 50-75

[020] (F-Skirt) HEMINGRAY-660
(R-Skirt) MADE IN U.S.A./[Numbers and dots] CB
Clear, Straw 3-5

[030] (F-Skirt) HEMINGRAY-660
(R-Skirt) MADE IN U.S.A./[Numbers and dots] SB
Clear 3-5
Ice Blue 10-15
Olive Amber 20-30

[040] (F-Skirt) HEMINGRAY-660/LOWEX
(R-Skirt) MADE IN U.S.A./[Numbers and dots] SB
Clear, Ice Green 5-10
Olive Amber 20-30
Honey Amber 30-40
Clear w/ Treated Top,
Ice Green w/ Treated Top 200-250

[050] (F-Skirt) HEMINGRAY-660/LOWEX/
['[Numbers]' blotted out] (R-Skirt) MADE IN U.S.A./[Numbers and dots] SB
Honey Amber 30-40

[070] (F-Skirt) HEMINGRAY-660/[Numbers and dots] (R-Skirt) MADE IN U.S.A./[Number] SB
Olive Amber 20-30
Honey Amber, Yellow Amber 30-40

NO EMBOSSING

[010] [No embossing] SB
Clear 15-20

CD 220

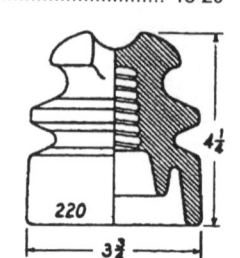

HEMINGRAY

[010] (F-Skirt) HEMINGRAY-67/[Numbers and dots] (R-Skirt) MADE IN U.S.A./[Number] SB
Light Lemon 75-100
Light Golden Amber 200-250

[020] (F-Skirt) HEMINGRAY-670
(R-Skirt) MADE IN U.S.A./[Numbers and dots] CB
Clear, Off Clear 5-10
Ice Aqua 10-15

[030] (F-Skirt) HEMINGRAY-670
(R-Skirt) MADE IN U.S.A./[Numbers and dots] SB
Clear 5-10

[040] (F-Skirt) HEMINGRAY-670
(R-Skirt) MADE IN U.S.A./[Numbers] {Mold variation has a smaller ridge and a saddle groove running from front to rear} CB
Clear 10-15

[050] (F-Skirt) HEMINGRAY-670/LOWEX
(R-Skirt) MADE IN U.S.A./[Numbers and dots] SB
Ice Green 20-30
Ice Green w/ Treated Top 200-250

[060] (F-Skirt) HEMINGRAY-670/LOWEX/
REG. U.S. PAT. OFF. (R-Skirt) MADE IN U.S.A./[Numbers and dots] CB
Ice Blue 10-15
Olive Amber 125-150

[070] (F-Skirt) HEMINGRAY-670/LOWEX/
REG. U.S. PAT. OFF. (R-Skirt) MADE IN U.S.A./[Numbers and dots] SB
Off Clear 5-10
Ice Green 10-15
Olive Amber 125-150

[080] (F-Skirt) HEMINGRAY-670/[Numbers and dots] (R-Skirt) MADE IN U.S.A./[Number] SB
Clear 5-10
Olive Amber 125-150
Dark Orange Amber, Honey Amber 150-175

CD 226

[090] (F-Skirt) HEMINGRAY-670/['LOWEX/ REG. U.S. PAT. OFF.' blotted out]
(R-Skirt) MADE IN U.S.A./[Numbers and dots] CB
Clear, Straw 5-10
Blue Tint, Green Tint 10-15

[100] (F-Skirt) HEMINGRAY-670/['LOWEX' blotted out] (R-Skirt) MADE IN U.S.A./[Numbers and dots] CB
Clear 5-10
Blue Tint, Ice Green 10-15

[110] (F-Skirt) HEMINGRAY-670/['LOWEX' blotted out]/REG. U.S. PAT. OFF.
(R-Skirt) MADE IN U.S.A./[Number and dots] CB
Clear, Straw 5-10
Blue Tint 10-15
Clear w/ Treated Top 200-250

CD 221

HEMINGRAY

[010] (F-Skirt) HEMINGRAY-68
(R-Skirt) MADE IN U.S.A./[Number and dots] SB
Golden Amber 400-500

[020] (F-Skirt) HEMINGRAY-68/[Numbers and dots] (R-Skirt) MADE IN U.S.A./[Number] SB
Light Lemon 75-100
Olive Amber 300-350

[030] (F-Skirt) HEMINGRAY-68/[Numbers and dots] (R-Skirt) MADE IN U.S.A./[Number] {Aluminum coated top} SB
Green Tint 200-250

[040] (F-Skirt) HEMINGRAY-68/['LOWEX/ REG. U.S. PAT. OFF.' blotted out]
(R-Skirt) MADE IN U.S.A./[Numbers and dots] CB
Light Straw 75-100

[050] (F-Skirt) HEMINGRAY-680
(R-Skirt) MADE IN U.S.A./[Numbers and dots] CB
Clear, Green Tint, Straw 5-10
Ice Green 10-15

[060] (F-Skirt) HEMINGRAY-680
(R-Skirt) MADE IN U.S.A./[Numbers and dots] SB
Clear 5-10

[070] (F-Skirt) HEMINGRAY-680/LOWEX
(R-Skirt) MADE IN U.S.A./[Numbers and dots] SB
Clear 10-15
Ice Green 15-20
Ice Green w/ Treated Top 200-250

[080] (F-Skirt) HEMINGRAY-680/LOWEX/ REG. U.S. PAT. OFF. (R-Skirt) MADE IN U.S.A./[Numbers and dots] SB
Clear 10-15
Ice Green 15-20
Clear w/ Treated Top 200-250

[090] (F-Skirt) HEMINGRAY-680/['LOWEX/ REG. U.S. PAT. OFF.' blotted out]
(R-Skirt) MADE IN U.S.A./[Numbers and dots] CB
Clear 5-10

[100] (F-Skirt) HEMINGRAY-680/['LOWEX/ REG. U.S. PAT. OFF.' blotted out]
(R-Skirt) MADE IN U.S.A./[Numbers and dots] SB
Clear 5-10

NO NAME

[010] (F-Skirt) 68 {Aluminum coated top} {Hemingray product} SB
Clear 200-250

WHITALL TATUM ▽

[010] (F-Skirt) WHITALL TATUM CO. NO. 514/ [Number] (R-Skirt) MADE IN U.S.A. ▽ SB
Dark Red Amber 2,000-2,500

CD 225

BROOKFIELD

[010] (F-Skirt) BROOKFIELD SB
Dark Aqua 400-500

CD 226

NO NAME

[010] (F-Skirt) № 115 SB
Aqua, Dark Aqua 30-40
Green 100-125

135

CD 226.3

CD 226.3

B

[010] (F-Skirt) B SB
Dark Emerald Green,
Emerald Green Blackglass 1,500-1,750

CD 228

BROOKFIELD

[010] (F-Skirt) BROOKFIELD SB
Dark Aqua 500-600

CD 228.5

ARMSTRONG Ⓐ

[010] (F-Skirt) Ⓐ (R-Skirt) 51-C2 SB
Clear 5,000-7,500

CD 229.6

NO EMBOSSING

[010] [No embossing] {Fry glass} SB
Black Opalescent,
Emerald Green Blackglass 2,000-2,500

CD 230

HEMINGRAY

[010] (F-Skirt) HEMINGRAY (R-Skirt) 512/
[Number]-MADE IN U.S.A.-[Number and dots]
CB
Clear, Light Straw 1-2

[020] (F-Skirt) HEMINGRAY (R-Skirt) 512/
[Number]-MADE IN U.S.A.-[Number and dots]
SB
Clear 1-2

[030] (F-Skirt) HEMINGRAY/MADE IN U.S.A.
(R-Skirt) D-512 SB
Green Tint, Ice Blue 3-5
Honey Amber, Light Citrine 30-40
Carnival 40-50
Amber Blackglass 100-125

[040] (F-Skirt) HEMINGRAY/MADE IN U.S.A.
(R-Skirt) LOWEX/D-512 SB
Honey Amber 30-40

[050] (F-Skirt) HEMINGRAY/MADE IN U.S.A.
(R-Skirt) [Number]/D-512 SB
Olive Amber 10-15
Dark Orange Amber 20-30
Yellow Amber 30-40
Carnival 40-50

[060] (F-Skirt) [Numbers and dots]/
HEMINGRAY/MADE IN U.S.A. (R-Skirt) D-512
SB
Blue Tint, Ice Blue 3-5
Olive Amber 10-15
Orange Amber 20-30

[070] (F-Skirt) [Numbers and dots]/
HEMINGRAY/MADE IN U.S.A.
(R-Skirt) LOWEX/D-512 SB
Ice Green 3-5
Olive Amber 10-15
Honey Amber, Whiskey Amber,
Yellow Amber 30-40

[080] (F-Skirt) [Numbers and dots]/
HEMINGRAY/MADE IN U.S.A.
(R-Skirt) LOWEX/D-512/REG. U.S. PAT. OFF.
SB
Off Clear 1-2
Ice Green 3-5
Lemon 10-15

[090] (F-Skirt) [Numbers and dots]/
HEMINGRAY/MADE IN U.S.A.
(R-Skirt) [Number]/D-512 SB
Clear, Lemon Tint 1-2
Blue Tint, Ice Blue, Light Green 3-5
Dark Olive Amber 10-15
Dark Orange Amber,
Light Olive Amber, Orange Amber 20-30
Golden Amber, Honey Amber 30-40
Carnival, Red Amber 40-50
Dark Carnival 50-75

CD 232

LOWEX

[010] (F-Skirt) LOWEX/REG. U.S. PAT. OFF.
(R-Skirt) 512/[Number]-MADE IN
U.S.A.-[Number] SB
Clear, Ice Green, Light Lemon 3-5
Dark Olive Amber 10-15
Orange Amber 20-30
Honey Amber 30-40

NO NAME

[010] (F-Skirt) 512/[Number]-MADE IN
U.S.A.-[Number and dots] {Hemingray product}
SB
Clear 5-10

CD 230.1

HEMINGRAY

[010] (F-Skirt) HEMINGRAY/MADE IN U.S.A.
(R-Skirt) D-512 SB
Aqua, Green Aqua 30-40
Carnival coated Aqua 200-250

CD 230.2

ARMSTRONG Ⓐ

[010] (F-Skirt) *Armstrong's* 51-C1A
(R-Skirt) MADE IN U.S.A. Ⓐ [Numbers] SB
Dark Red Amber, Root Beer Amber 30-40

CD 231

HEMINGRAY

[010] (F-Skirt) HEMINGRAY-820
(R-Skirt) MADE IN U.S.A./[Numbers and dots]
CB
Clear 20-30

[020] (F-Skirt) HEMINGRAY-820/TEMPERED
(R-Skirt) MADE IN U.S.A./[Numbers and dots]
CB
Clear 20-30

KIMBLE

[010] (F-Skirt) KIMBLE-820/TEMPERED
(R-Skirt) MADE IN U.S.A./[Numbers and dots]
CB
Clear, Light Pink, Straw 3-5
Peach 5-10

[020] (F-Skirt) KIMBLE-820/['HEMINGRAY'
blotted out]/TEMPERED (R-Skirt) MADE IN
U.S.A./[Numbers and dots] CB
Clear, Light Pink, Straw 3-5
Peach 5-10

CD 231.2

KIMBLE

[010] (F-Skirt) KIMBLE-820/TEMPERED
(R-Skirt) MADE IN U.S.A./[Numbers] CD
Clear, Light Straw 3-5
Ice Aqua 10-15

CD 232

HEMINGRAY

[010] (F-Skirt) HEMINGRAY-513
(R-Skirt) MADE IN U.S.A./[Numbers and dots]
CB
Clear 3-5

[020] (F-Skirt) HEMINGRAY/MADE IN U.S.A.
(R-Skirt) D-513 SB
Ice Aqua 10-15

[030] (F-Skirt) [Numbers and dots]/
HEMINGRAY/MADE IN U.S.A. (R-Skirt) D-513
SB
Ice Aqua 10-15

[040] (F-Skirt) [Numbers and dots]/
HEMINGRAY/MADE IN U.S.A.
(R-Skirt) [Number]/D-513 CB
Clear, Light Green Straw, Light Straw 3-5

[050] (F-Skirt) [Numbers and dots]/
HEMINGRAY/MADE IN U.S.A.
(R-Skirt) [Number]/D-513 SB
Clear 3-5
Pink Tint 5-10
Ice Aqua, Ice Blue 10-15
Carnival, Honey Amber 300-350

CD 232.1

CD 232.1

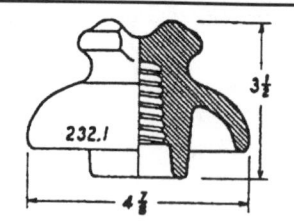

HEMINGRAY

[010] (F-Skirt) HEMINGRAY/MADE IN U.S.A. SB
Aqua, Dark Aqua 75-100

CD 233

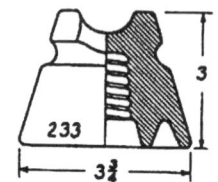

HEMINGRAY

[005] (F-Skirt) HEMINGRAY (R-Skirt) №42 {Note '№ 42'} {Specimen only} SDP
Aqua 500-600

[010] (F-Skirt) HEMINGRAY-E.3. (R-Skirt) MADE IN U.S.A. SB
Clear, Light Lemon, Off Clear 75-100
Lemon 100-125
Aqua 150-175

NO EMBOSSING

[010] [No embossing] SB
Dark Aqua 700-800

PYREX

[010] (F-Skirt) PYREX REG. U.S. PAT. OFF. (R-Skirt) 61/CORNING MADE IN U.S.A. SB
Clear, Straw 1-2
Lemon 3-5
Carnival 75-100

[020] (F-Skirt) PYREX REG. U.S. PAT. OFF. (R-Skirt) 661/CORNING MADE IN U.S.A. SB
Clear 1-2
Lemon 3-5
Carnival 75-100

[030] (F-Skirt) PYREX REG. U.S. PAT. OFF. (R-Skirt) CORNING MADE IN U.S.A. SB
Clear, Straw 1-2

CD 233.2

HEMINGRAY

CD 233.2
$3\ ^3/_4$ x 3

[010] (F-Skirt) HEMINGRAY-511 (R-Skirt) [Numbers and dots]/MADE IN U.S.A. SB
Clear 125-150

[020] (F-Skirt) HEMINGRAY-511/LOWEX (R-Skirt) [Numbers]/MADE IN U.S.A. SB
Clear, Green Tint 150-175

CD 234

PYREX

[010] (F-Skirt) PYREX T.M. REG. U.S. PAT. OFF. (R-Skirt) [Number or letter]/MADE IN U.S.A. 63 SB
Clear, Light Yellow, Straw 1-2
Carnival, Dark Carnival 30-40

CD 235

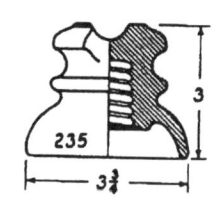

PYREX

[010] (F-Skirt) CORNING PYREX T.M. REG. U.S. (R-Skirt) PAT. OFF. MADE IN U.S.A. 662 SB
Clear 3-5
Carnival 40-50

[015] (F-Skirt) CORNING PYREX T.M. REG. U.S. (R-Skirt) [Number]/PAT. OFF. MADE IN U.S.A. 662 {Lettering is large and may be hand scribed} SB
Clear 3-5
Carnival 40-50

[020] (F-Skirt) CORNING PYREX T.M. REG. U.S. PAT. (R-Skirt) OFF. MADE IN U.S.A. 62 SB
Clear 3-5
Carnival 30-40

[030] (F-Skirt) CORNING PYREX T.M. REG. U.S. PAT. (R-Skirt) OFF. MADE IN U.S.A. 662 SB
Clear 1-2
Carnival 30-40

[040] (F-Skirt) CORNING PYREX T.M. REG. U.S. PAT. (R-Skirt) [Number]/OFF. MADE IN U.S.A. 662 SB
Clear 1-2
Carnival 30-40

[050] (F-Skirt) ϽORNING PYREX T.M. REG. U.S. PAT. {Note backwards letter} (R-Skirt) OFF. MADE IN U.S.A. 62 SB
Carnival 30-40

[060] (F-Skirt) ϽRNING PYREX T.M. REG. U.S. PAT. OFF. {Note backwards letter} {Note spelling} (R-Skirt) MADE IN U.S.A. 662 SB
Carnival 30-40

[065] (F-Skirt) ϽRNING PYREX T.M. REG. U.S. PAT. {Note backwards letter} {Note spelling} (R-Skirt) [Number]/OFF. MADE IN U.S.A. 662 SB
Carnival 30-40

[070] (F-Skirt) [Number]/CORNING PYREX T.M. REG. U.S. PAT. (R-Skirt) OFF. MADE IN U.S.A. 662 SB
Clear 1-2

CD 236

BROOKFIELD

[010] (F-Skirt) BROOKFIELD SB
Dark Aqua 500-600

CD 237

HEMINGRAY

[010] (F-Skirt) HEMINGRAY-72 (R-Skirt) MADE IN U.S.A. RDP
Clear 5-10
Ice Aqua, Ice Blue 10-15

[020] (F-Skirt) HEMINGRAY-72.A/[Numbers and dots] (R-Skirt) MADE IN U.S.A./[Number] {Wide saddle groove} SB
Clear 5-10

[030] (F-Skirt) HEMINGRAY-72.A/[Numbers] (R-Skirt) MADE IN U.S.A. RDP
Ice Blue 10-15

[040] (F-Skirt) HEMINGRAY-72.A/[Numbers] (R-Skirt) MADE IN U.S.A. SB
Clear 5-10

[050] (F-Skirt) HEMINGRAY-72/[Numbers and dots] (R-Skirt) MADE IN U.S.A. SB
Clear 3-5
Ice Blue 5-10

[060] (F-Skirt) HEMINGRAY-72/[Numbers and dots] (R-Skirt) MADE IN U.S.A./[Number] SB
Clear 3-5
Ice Blue 5-10

[070] (F-Skirt) HEMINGRAY-720 (R-Skirt) MADE IN U.S.A./[Numbers and dots] CB
Clear 3-5

[080] (F-Skirt) HEMINGRAY-720/LOWEX (R-Skirt) MADE IN U.S.A./[Numbers and dots] SB
Clear 15-20

[090] (F-Skirt) HEMINGRAY-720/[Numbers] (R-Skirt) MADE IN U.S.A./[Numbers] SB
Clear 3-5
Ice Blue 5-10

[100] (F-Skirt) HEMINGRAY-720/['LOWEX' blotted out] (R-Skirt) MADE IN U.S.A./[Numbers and dots] CB
Clear 3-5

CD 238

HEMINGRAY

[010] (F-Skirt) HEMINGRAY-514 (R-Skirt) MADE IN U.S.A./[Numbers and dots] CB
Clear, Green Tint, Light Straw 3-5
Honey Amber 300-350

[020] (F-Skirt) HEMINGRAY-514 (R-Skirt) MADE IN U.S.A./[Number] SB
Clear 3-5
Light Honey Amber 300-350

CD 238

[030] (F-Skirt) HEMINGRAY/MADE IN U.S.A./ ['LOWEX/REG. IN U.S. PAT. OFF.' blotted out] (R-Skirt) D-514 CB
Off Clear 3-5

[040] (F-Skirt) [Numbers and dots]/ HEMINGRAY/MADE IN U S A {Note no periods} (R-Skirt) [Number]/D-514 SB
Ice Blue 10-15

[050] (F-Skirt) [Numbers and dots]/ HEMINGRAY/MADE IN U.S.A. (R-Skirt) D-514 CB
Clear, Green Tint 3-5
Ice Blue 10-15

[060] (F-Skirt) [Numbers and dots]/ HEMINGRAY/MADE IN U.S.A. (R-Skirt) D-514 SB
Clear 3-5
Green Tint 5-10
Honey Amber 300-350

[070] (F-Skirt) [Numbers and dots]/ HEMINGRAY/MADE IN U.S.A. (R-Skirt) [Number]/D-514 CB
Clear 3-5
Green Tint 5-10

[080] (F-Skirt) [Numbers and dots]/ HEMINGRAY/MADE IN U.S.A. (R-Skirt) [Number]/D-514 SB
Off Clear 5-10
Ice Blue 10-15
Honey Amber 300-350

[090] (F-Skirt) [Numbers and dots]/ HEMINGRAY/MADE IN U.S.A./LOWEX/REG. IN U.S. PAT. OFF. (R-Skirt) [Number]/D-514 SB
Clear 15-20
Green Tint 20-30

CD 238.1

HEMINGRAY

CD 238.1
$5\ ^{1}/_{2}\ \times\ 3\ ^{5}/_{8}$

[010] (F-Skirt) [Numbers]/HEMINGRAY/MADE IN U.S.A. (R-Skirt) D-514 SB
Ice Blue 200-250

CD 238.2

ARMSTRONG Ⓐ

CD 238.2
$5\ ^{1}/_{2}\ \times\ 4\ ^{1}/_{8}$

[010] (F-Skirt) Ⓐ/[Numbers] {Comes in two dome shapes} SB
Clear 5,000-7,500

CD 239

HEMINGRAY

[010] (F-Skirt) HEMINGRAY-830/TEMPERED (R-Skirt) MADE IN U.S.A./[Numbers and dots] CB
Clear, Off Clear 5-10

[020] (F-Skirt) HEMINGRAY/['KIMBLE-830' blotted out]/TEMPERED (R-Skirt) MADE IN U.S.A./[Number and dots] SB
Clear w/ Light Carnival Top 50-75

[030] (F-Skirt) KIMBLE/HEMINGRAY-830 [Letter]/TEMPERED (R-Skirt) MADE IN U.S.A./[Numbers and dots] CB
Clear 5-10

KIMBLE

[010] (F-Skirt) KIMBLE [Letter]/830/ TEMPERED (R-Skirt) MADE IN U.S.A./ [Numbers and dots] CB
Clear 3-5

[020] (F-Skirt) KIMBLE-830 (R-Skirt) MADE IN U.S.A./[Numbers] SB
Clear w/ Light Carnival Top 50-75

[025] (F-Skirt) KIMBLE-830 (R-Skirt) MADE IN U.S.A./[Numbers] {Square shoulder variant} SB
Clear 75-100

CD 241

[030] (F-Skirt) KIMBLE-830/TEMPERED (R-Skirt) MADE IN U.S.A./[Numbers and dots] CB
Clear, Green Tint, Light Green, Straw 3-5
Peach 10-15

[040] (F-Skirt) KIMBLE-830/['HEMINGRAY' blotted out]/TEMPERED (R-Skirt) MADE IN U.S.A./[Numbers and dots] CB
Light Straw 3-5
Peach 10-15
Clear w/ Light Carnival Top,
Light Straw w/ Treated Top 50-75

CD 239.2

CD 239.2
7 x 5

KIMBLE

[010] (F-Skirt) KIMBLE-850-1 (R-Skirt) MADE IN U.S.A./[Numbers] CB
Clear 5,000-7,500

CD 240

PYREX

[010] (F-Skirt) CORNING PYREX REG. U.S. PAT. OFF. (R-Skirt) MADE IN U.S.A. - 131 [Number] SB
Clear, Yellow Tint 15-20

CD 240.2

NO EMBOSSING

[010] [No embossing] SB
Clear 2,000-2,500

[020] [No embossing] {Metal cap covers the insulator above the wire groove} SB
Clear 2,000-2,500

NO NAME

[010] (F-Skirt) 1025 {Metal cap covers the insulator above the wire groove} SB
Clear 2,000-2,500

CD 240.5

PYREX

[010] (F-Skirt) CORNING PYREX REG. U.S. PAT. OFF. (R-Skirt) MADE IN U.S.A. 161 SB
Clear, Light Lemon 15-20

[020] (Skirt) CORNING PYREX T.M. REG. U.S. PAT. OFF. MADE IN U.S.A. 161 [Number] {Embossed all the way around the skirt} SB
Clear 15-20

[030] (Skirt) CORNING PYREX T.M. REG. U.S. PAT. OFF. MADE IN U.S.A. 161 {Embossed all the way around the skirt} SB
Clear 15-20

CD 241

HEMINGRAY

[010] (F-Skirt) HEMINGRAY-23 (R-Skirt) MADE IN U.S.A. RDP
Aqua, Blue Aqua 15-20
Hemingray Blue 20-30
Honey Amber 50-75

[020] (F-Skirt) HEMINGRAY-23 (R-Skirt) MADE IN U.S.A. SDP
Aqua 15-20
Hemingray Blue 20-30
Honey Amber, Red Amber,
Root Beer Amber 50-75
Orange Amber, Yellow Amber 75-100

LOCKE

[010] (F-Skirt) PAT'D BY F.M. LOCKE/VICTOR/N.Y. (R-Skirt) MARCH 31, 1914 FEB 2, 1915/OCT. 12, 1915 {Embossing is upside down} SB
Aqua, Clear, Straw 2,500-3,000
White Opalescent 4,000-4,500

141

CD 241.2

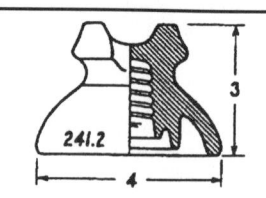

CD 241.2

HEMINGRAY

[010] (F-Skirt) HEMINGRAY-24 (R-Skirt) MADE IN U.S.A. SDP
Aqua, Hemingray Blue 75-100
Golden Amber 1,250-1,500

CD 242

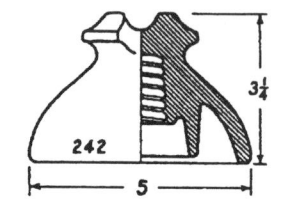

HEMINGRAY

[010] (F-Skirt) HEMINGRAY-66 (R-Skirt) MADE IN U.S.A. SDP
Aqua, Hemingray Blue 250-300

CD 243

HEMINGRAY

[010] (F-Skirt) HEMINGRAY-88 (R-Skirt) MADE IN U.S.A. SDP
Aqua, Hemingray Blue 2,000-2,500

CD 244

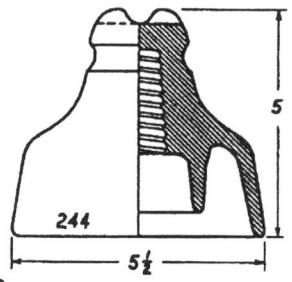

NO EMBOSSING

[010] [No embossing] SB
Purple 4,500-5,000

CD 244.5

GREELEY

CD 244.5
$5\ ^3/_2\ \times\ 4$

[010] (F-Skirt) GREELEY/N.Y. (R-Skirt) L & S/ PATTERN SB
Aqua 10,000-15,000

CD 245

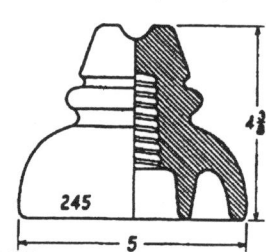

NO NAME

[010] (F-Skirt) 9200 SB
Aqua, Light Blue 100-125
Green Aqua 150-175
Green 300-350

T-H.

[010] (F-Skirt) T-H (R-Skirt) 9200 SB
Aqua, Light Aqua, Light Blue 100-125
Light Green 200-250

CD 247

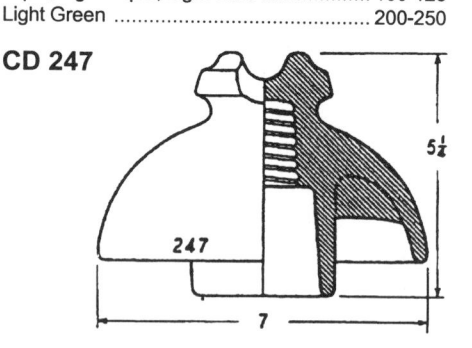

NO EMBOSSING

[010] [No embossing] SB
Aqua 400-500
Blue 600-700
Light Green, Yellow Green 800-1,000

CD 250.2

CD 248/311/311

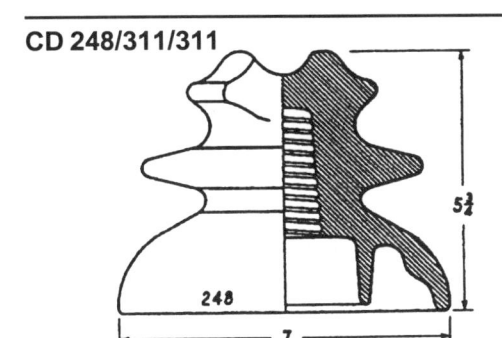

(CD 248 is typically installed with two CD 311 sleeves)

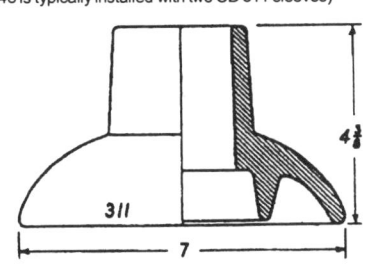

HEMINGRAY

[010] (F-Skirt) HEMINGRAY (R-Skirt) Nº 79 {1 5/8" pinhole} {CD 311 sleeve is unembossed} SDP
Aqua, Blue Aqua 75-100
Hemingray Blue 100-125
Aqua Top w/ Honey Amber Sleeves,
Hemingray Blue Top
w/ Honey Amber Sleeves 600-700

NO EMBOSSING

[010] [No embossing] {Pyrex product} {See CD 311 for sleeve embossing} SB
Straw 50-75

PYREX

[010] (F-Skirt) CORNING PYREX REG. U.S. PAT. OFF. (R-Skirt) MADE IN U.S.A. {See CD 311 for sleeve embossing} SB
Clear, Straw 50-75

[015] (F-Skirt) CORNING PYREX T. M. REG. U.S. PAT. OFF. MADE IN U.S.A. {See CD 311 for sleeve embossing} SB
Clear, Yellow Tint 50-75

[020] (Skirt) PYREX PAT. 5-27-19. MADE BY CORNING GLASS WORKS, CORNING N.Y. U.S.A. {Embossed all the way around the skirt} {See CD 311 for sleeve embossing} SB
Clear, Straw, Yellow Tint 50-75

CD 249

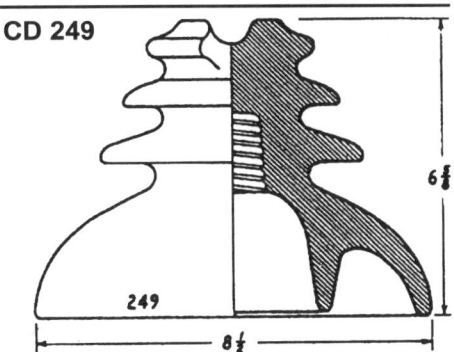

HEMINGRAY

[005] (F-Skirt) HEMINGRAY (R-Skirt) No 0, PROVO TYPE/PATENTED APRIL 25 1899 {1 3/8" pinhole} SB
Aqua 150-175

CD 250

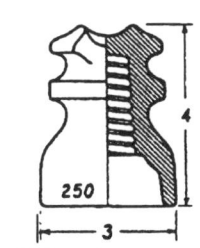

N.E.G.M.CO.

[010] (F-Skirt) N.E.G.M.CO. {Note no inner skirt} SB
Aqua 1,000-1,250

CD 250.2

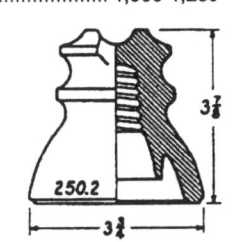

ERICSSON

[010] (F-Skirt) TELEFONOS ERICSSON-LD-1 SB
Aqua 15-20
Celery Green, Green 30-40
Olive Green 200-250

CD 251

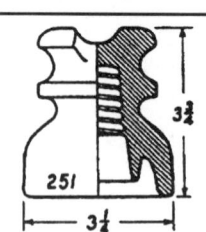

CD 251

HEMINGRAY

[010] (F-Skirt) HEMINGRAY-61
(R-Skirt) MADE IN U.S.A. RDP
Clear 5-10
Aqua 10-15
Ice Blue 15-20

[020] (F-Skirt) HEMINGRAY-61
(R-Skirt) MADE IN U.S.A. SB
Clear 5-10

[030] (F-Skirt) HEMINGRAY-61
(R-Skirt) MADE IN U.S.A./[Numbers and dots]
CB
Clear 5-10
Ice Blue 10-15

[040] (F-Skirt) HEMINGRAY-61
(R-Skirt) MADE IN U.S.A./[Numbers and dots]
SB
Clear 5-10
Aqua, Blue Tint, Ice Blue 10-15

[050] (F-Skirt) HEMINGRAY-61/[Numbers and dots] (R-Skirt) MADE IN U.S.A./[Number] RDP
Clear 5-10

[060] (F-Skirt) HEMINGRAY-61/['[Numbers]' blotted out] (R-Skirt) MADE IN U.S.A./
[Numbers and dots] SB
Clear 5-10

[065] (F-Skirt) HEMINGRAY/№ 1 CABLE
(R-Skirt) PATENTED/MAY 2 1893 RDP
Green Aqua 20-30

[070] (F-Skirt) HEMINGRAY/№ 1 CABLE
(R-Skirt) PATENTED/MAY 2 1893 SDP
Aqua, Light Aqua 10-15

[080] (F-Skirt) MADE IN U.S.A./HEMINGRAY/
№ 61 CABLE (R-Skirt) PATENTED/MAY 2
1893 SDP
Aqua 15-20
Hemingray Blue 20-30

[090] (F-Skirt) № 61/HEMINGRAY/MADE IN
U.S.A. (R-Skirt) PATENTED/MAY 2 1893 SDP
Aqua 15-20
Hemingray Blue 20-30

LYNCHBURG Ⓛ

[010] (F-Skirt) Ⓛ/LYNCHBURG
(R-Skirt) [Number]/MADE IN NO.1 U.S.A.
SDP
Light Aqua 40-50

[015] (F-Skirt) Ⓛ/LYNCHBURG
(R-Skirt) [Number]/MADE IN NO.1 U.S.A./
['PAT JUNE 17 1890' blotted out] SB
Light Aqua 40-50

[020] (F-Skirt) Ⓛ/LYNCHBURG/
['N.E.G.M.CO.' blotted out] (R-Skirt) [Number]/
MADE IN NO.1 U.S.A./['PAT JUNE 17 1890'
blotted out] SDP
Light Aqua 40-50
Aqua w/ Milk Swirls 100-125

N.E.G.M.CO.

[010] (F-Skirt) N.E.G.M.CO. SB
Aqua 15-20
Ice Blue 20-30
Green 50-75
Sapphire Blue 250-300

[020] (F-Skirt) N.E.G.M.CO. (R-Skirt) "PAT
JUNE 17 1890" SB
Aqua, Light Aqua 15-20
Ice Aqua, Ice Blue 20-30
Green 30-40
Dark Blue, Yellow Green 50-75
Dark Yellow Green 75-100
Olive Green 150-175

CD 252

CABLE

[005] (F-Skirt) NO.2 CABLE RB
Aqua, Dark Aqua 5-10
Dark Green 20-30
7-up Green, Dark Olive Green,
Dark Yellow Green, Milky Aqua 75-100
Dark Olive Amber 100-125
Orange Amber 300-350

[020] (F-Skirt) NO.2 CABLE (R-Skirt) CABLE
{Large, faint, freehand embossing} SB
Aqua 20-30

CD 252

[010] (F-Skirt) NO.2 CABLE {Canadian style} SB
Light Aqua 5-10
Ice Green, Light Blue 15-20
Light Steel Blue, Light Straw,
Peach, Sage Green 20-30
Dark Steel Blue 50-75

E.S.S.CO.

[010] (F-Skirt) E.S.S.Co./[Blotted out embossing] (R-Skirt) № 401 SB
Aqua 100-125

GAYNER

[010] (F-Skirt) GAYNER (R-Skirt) NO.620/[Number] SDP
Aqua 40-50
Blue Aqua 50-75

HEMINGRAY

[010] (F-Skirt) HEMINGAY-62 {Note spelling} (R-Skirt) MADE IN U.S.A. RDP
Aqua 15-20
Hemingray Blue, Ice Blue 20-30

[020] (F-Skirt) HEMINGRAY-62 (R-Skirt) MADE IN U.S.A. RDP
Aqua, Green Aqua 5-10
Hemingray Blue, Ice Blue 10-15
Carnival 350-400

[030] (F-Skirt) HEMINGRAY-62 (R-Skirt) MADE IN U.S.A./[Numbers and dots] CB
Clear 3-5
Green Tint 5-10

[040] (F-Skirt) HEMINGRAY-62 (R-Skirt) MADE IN U.S.A./[Numbers and dots] SB
Clear 3-5
Light Pink 10-15
Flashed Red Amber,
Flashed Yellow Amber 40-50
Olive Amber 75-100

[050] (F-Skirt) HEMINGRAY-62 (R-Skirt) ['HEMINGRAY' blotted out]/MADE IN U.S.A. RDP
Ice Blue 10-15

[060] (F-Skirt) HEMINGRAY-62/LOWEX (R-Skirt) MADE IN U.S.A./[Numbers and dots] SB
Flashed Amber 40-50
Olive Amber 75-100

[070] (F-Skirt) HEMINGRAY-62/LOWEX/REG. IN U.S. PAT. OFF. (R-Skirt) MADE IN U.S.A./[Numbers and dots] SB
Clear 5-10
Green Tint, Lemon Tint, Pink Tint 10-15
Olive Amber 75-100

[080] (F-Skirt) HEMINGRAY-62/LOWEX/REG. U.S. PAT. OFF. (R-Skirt) MADE IN U.S.A./[Numbers and dots] SB
Clear 5-10
Light Pink 10-15

[090] (F-Skirt) HEMINGRAY-62/[Numbers and dots] (R-Skirt) MADE IN U.S.A./[Number] RDP
Clear 3-5
Blue Tint, Green Tint 5-10

[100] (F-Skirt) HEMINGRAY-62/[Numbers and dots] (R-Skirt) MADE IN U.S.A./[Number] SB
Green Tint, Ice Blue, Light Lemon 5-10

[103] (F-Skirt) HEMINGRAY-62/[Numbers and dots] (R-Skirt) ['HEMINGRAY' blotted out]/MADE IN U.S.A./[Number] SB
Ice Blue 10-15

[107] (F-Skirt) HEMINGRAY-62/['LOWEX/REG. IN U.S. PAT. OFF.' blotted out] (R-Skirt) MADE IN U.S.A./[Numbers and dots] CB
Clear 3-5
Smoke 5-10

[110] (F-Skirt) HEMINGRAY-62/['LOWEX/REG. U.S. PAT. OFF.' blotted out] (R-Skirt) MADE IN U.S.A./[Numbers and dots] SB
Clear 3-5
Pink Tint 10-15
Flashed Amber 40-50

[120] (F-Skirt) HEMINGRAY-62/['LOWEX' blotted out] (R-Skirt) MADE IN U.S.A./[Number and dots] CB
Clear, Smoke 3-5

[130] (F-Skirt) HEMINGRAY-62/['LOWEX' blotted out] (R-Skirt) MADE IN U.S.A./[Numbers and dots] SB
Clear 3-5

[140] (F-Skirt) HEMINGRAY/LOWEX (R-Skirt) MADE IN U.S.A. SB
Olive Amber 75-100

[150] (F-Skirt) HEMINGRAY/№ 2 CABLE (R-Skirt) PAT.MAY 2ND 1893 SDP
Aqua 5-10
Blue Aqua 10-15
Green 40-50

145

CD 252

[160] (F-Skirt) HEMINGRAY/N⁰ 2 CABLE (R-Skirt) ['KNOWLES CABLE' blotted out]/ PAT.MAY 2ND 1893 SDP
Aqua, Blue Aqua, Dark Aqua 5-10
Green 40-50

[170] (F-Skirt) HEMINGRAY/N⁰ 2 CABLE (R-Skirt) ['KNOWLES CABLE' blotted out]/ PAT.MAY 2ND 1893/['INSULATOR' blotted out] SDP
Blue Aqua 5-10

[180] (F-Skirt) HEMINGRAY/N⁰ 2 CABLE/ ['KNOWLES CABLE/INSULATOR' blotted out] (R-Skirt) PAT.MAY 2ND 1893 SDP
Aqua 5-10

[190] (F-Skirt) HEMINGRAY/N⁰ 62 CABLE (R-Skirt) MADE IN U.S.A./KNOWLES CABLE/ PAT.MAY 2ND 1893 SDP
Dark Aqua 5-10

[200] (F-Skirt) HEMINGRAY/N⁰ 62 CABLE (R-Skirt) MADE IN U.S.A./PAT.MAY 2ND 1893 RDP
Ice Blue 15-20

[210] (F-Skirt) HEMINGRAY/N⁰ 62 CABLE (R-Skirt) MADE IN U.S.A./PAT.MAY 2ND 1893 SDP
Hemingray Blue 10-15

[220] (F-Skirt) HEMINGRAY/N⁰ 62 CABLE (R-Skirt) MADE IN U.S.A./PAT.MAY 2ND 1893/ ['KNOWLES CABLE/INSULATOR' blotted out] SDP
Aqua 5-10
Hemingray Blue 10-15
Ice Blue 15-20

[230] (F-Skirt) HEMINGRAY/N⁰ 62 CABLE (R-Skirt) MADE IN U.S.A./PAT.MAY 2ND 1893/ ['KNOWLES CABLE' blotted out] RDP
Aqua 5-10
Ice Aqua 15-20

[240] (F-Skirt) HEMINGRAY/N⁰ 62 CABLE (R-Skirt) PAT.MAY 2 1893 SDP
Aqua 5-10

[250] (F-Skirt) HEMINGRAY/N⁰ 62 CABLE (R-Skirt) PAT.MAY 2ND 1893 SDP
Aqua 5-10

[260] (F-Skirt) HEMINGRAY/N⁰ 62 CABLE (R-Skirt) ['KNOWLES CABLE' blotted out]/ PAT.MAY 2ND 1893/['INSULATOR' blotted out] SDP
Aqua 5-10

[270] (F-Skirt) HEMINGRAY/N⁰ 62 CABLE/ ['KNOWLES CABLE/INSULATOR' blotted out] (R-Skirt) MADE IN U.S.A./PAT.MAY 2ND 1893 SDP
Aqua, Blue 5-10
Hemingray Blue 10-15
Ice Blue 15-20

[280] (F-Skirt) ['MADE IN U.S.A.' blotted out]/ HEMINGRAY-62/['LOWEX/REG. IN U.S. PAT. OFF.' blotted out] (R-Skirt) MADE IN U.S.A./ [Numbers and dots] CB
Clear 3-5

KNOWLES

[010] (F-Skirt) KNOWLES CABLE/ INSULATOR SB
Aqua, Light Aqua 10-15
Blue, Ice Aqua, Ice Green, Light Green 20-30
Yellow Green 100-125

[020] (F-Skirt) KNOWLES/NO.2 CABLE (R-Skirt) ✉ SB
Aqua 20-30
Yellow Green 100-125

[030] (F-Skirt) KNOWLES/NO.2 CABLE {Slug embossing} (R-Skirt) (Arc)PATENTED/✉/ JUNE.17. 1890 SB
Aqua, Blue Aqua 10-15
Light Blue 15-20
Blue 20-30
Apple Green, Green 50-75
Yellow Green 100-125
Emerald Green 175-200
Dark Yellow Green 200-250
Dark Yellow Olive Green 350-400

LYNCHBURG Ⓛ

[010] (F-Skirt) LYNCHBURG (R-Skirt) N⁰ 2 CABLE/MADE IИ U.S.A. {Note backwards letter} SDP
Light Aqua 15-20

[020] (F-Skirt) Ⓛ/LYNCHBURG {Note backwards letter} (R-Skirt) N⁰ 2 CABLE/MADE IN U.Ƨ.A. {Note backwards letter} SDP
Aqua 15-20

[030] (F-Skirt) Ⓛ/LYNCHBURG (R-Skirt) N⁰ 2 CABLE/MADE IN U.S.A. SDP
Aqua 15-20
Sage Green 20-30
Aqua w/ Milk Swirls 75-100

[040] (F-Skirt) Ⓛ/LYNCHBURG (R-Skirt) N⁰ 2 CABLE/MADE IИ U.S.A. {Note backwards letter} SDP
Aqua 15-20
Sage Green 20-30

CD 254

[050] (F-Skirt) Ⓛ/LYИCHBURG {Note backwards letter} (R-Skirt) № 2 CABLE/MADE IИ U.S.A. {Note backwards letter} SDP
Aqua 15-20
Celery Green 20-30

[060] (F-Skirt) Ⓛ/LYNCHBURG/MADE IN U.S.A. (R-Skirt) № 62 CABLE {Note '№ 62' style number} SDP
Aqua 50-75

M.& E.CO.

[010] (F-Skirt) M & E Co./PHILA {Slug embossing} (R-Skirt) ⌘ SB
Green 100-125
Light Yellow Green 200-250
Dark Yellow Green 300-350

[020] (F-Skirt) THE M & E Co. (R-Skirt) № 401 SB
Aqua, Light Green Aqua 100-125

[030] (F-Skirt) THE/M & E Co./PHILADELPHIA (R-Skirt) KNOWLES CABLE/INSULATOR {Can be 1/2" taller than most CD 252's} SB
Aqua, Green Aqua, Light Aqua 40-50
Green 350-400
Yellow Green 400-500

MAYDWELL

[010] (F-Skirt) MAYDWELL-62 (R-Skirt) U.S.A. SDP
Light Green, Light Straw, Off Clear 10-15
Straw, Yellow Green Tint 15-20
Gingerale, Pink, Purple Tint 20-30

[020] (F-Skirt) MAYDWELL-62/['MᶜLAUGHLIN' blotted out] (R-Skirt) U.S.A. RDP
Blue Tint, Yellow Green Tint 20-30

MᶜLAUGHLIN

[010] (F-Skirt) MᶜLAUGHLIN-62 RDP
Light Green 10-15
Apple Green 30-40
Delft Blue 50-75
Emerald Green 250-300

[020] (F-Skirt) MᶜLAUGHLIN-62 SDP
Aqua, Ice Green 10-15
Blue Aqua, Light Blue, Sage Green 15-20
Steel Blue 20-30
Apple Green, Blue Purple 30-40
Delft Blue, Light Cornflower Blue 50-75

NO EMBOSSING

[010] [No embossing] {Knowles style} SB
Brooke's Blue 75-100

[020] [No embossing] {NO. 2 CABLE style} RB
Olive Green 75-100

PRISM ⌘

[010] (F-Skirt) ⌘ SB
Aqua 50-75

[020] (F-Skirt) ⌘ (R-Skirt) NO.2 SB
Aqua, Blue Aqua, Ice Blue,
Light Green Aqua 20-30
Green 50-75

CD 253

KNOWLES

[010] (F-Skirt) KNOWLES CABLE/INSULATOR SB
Light Aqua 40-50
Blue 50-75
Green 100-125
Yellow Green 200-250
Milky Aqua 250-300

[020] (F-Skirt) KNOWLES CABLE/INSULATOR/"PATENTED.JUNE. 17. 1890" SB
Aqua, Green Aqua 40-50
Blue 50-75
Dark Blue 75-100
Green 100-125
Yellow Green 200-250
Dark Yellow Green 300-350

CD 254

CABLE

[010] (F-Skirt) NO.3 CABLE SB
Aqua 20-30
Green 40-50
Dark Green, Light Blue Aqua 50-75
Dark Yellow Green 200-250

[020] (F-Skirt) NO.3 CABLE/[Blotted out embossing] SB
Emerald Green 175-200

147

CD 254

HEMINGRAY

[010] (F-Skirt) HEMINGRAY (R-Skirt) № 3 CABLE SB
Aqua, Dark Aqua 30-40
Green 75-100

[020] (F-Skirt) HEMINGRAY (R-Skirt) № 3 CABLE SDP
Aqua, Blue, Dark Aqua 30-40
Blue/Green Two Tone 50-75

[030] (F-Skirt) HEMINGRAY-63 (R-Skirt) MADE IN U.S.A. SB
Clear, Ice Aqua, Ice Blue 200-250

[040] (F-Skirt) HEMINGRAY-63 (R-Skirt) MADE IN U.S.A. SDP
Aqua 200-250

[050] (F-Skirt) HEMINGRAY-63/[Number] (R-Skirt) MADE IN U.S.A./[Number] SB
Clear, Off Clear 200-250

[060] (F-Skirt) HEMINGRAY/MADE IN U.S.A. (R-Skirt) № 63 CABLE SDP
Hemingray Blue 300-350

M.& E.CO.

[010] (F-Skirt) THE M & E Co. (R-Skirt) № 58 SB
Aqua, Blue Aqua, Light Aqua 200-250

NO EMBOSSING

[010] [No embossing] SB
Dark Aqua 30-40

CD 256

MANHATTAN

[010] (F-Skirt) MANHATTAN SB
Aqua, Blue 40-50

[020] (F-Skirt) MANHATTAN (R-Skirt) PATENTED JUNE 17 1890 SB
Aqua, Blue, Green Aqua 40-50
Ice Blue, Light Green 75-100

[030] (F-Skirt) MANHATTAN (R-Skirt) "PATENTED JUNE 17 1890" SB
Aqua, Blue, Light Aqua 40-50
Green 75-100
Yellow Green 125-150
Yellow Olive Green 250-300
Olive Green 300-350

CD 257

HEMINGRAY

[010] (F-Skirt) HEMINGRAY (R-Skirt) MADE IN U.S.A./№ 60 RDP
Blue Aqua 20-30

[020] (F-Skirt) HEMINGRAY (R-Skirt) MADE IN U.S.A./№ 60 SDP
Hemingray Blue 20-30

[030] (F-Skirt) HEMINGRAY (R-Skirt) № 60 SDP
Aqua, Green Aqua 15-20
Light Green 40-50
Emerald Green 100-125

[040] (F-Skirt) HEMINGRAY (R-Skirt) PATENT/JUNE 17 1890 MAY 2 1893 SDP
Aqua, Blue Aqua, Light Aqua 15-20
Green 50-75
Emerald Green 125-150
Milky Aqua, Milky Blue Aqua,
Milky Green Aqua 500-600
Light Electric Blue 800-1,000
Electric Blue 1,000-1,250
Electric Blue w/ Milk Swirls 1,250-1,500
Dark Electric Blue 1,500-1,750

[050] (F-Skirt) HEMINGRAY (R-Skirt) PATENT/JUNE 17 1890 MAY 2 1893 {Wide groove} SDP
Aqua, Blue Aqua, Light Aqua 15-20
Green 50-75
Emerald Green, Forest Green 125-150

[060] (F-Skirt) HEMINGRAY (R-Skirt) PATENT/JUNE 17 1890 MAY 2 1893 {'0' is engraved over a '3'} SDP
Blue Aqua 15-20
Electric Blue 1,000-1,250

[070] (F-Skirt) HEMINGRAY (R-Skirt) PATENT/JUNE 17 1893 MAY 2 1893 {Note 'JUNE 17 1893' date} SDP
Blue Aqua 20-30
Ice Aqua 50-75
Light Jade Green Milk 300-350

CD 257

[080] (F-Skirt) HEMINGRAY
(R-Skirt) PATENT/JUNE 17 1893 MAY 2 1893
{Note 'JUNE 17 1893' date} {Wide groove}
SDP
Aqua 20-30
Green 50-75
Green/Aqua Two Tone 75-100
Emerald Green 125-150

[090] (F-Skirt) HEMINGRAY-60
(R-Skirt) MADE IN U.S.A. RDP
Aqua, Green Tint, Hemingray Blue, Ice Blue 15-20

[100] (F-Skirt) HEMINGRAY-60
(R-Skirt) MADE IN U.S.A. SDP
Aqua, Blue 15-20
Hemingray Blue 20-30

[110] (F-Skirt) HEMINGRAY-60
(R-Skirt) MADE IN U.S.A. {Wide groove} RDP
Clear, Straw 10-15
Aqua, Hemingray Blue, Ice Blue 15-20

[120] (F-Skirt) HEMINGRAY-60/[Numbers and dots] (R-Skirt) MADE IN U.S.A./[Number] RDP
Clear, Light Straw, Off Clear 10-15
Blue Tint, Green Tint, Ice Blue, Light Pink .. 15-20

[130] (F-Skirt) HEMINGRAY-60/[Numbers and dots] (R-Skirt) MADE IN U.S.A./[Number] {Wide groove} RDP
Clear 10-15

[140] (F-Skirt) HEMINGRAY-63 {Note '63' style number} (R-Skirt) MADE IN U.S.A. {Wide groove} RDP
Aqua, Clear, Hemingray Blue 20-30

[150] (F-Skirt) HEMINGRAY/№ 60-A
(R-Skirt) MADE IN U.S.A./JUNE 17 1890 MAY 2 1893 SDP
Aqua 20-30

[160] (F-Skirt) HEMINGRAY/№ 60-A
(R-Skirt) MADE IN U.S.A./JUNE 17 1890 MAY 2 1893 {Wide groove} SDP
Hemingray Blue 20-30

[170] (F-Skirt) HEMINGRAY/№ 60-A
(R-Skirt) MADE IN U.S.A./['PATENT' blotted out]/JUNE 17 1890 MAY 2 1893 {Wide groove} SDP
Hemingray Blue 20-30

[180] (F-Skirt) HEMINGRAY/№ 60-A/[Blotted out embossing] (R-Skirt) MADE IN U.S.A./MAY 2 1893 {Wide groove} SDP
Hemingray Blue 20-30

[190] (F-Skirt) HEMINGRAY/№ 63-A {Note '№ 63-A' style number} (R-Skirt) MADE IN U.S.A./JUNE 17 1890 MAY 2 1893 {Wide groove} SDP
Hemingray Blue 30-40

[200] (F-Skirt) HEMINGRAY/[Numbers and dots] (R-Skirt) MADE IN U.S.A./№ 60/[Number] RDP
Clear, Light Pink, Off Clear 10-15
Ice Aqua 15-20

[205] (F-Skirt) HEMINGRAY/['JUNE 17 1890' blotted out] (R-Skirt) PATENT/['HEMINGRAY' blotted out]/JUNE 17 1890 MAY 2 1893 SDP
Aqua 20-30

[210] (F-Skirt) HEMINGRAY/['PATENT' blotted out]/№ 60A/['JUNE 17 1890' blotted out] (R-Skirt) MADE IN U.S.A./['PATENT/MAY 2 1893' blotted out] {Wide groove} SDP
Hemingray Blue 20-30

[220] (F-Skirt) MADE IN U.S.A./HEMINGRAY/ № 60-A (R-Skirt) PATENT/JUNE 17 1890 MAY 2 1893 {Wide groove} SDP
Hemingray Blue 20-30

[230] (F-Skirt) MADE IN U.S.A./HEMINGRAY/ № 60-A (R-Skirt) PATENT/JUNE 17 1893 MAY 2 1893 {Note 'JUNE 17 1893' date} {Wide groove} SDP
Hemingray Blue 20-30

[240] (F-Skirt) № 60-A/HEMINGRAY
(R-Skirt) MADE IN U.S.A./['PATENT' blotted out]/JUNE 17 1890 MAY 2 1893 {Wide groove} SDP
Aqua, Hemingray Blue 20-30

[250] (F-Skirt) № 60-A/HEMINGRAY
(R-Skirt) MADE IN U.S.A./['PATENT' blotted out]/JUNE 17 1893 MAY 2 1893 {Note 'JUNE 17 1893' date} {Wide groove} SDP
Aqua, Hemingray Blue 20-30

PATENT - OTHER

[010] (F-Skirt) PATENT/JUNE 17 1890
(R-Skirt) PATENT/MAY 2 1893 {Hemingray product} SDP
Aqua, Green Aqua 20-30
Blue, Ice Green, Light Aqua 30-40
Ice Aqua 40-50
Hemingray Blue, Ice Blue 50-75
Green 100-125
Clear 200-250
Ice Blue/Purple Two Tone 1,250-1,500

149

CD 257

[020] (F-Skirt) PATENT/JUNE 17 1890
(R-Skirt) PATENT/MAY 2 1893 {Wide groove}
{Hemingray product} SDP
Aqua 40-50
Light Aqua 50-75

CD 258

CABLE

[010] (F-Skirt) CABLE SDP
Blue Aqua, Dark Aqua 175-200

[020] (F-Skirt) CABLE {Comes with and without threading on the inside of the outer skirt} SB
Aqua, Dark Aqua 20-30
Blue 50-75
Green 75-100
Emerald Green 400-500

CD 259

CABLE

[010] (Base) CABLE INSULATOR PATENTS APP'D FOR {Inside of outer skirt is threaded} BE
Aqua 800-1,000

[015] (F-Skirt) CABLE SB
Dark Green Aqua 40-50

[020] (F-Skirt) CABLE {Inside of outer skirt is threaded} SB
Aqua, Dark Aqua 10-15
Light Aqua, Light Blue 20-30
Green 100-125
Emerald Green 250-300
Dark Yellow Green 600-700

[030] (F-Skirt) CABLE {Inside of outer skirt is threaded} SDP
Dark Aqua 40-50
Green 175-200

OAKMAN MFG.CO.

[010] (F-Skirt) CABLE (Base) OAKMAN M'F'G.CO. BOSTON/PAT'D. JUNE 17, 1890, AUG. 19, 1890. {Inside of outer skirt is threaded} BE
Aqua 20-30
Light Aqua, Light Blue 30-40
Celery Green, Light Green 75-100
Milky Aqua 200-250
Lime Green 300-350

PATENT - OTHER

[010] (Base) PAT'D JUNE 17, 1890, AUG. 19, 1890 BE
Light Aqua, Light Blue 250-300

CD 260

CABLE

[010] (F-Skirt) CABLE SB
Aqua, Dark Aqua 200-250
Green 350-400

CALIFORNIA

[010] (F-Skirt) CALIFORNIA SB
Sage Green 50-75
Green 300-350
Smoke 400-500
Peach 500-600
Purple 600-700
Gingerale, Yellow 700-800
Purple/Peach Two Tone,
Purple/Yellow Two Tone 1,000-1,250

[020] (F-Skirt) CALIFORNIA {Pinch eared} SB
Aqua, Green Aqua 175-200

HAWLEY ⌗

[010] (F-Skirt) ⌗ (R-Skirt) HAWLEY, PA./U.S.A. SB
Aqua 1,750-2,000

NO EMBOSSING

[010] [No embossing] SB
Aqua 200-250

STAR ✯

[010] (F-Skirt) ✯ SB
Blue Aqua, Green Aqua 100-125

CD 264

[020] (F-Skirt) ✱ (R-Skirt) "PATENTED.JUNE. 17. 1890" SB
Aqua, Light Aqua 100-125
Light Blue 125-150
Light Green 175-200
Green, Lime Green 200-250
Yellow Green 300-350
Dark Yellow Green 400-500

CD 262

COLUMBIA

[010] (F-Skirt) № 2 COLUMBIA (R-Skirt) PAT'D MAY 12TH 1891 SB
Aqua, Dark Aqua, Green Aqua 125-150
Light Blue Aqua 175-200
Green 250-300

CD 263

COLUMBIA

[010] (F-Skirt) COLUMBIA SB
Light Aqua 200-250

[020] (F-Skirt) COLUMBIA {Inside of outer skirt is threaded} SB
Light Aqua 125-150
Ice Aqua 150-175

[030] (F-Skirt) PAT'D MAY 12 1891/ COLUMBIA SB
Dark Aqua 125-150
Blue 150-175
Green 350-400
Yellow Green 500-600

[040] (F-Skirt) PAT'D MAY 12 1891/ COLUMBIA {Inside of outer skirt is threaded} SB
Dark Aqua, Green Aqua 100-125
Light Aqua, Light Blue 150-175
Dark Yellow Green 500-600

[050] (F-Skirt) PAT'D MAY 12 1891/ COLUMBIA {Wide ear} {Inside of outer skirt is threaded} SB
Aqua, Dark Aqua 175-200
Green 400-500

[060] (F-Skirt) PAT'D MAY 12TH 1891/ COLUMBIA {Inside of outer skirt is threaded} SB
Aqua 100-125

HEMINGRAY

[010] (F-Skirt) HEMINGRAY (R-Skirt) PAT'D. MAY 12 1891 SDP
Aqua 125-150

[020] (F-Skirt) HEMINGRAY (R-Skirt) PAT'D. MAY 12 1891/COLUMBIA SB
Aqua 125-150
Hemingray Blue 150-175
Light Green 200-250
Green 350-400

[030] (F-Skirt) HEMINGRAY (R-Skirt) PAT'D. MAY 12 1891/COLUMBIA SDP
Aqua, Blue Aqua 125-150
Hemingray Blue 150-175
Hemingray Blue w/ Milk Swirls 200-250

NO EMBOSSING

[010] [No embossing] SB
Light Blue 600-700

OAKMAN MFG.CO.

[010] (F-Skirt) COLUMBIA (Base) PAT.JUNE 17, 1890. AUG. 19, 1890. MAY 12, 1891./ OAKMAN M'F'G CO. BOSTON {Inside of outer skirt is threaded} BE
Light Aqua, Light Blue 300-350
Light Lime Green 400-500

PATENT - OTHER

[010] (F-Skirt) "PATENTED.JUNE. 17. 1890" SB
Blue Aqua 250-300
Light Blue, Light Green 300-350
Yellow Green 1,750-2,000

CD 264

COLUMBIA

[010] (F-Skirt) COLUMBIA SB
Light Aqua, Light Blue 400-500

CD 265

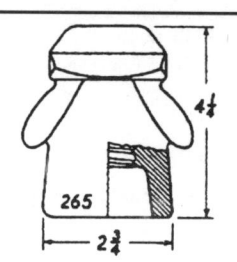

CD 265

FISHER

[010] (F-Skirt) FISHER/(Arc)ELECTRIC RAILWAY SYSTEM/BUILT BY/DETROIT ELECTRICAL/WORKS SB
Aqua, Ice Blue,
Light Yellow Green, Lime Green 4,500-5,000

NO EMBOSSING

[010] [No embossing] SB
Light Green 5,000-7,500

CD 266

PATENT - OTHER

[010] (F-Skirt) "PATENTED.JUNE. 17. 1890" SB
Aqua, Dark Aqua 1,000-1,250
Green 1,250-1,500

CD 267

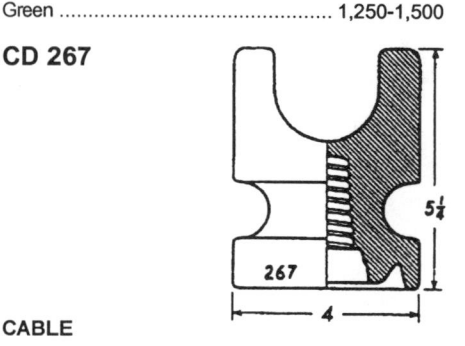

CABLE

[010] (F-Skirt) Nº 4 CABLE SB
Aqua 75-100
Dark Aqua 100-125
Blue 125-150
Green 300-350
Dark Yellow Green, Emerald Green 500-600

N.E.G.M.CO.

[010] (F-Skirt) N.E.G.M. SB
Aqua 400-500

[020] (F-Skirt) N.E.G.M.CO.
(R-Skirt) "PATENTED.JUNE. 17. 1890" SB
Aqua, Blue 100-125
Light Green 200-250
Green 350-400
Yellow Green 400-500

NO EMBOSSING

[010] [No embossing] SB
Aqua 75-100
Light Blue 100-125

CD 267.5

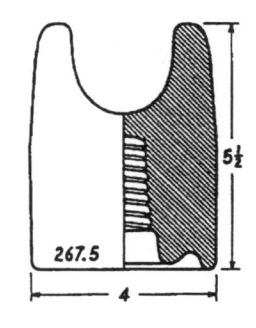

N.E.G.M.CO.

[010] (F-Skirt) N.E.G.M.CO. SB
Blue, Blue Aqua 75-100
Light Green 100-125
Green 125-150
Dark Green 150-175
Emerald Green 175-200

CD 268

CROWN

[010] (F-Skirt) THE CROWN/PAT ALLOWED SB
Aqua, Dark Aqua, Dark Green 15,000-20,000

152

CD 272

CD 268.5

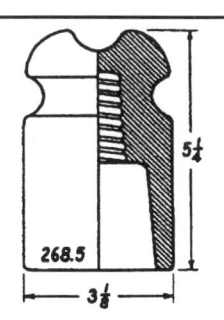

STANDARD GLASS INSULATOR CO.

[010] (F-Skirt) PATENTED/AUG. 8, 1893 (Base) THE STANDARD GLASS INSULATOR CO./BOSTON MASS BE
Aqua 3,500-4,000

CD 269

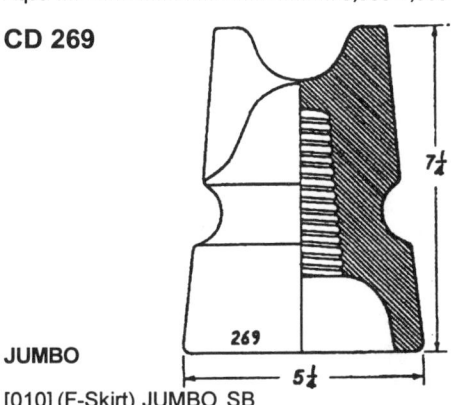

JUMBO

[010] (F-Skirt) JUMBO SB
Aqua, Dark Aqua 400-500
Green 800-1,000

[020] (F-Skirt) JUMBO
(R-Skirt) ['BROOKFIELD' blotted out] SB
Aqua 400-500

OAKMAN MFG.CO.

[010] (F-Skirt) JUMBO (R-Skirt) PAT'D JUNE 17, 1890, AUG 19, 1890 (Base) OAKMAN M'F'G.CO. BOSTON BE
Light Aqua, Light Blue Aqua 400-500
Ice Aqua 600-700
Green 800-1,000

CD 270

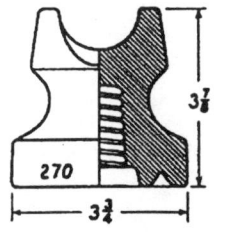

NO EMBOSSING

[010] [No embossing] SB
Aqua 350-400
Green 400-500

CD 271

HEMINGRAY

CD 271
5 x 4 $^{3}/_{8}$

[010] (F-Skirt) HEMINGRAY ---
(R-Skirt) [Unknown] {Specimen only} SDP
Hemingray Blue 500-600

CD 272

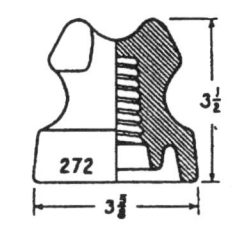

ARMSTRONG Ⓐ

[010] (F-Skirt) *Armstrong's* 511A
(R-Skirt) MADE IN U.S.A. Ⓐ [Numbers] CB
Root Beer Amber 5-10
Dark Red Amber 10-15

[020] (F-Skirt) *Armstrong's* 511A
(R-Skirt) MADE IN U.S.A. Ⓐ [Numbers] SB
Root Beer Amber 5-10
Dark Red Amber 10-15

NO EMBOSSING

[010] [No embossing] SB
Aqua, Green Aqua 20-30

WHITALL TATUM 〖W〗

[010] (F-Skirt) WHITALL TATUM CO. NO. 511A/[Number] (R-Skirt) MADE IN U.S.A. 〖W〗/[Number] SB
Root Beer Amber 5-10
Dark Red Amber 15-20

[020] (F-Skirt) WHITALL TATUM CO. NO 62/[Number] {Note 'NO 62' style number}
(R-Skirt) MADE IN U.S.A. SB
Dark Red Amber 250-300

[025] (F-Skirt) WHITALL TATUM CO./[Number] (R-Skirt) MADE IN U.S.A. 〖W〗/[Number] SB
Root Beer Amber 5-10

[030] (F-Skirt) WHITALL TATUM NO. 511A/[Number] (R-Skirt) MADE IN U.S.A. SB
Root Beer Amber 5-10

CD 272

[040] (F-Skirt) WHITALL TATUM № 511A/
[Number] (R-Skirt) MADE IN U.S.A. ▽ SB
Clear, Green Tint, Root Beer Amber,
Smoke, Straw 5-10

[050] (F-Skirt) WHITALL TATUM № 511A/
[Number] (R-Skirt) MADE IN U.S.A. ▽ /
[Number] SB
Clear, Root Beer Amber 5-10

CD 273

NO EMBOSSING

[010] [No embossing] SB
Dark Aqua 500-600

CD 275

LOCKE

[010] (F-Skirt) F.M. LOCKE VICTOR N.Y.
(R-Skirt) № 21 SB
Aqua, Green Aqua, Light Aqua 75-100

[020] (F-Skirt) LOCKE VICTOR N.Y.
(R-Skirt) № 21 SB
Aqua, Dark Aqua 75-100
Green 150-175

NO NAME

[010] (F-Skirt) № 21 SB
Dark Aqua 75-100
Green 150-175
Emerald Green 400-500

CD 280

HEMINGRAY

[010] (F-Skirt) HEMINGRAY (R-Skirt) № 100
H.V. SDP
Aqua 400-500

[020] (F-Skirt) HEMINGRAY/PATENT MAY 2
1893 (R-Skirt) HIGH VOLTAGE/TRIPLE
PETTICOAT/№ 1 SDP
Aqua, Blue Aqua, Green Aqua, Light Aqua .. 15-20
Ice Aqua 20-30
Green 100-125
Ice Aqua w/ Purple Swirls 800-1,000

[025] (F-Skirt) HEMINGRAY/PATENT MAY 2
1893 (R-Skirt) HIGH VOLTAGE/TRIPLE
PETTICOAT/['TRIPPLE' blotted out]/№ 1
SDP
Aqua 15-20

[030] (F-Skirt) HEMINGRAY/PATENT MAY 2
189Ɛ {'9' is engraved over a 'Ɛ'} {Note
backwards numbers} (R-Skirt) HIGH
VOLTAGE/TRIPLE PETTICOAT/No.1. SDP
Aqua 15-20

NO NAME

[010] (F-Skirt) № 135 SB
Green 100-125
Emerald Green 150-175

[020] (F-Skirt) № 180 SB
Dark Yellow Green, Emerald Green 400-500

PRISM ▭

[010] (F-Skirt) ▭ SB
Aqua 40-50
Green 75-100
Yellow Green 175-200
Dark Yellow Green 250-300

CD 281

HEMINGRAY

[010] (F-Skirt) HEMINGRAY-71
(R-Skirt) MADE IN U.S.A. RDP
Aqua, Blue Aqua, Dark Aqua 10-15
Hemingray Blue 15-20
Carnival 350-400
Root Beer Amber 2,500-3,000

[020] (F-Skirt) HEMINGRAY-71
(R-Skirt) MADE IN U.S.A. SDP
Hemingray Blue 15-20

[030] (F-Skirt) HEMINGRAY-71/[Numbers]
(R-Skirt) MADE IN U.S.A. RDP
Ice Blue 15-20

CD 282

[040] (F-Skirt) HEMINGRAY/№ 1 HIGH VOLTAGE (R-Skirt) PATENT MAY 2 1893 SDP
Dark Aqua, Green Aqua 10-15
Hemingray Blue 15-20

[050] (F-Skirt) HEMINGRAY/№ 1 HIGH VOLTAGE (R-Skirt) PATENTED MAY 2 1893 SDP
Aqua, Blue Aqua 10-15

[060] (F-Skirt) HEMINGRAY/№ 1 HIGH VOLTAGE/[Blot out under '1']
(R-Skirt) PATENT MAY 2 1893 SDP
Dark Aqua 10-15

[070] (F-Skirt) HEMINGRAY/№ 2 HIGH VOLTAGE (R-Skirt) PATENT MAY 2 1893 SDP
Aqua, Blue 10-15

[080] (F-Skirt) HEMINGRAY/№ 71 HIGH VOLTAGE (R-Skirt) PATENT MAY 2 1893/ MADE IN U.S.A. RDP
Hemingray Blue 15-20

[090] (F-Skirt) HEMINGRAY/№ 71 HIGH VOLTAGE (R-Skirt) PATENTED MAY 2 1893/ MADE IN U.S.A. SDP
Aqua 10-15
Hemingray Blue, Ice Blue 15-20

[100] (F-Skirt) HEMINGRAY/№ 71 HIGH VOLTAGE/[Blot out under '71']
(R-Skirt) PATENT MAY 2 1893/MADE IN U.S.A. SDP
Hemingray Blue 15-20

[110] (F-Skirt) HEMINGRAY/№ 71 HIGH VOLTAGE/[Blot out under '71']
(R-Skirt) PATENTED MAY 2 1893/MADE IN U.S.A. SDP
Hemingray Blue 15-20

LYNCHBURG Ⓛ

[010] (Dome) NO. 180 (F-Skirt) Ⓛ/ LYNCHBURG (R-Skirt) MADE IN U.S.A. SDP
Aqua 1,500-1,750
Green 1,750-2,000

[015] (Dome) NO. 180 (F-Skirt) Ⓛ/ LYNCHBURG (R-Skirt) [Number] MADE IN U.S.A. SDP
Green 1,750-2,000

CD 282

HEMINGRAY

[010] (F-Skirt) HEMINGRAY/№ 2 PROVO TYPE (R-Skirt) PATENT MAY 2 1893 SDP
Aqua 40-50
Light Hemingray Blue 75-100

[020] (F-Skirt) HEMINGRAY/№ 2 PROVO TYPE (R-Skirt) PATENT MAY 2 1893/ PATENTED APRIL 25 1899 SDP
Aqua, Dark Aqua 40-50
Hemingray Blue 50-75
Light Aqua 125-150
Olive Green 1,250-1,500

[030] (F-Skirt) HEMINGRAY/№ 2 "PROVO" TYPE (R-Skirt) PATENT MAY 2 1893 SDP
Light Aqua 125-150
Ice Green 250-300

[040] (F-Skirt) HEMINGRAY/№ 2 "PROVO" TYPE (R-Skirt) PATENT MAY 2 1893/ PATENTED APRIL 25 1899 SDP
Aqua 40-50
Hemingray Blue 50-75

KNOWLES

[010] (F-Skirt) KNOWLES 5 1/2 (R-Skirt) BOSTON SB
Aqua 200-250
Light Blue 250-300

PRISM ⊠

[010] (F-Skirt) ⊠ SB
Aqua, Blue Aqua 75-100
Lime Green 200-250
Yellow Green 300-350

155

CD 283

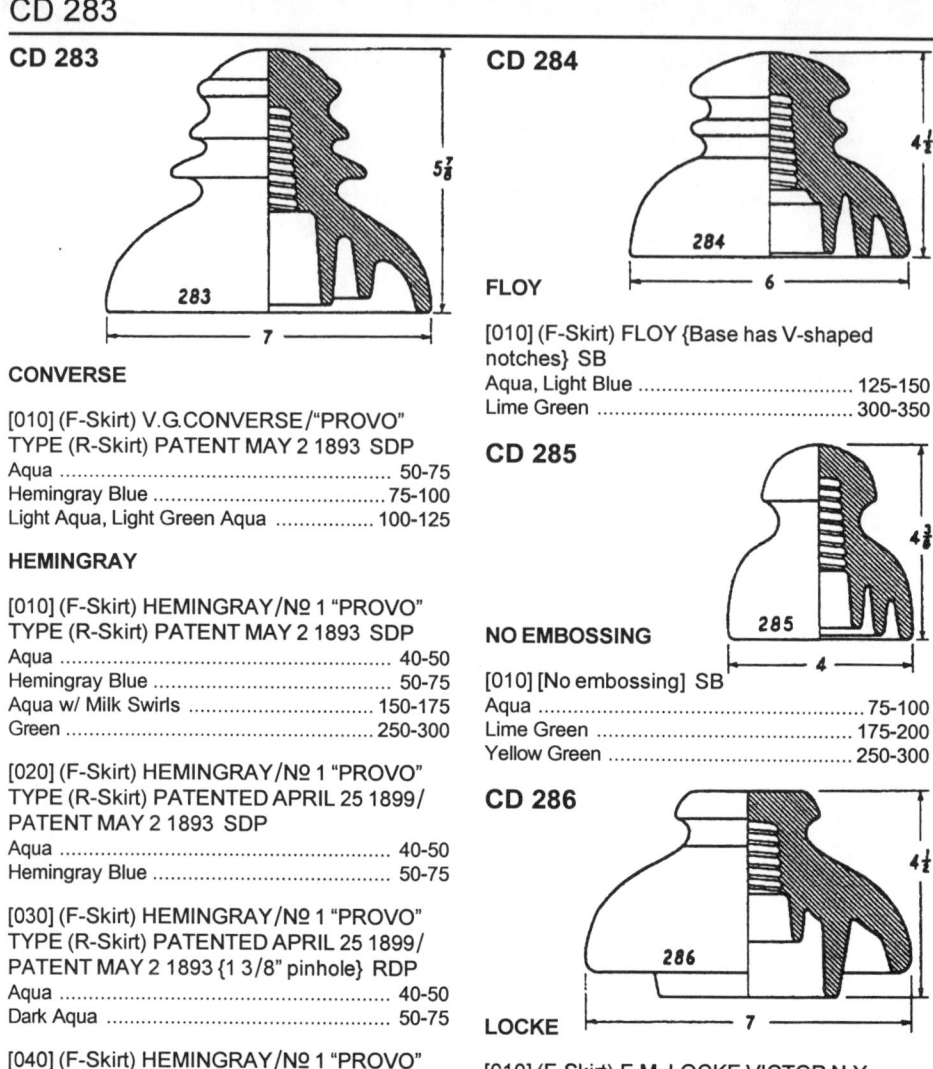

CD 283

CONVERSE

[010] (F-Skirt) V.G.CONVERSE/"PROVO" TYPE (R-Skirt) PATENT MAY 2 1893 SDP
Aqua 50-75
Hemingray Blue 75-100
Light Aqua, Light Green Aqua 100-125

HEMINGRAY

[010] (F-Skirt) HEMINGRAY/№ 1 "PROVO" TYPE (R-Skirt) PATENT MAY 2 1893 SDP
Aqua 40-50
Hemingray Blue 50-75
Aqua w/ Milk Swirls 150-175
Green 250-300

[020] (F-Skirt) HEMINGRAY/№ 1 "PROVO" TYPE (R-Skirt) PATENTED APRIL 25 1899/ PATENT MAY 2 1893 SDP
Aqua 40-50
Hemingray Blue 50-75

[030] (F-Skirt) HEMINGRAY/№ 1 "PROVO" TYPE (R-Skirt) PATENTED APRIL 25 1899/ PATENT MAY 2 1893 {1 3/8" pinhole} RDP
Aqua 40-50
Dark Aqua 50-75

[040] (F-Skirt) HEMINGRAY/№ 1 "PROVO" TYPE (R-Skirt) PATENTED APRIL 25 1899/ PATENT MAY 2 1893 {1 3/8" pinhole} SDP
Aqua 50-75

[050] (F-Skirt) HEMINGRAY/№ 1 "PROVO" TYPE (R-Skirt) PATENTED MAY 2 1893 SDP
Hemingray Blue 50-75

PRISM ▭

[010] (F-Skirt) ▭ SB
Aqua, Blue 75-100
Green, Lime Green 250-300
Yellow Olive Green 400-500

CD 284

FLOY

[010] (F-Skirt) FLOY {Base has V-shaped notches} SB
Aqua, Light Blue 125-150
Lime Green 300-350

CD 285

NO EMBOSSING

[010] [No embossing] SB
Aqua 75-100
Lime Green 175-200
Yellow Green 250-300

CD 286

LOCKE

[010] (F-Skirt) F.M. LOCKE VICTOR N.Y. (R-Skirt) PAT.MAY 22 94/" NOV. 24 96/" DEC. 15 96/" SEPT. 28 97/" JUNE 7 98 SB
Aqua, Light Blue 40-50
Light Green 175-200
Green 400-500
Yellow Olive Green 800-1,000

[020] (F-Skirt) F.M. LOCKE VICTOR N.Y. (R-Skirt) PAT.MAY 22 96/" NOV. 24 96/" DEC. 15 96/" SEPT. 28 97/" JUNE 7 98 {Note 'MAY 22 96' date} SB
Aqua 40-50

CD 287

CD 286.9

WESTINGHOUSE   CD 286.9   7 $\frac{1}{4}$ x 6 $\frac{1}{2}$

[010] (F-Skirt) WESTINGHOUSE ELECTRIC & M'F'G CO./PITTSBURG, PA.
(R-Skirt) "TELLURIDE" TYPE C. SB
Aqua 7,500-10,000

CD 287

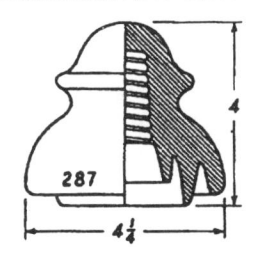

LOCKE

[010] (Dome) [Number] (F-Skirt) F.M. LACKE VICTOR N.Y./PAT'D MAY 22 1894 {Note spelling} (R-Skirt) № 15 SB
Light Aqua 15-20

[020] (Dome) [Number] (F-Skirt) F.M. LOCKE VICTOR N.Y. (R-Skirt) № 15 SB
Aqua 5-10
Emerald Green 30-40

[030] (Dome) [Number] (F-Skirt) F.M. LOCKE VICTOR N.Y./PAT'D MAY 22 1894 (R-Skirt) № 15 SB
Aqua, Light Aqua 5-10

[040] (Dome) [Number] (F-Skirt) F.M. LOCKE VICTOR N.Y./[Blotted out embossing] (R-Skirt) № 15 SB
Aqua 5-10
Dark Yellow Green 50-75

[050] (Dome) [Number] (F-Skirt) F.M. LOCKE VICTOR N.Y./['PAT'D MAY 22 1894' blotted out] (R-Skirt) № 15 SB
Aqua 5-10

[060] (Dome) [Number] (F-Skirt) LOCKE VICTOR N.Y. (R-Skirt) № 15 SB
Dark Aqua 5-10
Dark Yellow Green 50-75

[065] (Dome) [Number] (F-Skirt) ['F.M.' blotted out] LOCKE VICTOR N.Y. (R-Skirt) № 15 SB
Dark Yellow Green 50-75

[070] (F-Skirt) F.M. LOCKE & CO. VICTOR N.Y./PAT'D MAY 22 1894 SB
Aqua 5-10
Blue 10-15
Ice Blue 15-20
Green 20-30

[080] (F-Skirt) F.M. LOCKE VICTOR N.Y. (R-Skirt) № 15 SB
Aqua 5-10

[085] (F-Skirt) F.M. LOCKE VICTOR N.Y./ PAT'D MAY 22 1894 (R-Skirt) № 15 SB
Light Aqua 5-10

[090] (F-Skirt) F.M. LOCKE VICTOR N.Y./ PAT'D. MAY 22 1894 SB
Light Aqua 10-15

[100] (F-Skirt) LOCKE VICTOR N.Y. (R-Skirt) № 15 SB
Aqua 5-10
Dark Green 20-30

[110] (F-Skirt) ['F.M. LOCKE & CO. VICTOR N.Y./PAT'D MAY 22 1894' blotted out] (R-Skirt) № 135 SB
Dark Green Aqua 15-20

[120] (F-Skirt) ['F.M. LOCKE VICTOR N.Y.' blotted out]/PAT'D MAY 22 1894 (R-Skirt) № 15 SB
Emerald Green 30-40

NO NAME

[005] (Dome) [Number] (F-Skirt) № 135 SB
Dark Aqua 15-20
Green 30-40

[010] (Dome) [Number] (F-Skirt) № 15 SB
Aqua 5-10
Green 20-30

[020] (Dome) [Number] (F-Skirt) [Blotted out embossing] (R-Skirt) № 15 SB
Aqua, Dark Aqua 5-10
Olive Green 75-100
Olive Amber 150-175

[030] (F-Skirt) № 135 SB
Aqua 15-20
Green 30-40

[040] (F-Skirt) ['№ 135' blotted out] SB
Emerald Green 30-40

157

CD 287.1

CD 287.1

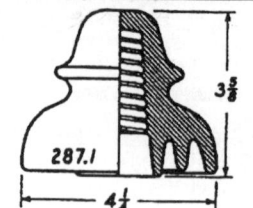

LOCKE

[010] (F-Skirt) F.M. LOCKE & Co VICTOR. N.Y./PATD MAY 22 1894 SB
Aqua, Light Aqua 20-30
Ice Aqua 30-40
Red Amber 3,000-3,500
Peacock Blue 3,500-4,000

[020] (F-Skirt) F.M. LOCKE ['& Co' blotted out] VICTOR. N.Y./PATD MAY 22 1894 SB
Light Aqua 20-30
Ice Aqua 30-40

CD 287.2

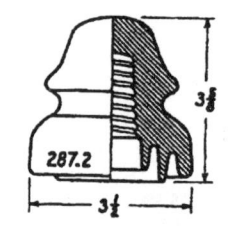

BROOKFIELD

[010] (F-Crown) (Arc)W. BROOKFIELD/NEW YORK. (F-Skirt) LOCKE PATENTS PENDING. {MLOD} SB
Aqua, Light Aqua 200-250

CD 288

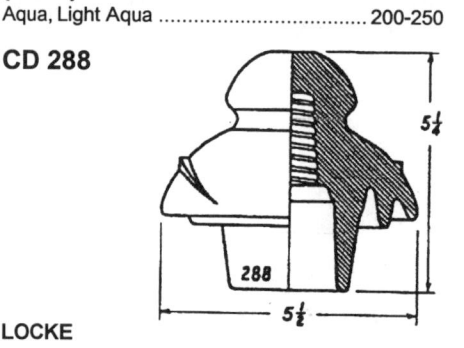

LOCKE

[005] (F-Umbrella/Above ridge) (Arc)FRED M. LOCKE VICTOR N.Y. (F-Umbrella/Below ridge) № 18 SB
Aqua 125-150

[010] (F-Umbrella/Above ridge) (Arc)FRED M. LOCKE VICTOR N.Y. (R-Umbrella/Above ridge) PAT.MAY 22-94/" NOV. 24-96 (R-Umbrella/Below ridge) SEPT. 28-97/& JUNE 7-98 SB
Aqua 125-150

[020] (F-Umbrella/Above ridge) (Arc)LOCKE VICTOR N.Y. (F-Umbrella/Below ridge) № 18 SB
Aqua 125-150

MERSHON

[010] (F-Umbrella) By R.D.MERSHON {One ridge} SB
Aqua 75-100

[020] (F-Umbrella) By R.D.MERSHON {Three ridges} SB
Aqua 300-350

[030] (F-Umbrella/Below ridge) PAT'D/By R.D. MERSHON {One ridge} SB
Light Aqua 75-100
Light Green 200-250

[040] (F-Umbrella/Below ridge) R.D.MERSHON {One ridge} SB
Aqua 75-100
Dark Aqua 125-150

NO EMBOSSING

[010] [No embossing] SB
Aqua, Light Aqua 50-75
Light Blue 75-100
Hemingray Blue, Ice Green 100-125

[020] [No embossing] {Denver style} SB
Light Aqua 150-175

CD 289

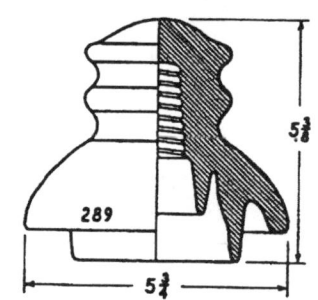

LOCKE

[010] (F-Skirt) FRED M. LOCKE VICTOR N.Y. SB
Aqua 40-50

[020] (F-Skirt) FRED M. LOCKE VICTOR N.Y. (R-Skirt) PAT.MAY 22 1894 SB
Aqua 40-50
Blue 50-75
Green 100-125

[030] (F-Skirt) FRED. M. LOCKE VICTOR N.Y. (R-Skirt) [Blotted out embossing] SB
Aqua, Dark Aqua 40-50

CD 292.5

NO EMBOSSING

[010] [No embossing] SB
Aqua, Dark Aqua 30-40

CD 289.9

CD 289.9
3 $\frac{1}{8}$ x 3 $\frac{7}{8}$

PATENT - OTHER

[010] (F-Skirt) PATD. MAY 7. 89 SB
Aqua 15,000-20,000

CD 291

HEMINGRAY

[010] (F-Skirt) HEMINGRAY/PATENT MAY 2 1893 SDP
Aqua 500-600

[020] (F-Skirt) HEMINGRAY/PATENT MAY 2 1893 (R-Skirt) HIGH VOLTAGE № 1/TRIPLE PETTICOAT SDP
Aqua 500-600

[030] (F-Skirt) HEMINGRAY/PATENT MAY 2 1893 (R-Skirt) HIGH VOLTAGE № 3/TRIPLE PETTICOAT SDP
Aqua 500-600

NO NAME

[010] (F-Skirt) № 5 SB
Aqua 600-700

[020] (F-Skirt) № 5 SDP
Aqua 700-800

CD 292

NO EMBOSSING

[010] [No embossing] SB
Aqua 30-40
Blue 50-75

PATENT - OTHER

[010] (F-Skirt) PATENTED.JUNE. 17 1890 SB
Green 50-75
Dark Yellow Green 175-200

[020] (F-Skirt) "PATENTED.JUNE. 17 1890" SB
Aqua 15-20
Light Blue 20-30
Green 50-75
Emerald Green 125-150
Yellow Green 150-175

PRISM ▭

[010] (F-Skirt) ▭ SB
Aqua 30-40
Green 75-100

CD 292.5

KNOWLES

[010] (F-Skirt) BOSTON (R-Skirt) "KNOWLES 6" SB
Aqua, Blue 250-300
Dark Green 400-500

[020] (F-Skirt) KNOWLES-6 (R-Skirt) BOSTON/PATENTED JUNE 17, 1890 SB
Aqua 200-250

[030] (F-Skirt) ▭/KNOWLES 6 (R-Skirt) BOSTON {No ribs on inner skirt} SB
Light Green 300-350

[040] (F-Skirt) ▭/"KNOWLES 6" (R-Skirt) [Blotted out embossing]/BOSTON SB
Green 350-400

[050] (F-Skirt) ▭/"KNOWLES 6"/ "PATENTED.JUNE. 17. 1890" (R-Skirt) BOSTON SB
Blue, Light Aqua 200-250

[060] (F-Skirt) ▭/"KNOWLES 6"/ "PATENTED.JUNE. 17. 1890" (R-Skirt) BOSTON {No ribs on inner skirt} SB
Blue, Light Aqua 200-250

159

CD 293

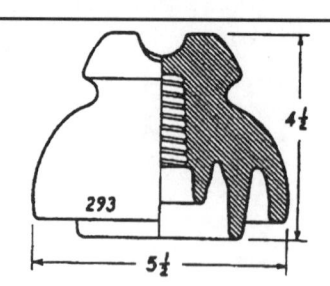

CD 293

LOCKE

[010] (F-Skirt) FRED M. LOCKE VICTOR N.Y./ PATD MAY 22-94 NOV. 24-96 DEC. 15-96 - SEP'T.28-97 (R-Skirt) OTHER PAT'S PENDING SB
Aqua 20-30
Light Blue 30-40

[020] (F-Skirt) FRED M. LOCKE VICTOR N.Y./ PAT'D MAY 22-94 NOV. 24-96 DEC. 15-96 (R-Skirt) SEP'T.28-97 OTHER PAT'S PENDING SB
Aqua, Blue Aqua 20-30
Light Green 75-100
Light Gray Purple 100-125
Light Purple 125-150
Light Purple w/ Milk Swirls 250-300

CD 293.1

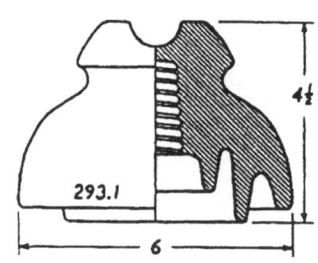

LOCKE

[010] (F-Skirt) FRED M LOCKE VICTOR N.Y./ PAT'D. MAY 22-94 NOV. 24-96 DEC. 15-96 SEPT. 28-97 (R-Skirt) OTHER PAT'S PENDING SB
Blue Aqua, Light Aqua 40-50
Light Green 75-100
Lime Green w/ Milk Swirls 175-200

CD 294

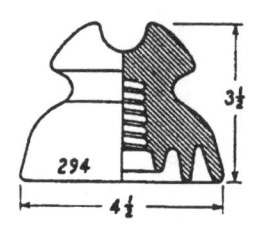

N.E.G.M.CO.

[010] (F-Skirt) N.E.G.M.CO. (R-Skirt) TRIPLE PETTICOAT SB
Aqua 50-75
Light Blue 75-100

CD 295

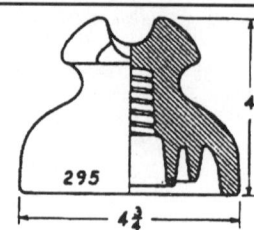

HEMINGRAY

[010] (F-Skirt) HEMINGRAY-72 (R-Skirt) MADE IN U.S.A. RDP
Aqua 5-10
Ice Blue 10-15
Dark Hemingray Blue 15-20

[020] (F-Skirt) HEMINGRAY-72 (R-Skirt) MADE IN U.S.A./[Number] SB
Aqua, Clear 5-10

[030] (F-Skirt) HEMINGRAY-72.A (R-Skirt) MADE IN U.S.A. {Wide saddle groove} RDP
Aqua, Blue Aqua, Dark Aqua 5-10

[040] (F-Skirt) HEMINGRAY-72.A/[Numbers and dots] (R-Skirt) MADE IN U.S.A./[Number] {Wide saddle groove} RDP
Aqua Tint, Clear 10-15

[050] (F-Skirt) HEMINGRAY-72.A/[Number] (R-Skirt) MADE IN U.S.A. {Wide saddle groove} RDP
Ice Aqua 15-20

[060] (F-Skirt) HEMINGRAY-72/[Numbers] (R-Skirt) MADE IN U.S.A. RDP
Ice Blue 10-15

[070] (F-Skirt) HEMINGRAY/PATENTED MAY 2 1893 (R-Skirt) HIGH VOLTAGE/MADE IN U.S.A./['TRIPLE PETTICOAT' blotted out]/№ 72/['4' blotted out] SDP
Hemingray Blue 15-20

[080] (F-Skirt) HEMINGRAY/PATENTED MAY 2 1893 (R-Skirt) HIGH VOLTAGE/TRIPLE PETTICOAT/№ 4 SDP
Aqua, Blue Aqua 5-10
Hemingray Blue 15-20

[085] (F-Skirt) HEMINGRAY/PATENTED MAY 2 1893 (R-Skirt) HIGH VOLTAGE/TRIPLE PETTICOAT/№ 4 {Tall variant; 1/2" extended outer skirt} SDP
Hemingray Blue 20-30

[090] (F-Skirt) HEMINGRAY/PATENTED MAY 2 1893 (R-Skirt) HIGH VOLTAGE/TRIPLE PETTICOAT/№ 72 SDP
Aqua, Blue Aqua 5-10
Hemingray Blue 15-20

CD 297

[100] (F-Skirt) HEMINGRAY/PATENTED MAY 2 1893 (R-Skirt) HIGH VOLTAGE/TRIPLE PETTICOAT/N⁰ 72/['4' blotted out] SDP
Aqua, Blue Aqua 5-10
Hemingray Blue 15-20

CD 296

LOCKE

[010] (F-Skirt) F.M. LOCK VICTOR N.Y. {Note spelling} (R-Skirt) N⁰ 20 SB
Aqua, Dark Aqua 20-30

[020] (F-Skirt) F.M. LOCKE VICTOR N.Y. (R-Skirt) N⁰ 20 SB
Aqua, Dark Aqua 10-15

[030] (F-Skirt) LOCKE VICTOR N.Y. SB
Aqua 10-15

[040] (F-Skirt) LOCKE VICTOR N.Y. (R-Skirt) N⁰ 20 SB
Aqua 10-15
Dark Green 30-40
Emerald Green 40-50
Dark Yellow Green 50-75

[050] (F-Skirt) ['F.M.' blotted out] LOCKE VICTOR N.Y. (R-Skirt) N⁰ 20 SB
Dark Aqua 10-15
Green 30-40
Emerald Green 40-50
Dark Yellow Green 50-75

NO NAME

[010] (F-Skirt) N⁰ 20 SB
Aqua 10-15
Green 30-40

[029] (F-Skirt) [Blotted out embossing] (R-Skirt) N⁰ 20 SB
Dark Aqua 10-15

PRISM ⌺

[010] (F-Skirt) ⌺ SB
Aqua 30-40
Blue 40-50
Green 75-100
Lime Green 100-125
Yellow Green 300-350

CD 297

LOCKE

[010] (F-Skirt) F.M. LOCKE VICTOR N.Y./ [Blotted out embossing] (R-Skirt) N⁰ 16 SB
Dark Aqua 10-15

[020] (F-Skirt) FRED M. LOCKE VICTOR N.Y. (R-Skirt) N⁰ 16/OTHER PATENTS PENDING SB
Aqua, Light Aqua 10-15

[030] (F-Skirt) FRED M. LOCKE VICTOR N.Y. (R-Skirt) N⁰ 16/[Blotted out embossing] SB
Aqua 10-15
Light Green 30-40
Emerald Green 50-75
Dark Olive Green 200-250

[035] (F-Skirt) FRED M. LOCKE VICTOR N.Y./ PAT'D MAY 22-94 NOV. 24 96 DEC. 15-98 SEPT. 28 97 {Note 'DEC. 15-98' date} (R-Skirt) No 16/OTHER PAT'S PENDING SB
Light Aqua 30-40

[040] (F-Skirt) FRED M. LOCKE VICTOR N.Y./ PAT'D MAY 22-94 NOV. 24-96 DEC. 15-98 SEPT. 28-97 {Note 'DEC. 15-98' date} (R-Skirt) OTHER PAT'S PENDING {Narrow neck} SB
Light Aqua, Light Blue 20-30
Light Green 40-50

[050] (F-Skirt) FRED M. LOCKE VICTOR N.Y./ [Blotted out embossing] (R-Skirt) N⁰ 16 SB
Green Aqua 10-15

[060] (F-Skirt) FRED. M. LOCKE VICTOR N.Y. (R-Skirt) N⁰ 16 SB
Aqua, Dark Aqua 10-15
Green 40-50

[070] (F-Skirt) LOCKE VICTOR N.Y. (R-Skirt) N⁰ 16 SB
Emerald Green 50-75

[080] (F-Skirt) ['FRED M.' blotted out] LOCKE VICTOR N.Y. (R-Skirt) N⁰ 16/[Blotted out embossing] SB
Dark Yellow Green 150-175

CD 297

NO EMBOSSING

[010] [No embossing] SB
Dark Aqua 10-15
Emerald Green 50-75

[020] [No embossing] {Denver style} SB
Light Aqua 200-250

NO NAME

[005] (F-Skirt) № 16 (R-Skirt) [Blotted out embossing] SB
Dark Green 50-75

[010] (F-Skirt) № 16/[Blotted out embossing] (R-Skirt) [Blotted out embossing] SB
Aqua, Dark Aqua 5-10
Green 40-50
Dark Green 50-75
Dark Yellow Green 125-150
Dark Olive Amber 400-500

CD 298

NO EMBOSSING

[010] [No embossing] SB
Aqua, Light Aqua 100-125
Blue, Ice Aqua 125-150

CD 299

BROOKFIELD

[010] (F-Skirt) BROOKFIELD SB
Dark Aqua 250-300
Emerald Green 500-600

[020] (Top of ear) BROOKFIELD SB
Dark Aqua 250-300
Dark Aqua/Emerald Green Two Tone 400-500
Emerald Green 500-600

CD 299.1

PRISM

[010] (F-Skirt) (R-Skirt) 4 1/2 SB
Light Aqua 300-350

CD 299.2

NO EMBOSSING

[010] [No embossing] {Fry glass} SB
Black Opalescent, White Opalescent . 5,000-7,500

CD 299.7

CD 299.7
$5\ ^1/_2\ \times\ 4\ ^1/_4$

LOWEX

[010] (F-Skirt) LOWEX/REG. U.S. PAT. OFF. (R-Skirt) D-514/[Number]-MADE IN U.S.A.-[Number] SB
Light Lemon 5,000-7,500

CD 300

BROOKFIELD

[005] (F-Skirt) BROOKFIELD SB
Dark Aqua 40-50

CD 301.5

[010] (F-Skirt) BROOKFIELD (R-Skirt) [Blotted out embossing] SB
Aqua 40-50

[020] (F-Skirt) No.17/BROOKFIELD SB
Aqua 40-50
Green 100-125

[030] (F-Skirt) No.17/BROOKFIELD (R-Skirt) [Blotted out embossing] SB
Dark Aqua 40-50
Green 100-125
Emerald Green 200-250

[040] (F-Skirt) ['No.17' blotted out]/BROOKFIELD (R-Skirt) [Blotted out embossing] SB
Dark Aqua 40-50
Emerald Green 200-250

LOCKE

[010] (F-Skirt) F.M. LOCKE VICTOR N.Y. (R-Skirt) No.17 SB
Aqua 30-40

[015] (F-Skirt) F.M. LOCKE VICTOR N.Y. (R-Skirt) PAT.MAY 22 96/" NOV. 24 96/" DEC. 15 96/" SEPT. 28 97/" JUNE 7 98 SB
Aqua 30-40

[020] (F-Skirt) F.M. LOCKE VICTOR N.Y. (R-Skirt) PAT.MAY 22, 94/" SEPT. 28, 97/" JUNE 7, 98 SB
Olive Green 500-600

[030] (F-Skirt) F.M. LOCKE VICTOR N.Y. PAT.MAY 22, 94/" NOV. 24 96/" DEC. 15 96/" SEPT 28 97/" JUNE 7 98 SB
Aqua 30-40
Olive Green 500-600

[040] (F-Skirt) No.17/F.M. LOCKE VICTOR N.Y. (R-Skirt) PAT.MAY 22 96/" NOV. 24 96/" DEC. 15 96/" SEPT. 28 97/" JUNE 7 98 SB
Aqua 30-40

[050] (F-Skirt) No.17/F.M. LOCKE VICTOR N.Y. (R-Skirt) PATENT/[Blotted out embossing] SB
Aqua 30-40
Green 100-125
Dark Green 175-200

[060] (F-Skirt) No.17/F.M. LOCKE VICTOR N.Y. (R-Skirt) PAT'D MAY 22, 94 MAY 24, 96/" SEPT.28, 97 SB
Aqua 30-40

[070] (F-Skirt) No.17/F.M. LOCKE VICTOR N.Y. (R-Skirt) [Blotted out embossing] SB
Aqua, Blue Aqua 30-40

[080] (Top of ear) 1 (F-Skirt) No.17/F.M. LOCKE VICTOR N.Y. (R-Skirt) [Blotted out embossing] SB
Dark Aqua 20-30
Green 100-125

CD 301

NO EMBOSSING

[010] [No embossing] SB
Aqua, Dark Green Aqua 600-700
Blue, Ice Blue 700-800
Green 800-1,000

CD 301.2

NO EMBOSSING

[010] [No embossing] {Fry glass} SB
Black Opalescent, Green Opalescent, White Opalescent 5,000-7,500

CD 301.5

CD 301.5
7 1/8 x 4 1/2

HEMINGRAY

[010] (F-Skirt) HEMINGRAY-515 (R-Skirt) MADE IN U.S.A./[Number and dots] {1 3/8" pinhole} SB
Clear 1,500-1,750

[020] (F-Skirt) HEMINGRAY-515/[Number] (R-Skirt) MADE IN U.S.A./[Number] {1 3/8" pinhole} SB
Honey Amber 2,000-2,500

CD 302

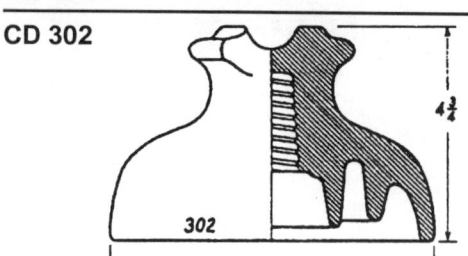

CD 302

HEMINGRAY

[020] (F-Skirt) HEMINGRAY/PATENT MAY 2ND 1893 (R-Skirt) MADE IN U.S.A./MUNCIE TYPE/N⁰ 75 RDP
Aqua 50-75
Hemingray Blue 75-100

[030] (F-Skirt) HEMINGRAY/PATENT MAY 2ND 1893 (R-Skirt) MUNCIE TYPE RDP
Aqua 30-40

[040] (F-Skirt) HEMINGRAY/PATENT MAY 2ND 1893 (R-Skirt) MUNCIE {1 3/8" pinhole} RDP
Aqua 40-50
Hemingray Blue 50-75

[050] (F-Skirt) HEMINGRAY/PATENT MAY 2ND 1893 (R-Skirt) OHIO RDP
Aqua, Blue Aqua 40-50

[060] (F-Skirt) HEMINGRAY/PATENT MAY 2ND 1893 (R-Skirt) OHIO {1 3/8" pinhole} RDP
Aqua 40-50

CD 303/310

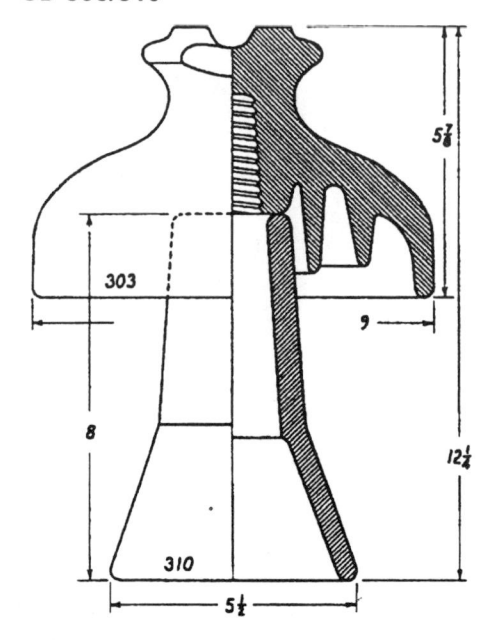

HEMINGRAY

[010] (F-Skirt) HEMINGRAY/PATENT MAY 2ND 1893 (R-Skirt) MUNCIE TYPE {CD 310 sleeve is unembossed} RDP
Aqua 75-100
Hemingray Blue 100-125
Green 300-350

[020] (F-Skirt) HEMINGRAY/PATENT MAY 2ND 1893 (R-Skirt) MUNCIE TYPE/N⁰ 76 {CD 310 sleeve is unembossed} RDP
Aqua, Blue Aqua 100-125
Hemingray Blue 125-150

[030] (F-Skirt) HEMINGRAY/PATENT MAY 2ND 1893 {CD 310 sleeve is unembossed} RDP
Aqua 75-100

CD 303.5

HEMINGRAY

CD 303.5 8 x 7 7/8

[010] (F-Skirt) HEMINGRAY/N⁰ 4 "PROVO" TYPE (R-Skirt) PATENT MAY 2 1893 SDP
Aqua 5,000-7,500

CD 304/310

HEMINGRAY

[010] (F-Skirt) HEMINGRAY/PATENT MAY 2 1893 {CD 310 sleeve is unembossed} SDP
Aqua, Green Aqua 100-125
Hemingray Blue 150-175
Aqua w/ Milk Swirls 250-300
Green 400-500

[020] (F-Skirt) HEMINGRAY/PATENTED MAY 2 1893 {CD 310 sleeve is unembossed} SDP
Aqua 100-125

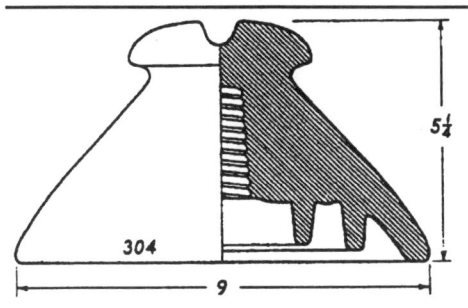

CD 309.5

(CD 310 sleeve shown with CD 303/310)

[030] (F-Skirt) [Blotted out embossing]/ HEMINGRAY/PATENTED MAY 2 1893 {CD 310 sleeve is unembossed} SDP
Aqua 100-125

NO EMBOSSING

[010] [No embossing] {Hemingray product} SDP
Aqua 125-150

CD 306

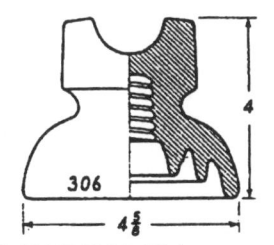

LYNCHBURG ⓛ

[010] (Top of ear) NO.181 (F-Skirt) ⓛ/ LYNCHBURG (R-Skirt) MADE IN U.S.A. RDP
Aqua 500-600

[020] (Top of ear) NO.181 (F-Skirt) ⓛ/ LYNCHBURG (R-Skirt) MADE IN U.S.A. SDP
Aqua 600-700

CD 307

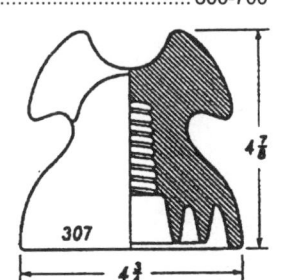

HEMINGRAY

[010] (F-Skirt) HEMINGRAY/PATENT MAY 2 1893 (R-Skirt) HIGH POTENTIAL/TRIPLE PETTICOAT CABLE SDP
Aqua 175-200
Hemingray Blue 200-250
Milky Aqua 400-500

CD 308

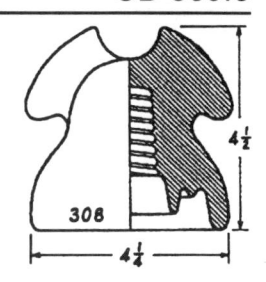

NO EMBOSSING

[010] [No embossing] SB
Light Aqua, Light Blue 50-75

[020] [No embossing] {Pennycuick style} SB
Light Apple Green 175-200

NO NAME

[010] (F-Skirt) № 100 SB
Light Aqua, Light Green Aqua 75-100
Light Blue 100-125
Dark Aqua 150-175
Green, Lime Green 200-250
Apple Green 300-350

CD 309

BEAL'S

CD 309
$5\ ^7/_8\ \times\ 5\ ^3/_4$

[010] (Top/Left ear) BEAL'S HIGH TENSION/ INSULATOR (Top/Right ear) PATENTED OCT. 10 1899 SB
Aqua 10,000-15,000

CD 309.5

(Illustration on next page)

WESTINGHOUSE

[010] (F-Skirt) WESTINGHOUSE ELECTRIC & M'F'G CO./PITTSBURG, PA. (R-Skirt) "TELLURIDE" TYPE B. SB
Aqua 10,000-15,000

165

CD 309.5

CD 309.5
7 3/8 x 10

CD 310

(CD 310 sleeve shown with CD 303/310)

NO EMBOSSING

[010] [No embossing] {Hemingray product} SB
Aqua, Blue Aqua, Green Aqua 15-20
Hemingray Blue 20-30
Green 75-100

CD 311

(CD 311 sleeve shown with CD 248/311/311)

NO EMBOSSING

[010] [No embossing] SB
Clear, Yellow Tint 10-15

[020] [No embossing] {Hemingray product} RDP
Aqua, Hemingray Blue 20-30

[030] [No embossing] {Hemingray product} SB
Honey Amber 250-300

[040] [No embossing] {Hemingray product} SDP
Aqua, Hemingray Blue 20-30

PYREX

[010] (F-Skirt) CORNING PYREX REG. U.S. PAT. OFF. (R-Skirt) MADE IN U.S.A. SB
Clear 15-20

[015] (F-Skirt) CORNING PYREX T. M. REG. U.S. PAT. OFF. MADE IN U.S.A. SB
Clear, Yellow Tint 10-15

[020] (F-Skirt) PYREX PAT. 5-27-19. MADE BY CORNING GLASS WORKS, CORNING N.Y. U.S.A. SB
Clear, Straw 15-20

CD 312

CD 312
7 1/2 x 11

NO EMBOSSING

[010] [No embossing] {Hemingray product} SB
Aqua 500-600

CD 313

CD 313
6 1/2 x 10 1/2

HEMINGRAY

[010] (F-Skirt) HEMINGRAY (R-Skirt) PATENT MAY 2 1893 SDP
Aqua 700-800

[020] (F-Skirt) "SECTION 2" (R-Skirt) HEMINGRAY PATENT MAY 2 1893 SDP
Aqua 700-800

CD 317

CD 313.1

CD 313.1
$4\,^7/_8 \times 8\,^1/_4$

NO EMBOSSING

[010] [No embossing] {Hemingray product} SB
Aqua 500-600

CD 314

CD 314
$7 \times 3\,^7/_8$

NO EMBOSSING

[010] [No embossing] {Hemingray product} SB
Aqua, Purple 400-500

CD 315

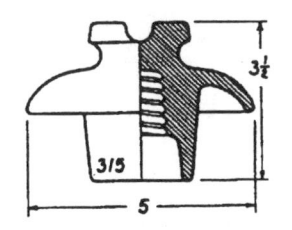

PRISM ⌸

[010] (F-Umbrella) (Arc)REGISTERED TRADE MARK/⌸ SB
Blue 125-150

[020] (F-Umbrella) (Arc)REGISTERED TRADE MARK/⌸ (R-Umbrella) PATENTED JUNE 17 1890 SB
Blue Aqua 125-150

[030] (F-Umbrella) ⌸ SB
Aqua 175-200
Blue 200-250
Lime Green 300-350

[040] (F-Umbrella) ⌸
(R-Umbrella) (Arc)PATENTED/JUNE 17, 1890 SB
Light Aqua 125-150

CD 316

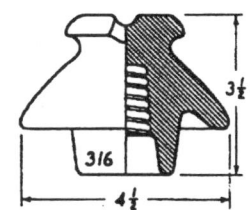

BROOKFIELD

[010] (F-Umbrella) BROOKFIELD SB
Aqua, Dark Aqua 1,500-1,750
Green 1,750-2,000

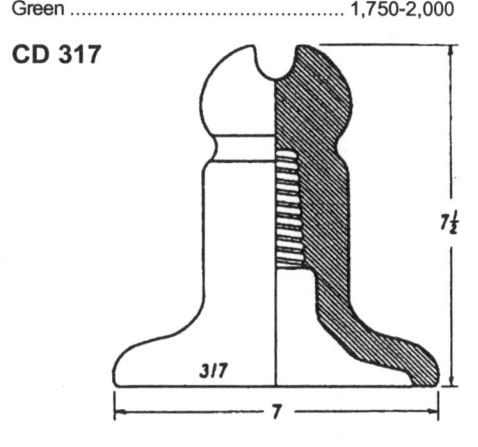

CD 317

CHAMBERS

[005] (F-Center) CHAMBERS/PAT AUG 14 1877 (R-Center) PATENT/DEC 19 1871 {'P' is engraved over a 'T'} SB
Aqua, Light Aqua 400-500

[010] (F-Center) CHAMBERS/PAT.AUG. 14-1877 (R-Center) PATENT/DEC 19 1871 SB
Aqua, Ice Green, Light Aqua 350-400
Blue, Light Green 400-500
Aqua w/ Milk Swirls, Dark Aqua 500-600
Lime Green 600-700
Teal Green 1,250-1,500
Off Clear 1,500-1,750
Cornflower Blue 2,000-2,500
Sapphire Blue 3,000-3,500
Light Purple 4,000-4,500

167

CD 317.5

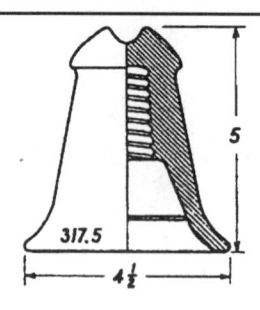

CD 317.5

CHAMBERS

[010] (F-Center) CHAMBERS/PAT AUG 14 1877 (R-Center) PATENT/DEC. 19. 1871 SB
Ice Green, Light Aqua 2,000-2,500

CD 317.7/314/314

CD 317.7
9 x 4 $^1/_2$
(See CD 314 for sleeve)

H.G.CO.

[010] (F-Skirt) H.G.CO./PATENT MAY 2. 1893 {Round drip points appear only on the inner skirt} {CD 314 sleeve is unembossed} RDP
Aqua 5,000-7,500

CD 317.8/313/313/313.1

PATENT - OTHER

CD 317.8
10 $^1/_2$ x 5 $^3/_8$
(See CD 313 and CD 313.1 for sleeve)

[010] (F-Skirt) "SECTION 1" (R-Skirt) PATENTED JUNE 10 1902 {See CD 313 and CD 313.1 for sleeve embossings} {Hemingray product} SB
Aqua 5,000-7,500

CD 317.9

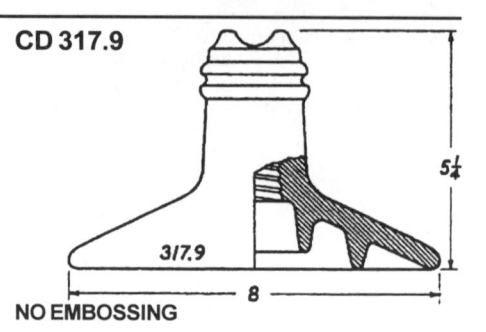

NO EMBOSSING

[010] [No embossing] SB
Aqua 5,000-7,500

[020] [No embossing] SDP
Aqua 5,000-7,500

CD 318

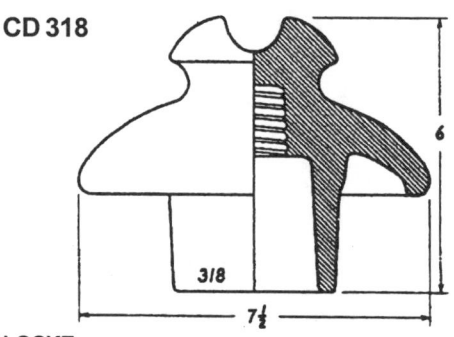

LOCKE

[010] (F-Umbrella) FRED M. LOCKE VICTOR N.Y. (R-Umbrella) PAT.MAY 22 96 " NOV 24 96 " DEC 15 96/ " SEPT 28 97 JUNE 7 98/Nº 19 SB
Light Blue 175-200

[020] (F-Umbrella) FRED M. LOCKE VICTOR N.Y. (R-Umbrella) [Blotted out embossing]/Nº 19 SB
Aqua, Dark Aqua, Light Green Aqua 150-175

CD 319

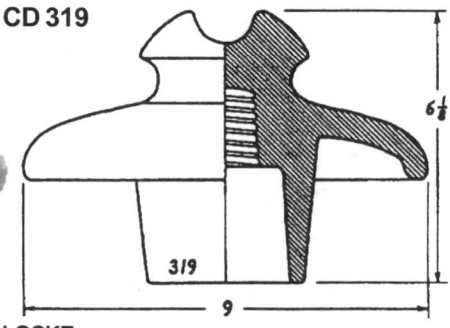

LOCKE

[010] (F-Umbrella) F.M. LOCKE VICTOR N.Y./Nº 23 SB
Aqua 30-40

CD 322

[020] (F-Umbrella) F.M. LOCKE VICTOR N.Y./ № 23 (R-Umbrella) PAT'D MAY 22 96 NOV 24 96 DEC 15 96/ " SEPT 28 97 JUNE 7 98 {Note 'MAY 22 96' date} SB
Aqua 75-100

[030] (F-Umbrella) F.M. LOCKE VICTOR N.Y./ № 23 (R-Umbrella) [Blotted out embossing] SB
Aqua 30-40
Blue 50-75
Green 200-250

[040] (F-Umbrella) FRED M. LOCKE VICTOR N.Y./№ 23 (R-Umbrella) [Blotted out embossing] SB
Aqua 30-40
Light Green 75-100
Green 200-250

CD 320

PYREX

[010] (F-Skirt) CORNING PYREX T.M. REG. U.S. PAT. (R-Skirt) OFF. MADE IN U.S.A. [Number]/171 SB
Clear, Straw 10-15
Carnival 50-75

[020] (Skirt) PYREX T.M. REG. U.S. PAT. OFF. MADE IN U.S.A. [Number]/171 {Embossed all the way around the skirt} SB
Clear, Straw 10-15
Carnival 50-75

CD 321

PATENT - OTHER

[010] (F-Umbrella) REGISTERED TRADE MARK (R-Umbrella) (Arc)PATENTED/JUNE 17, 1890 SB
Aqua, Blue 30-40
Green 75-100

PRISM

[010] (F-Skirt) REGISTERED/TRADE MARK (R-Skirt) (Arc)PATENTED/JUNE 17 1890 SB
Aqua 30-40

[020] (F-Skirt) SB
Aqua 30-40

[030] (F-Umbrella) (Arc)REGISTERED/ (Arc)TRADE MARK/ SB
Green Aqua 30-40
Light Blue 40-50
Blue 50-75

[040] (F-Umbrella) (Arc)REGISTERED/ (Arc)TRADE MARK/ (R-Umbrella) PATENTED JUNE 17, 1890 SB
Light Aqua 30-40
Light Blue 40-50
Blue 50-75
Green 75-100
Yellow Green 300-350

[050] (F-Umbrella) (R-Umbrella) (Arc)PATENTED/JUNE 17, 1890 SB
Aqua 30-40

CD 322

PYREX

[010] (Skirt) CORNING PYREX T.M. REG. U.S. PAT. OFF. MADE IN U.S.A. [Number]/233 {Embossed all the way around the skirt} {1" or 1 3/8" pinhole} SB
Clear, Light Straw 15-20
Carnival 100-125

CD 323

CD 323

PYREX

[010] (Skirt) CORNING PYREX T.M. REG. U.S. PAT. OFF. MADE IN U.S.A. [Letter or number]/271 {Embossed all the way around the skirt} {1 3/8" or 1 5/8" pinhole} SB
Clear, Green Straw, Light Straw 15-20
Carnival 100-125

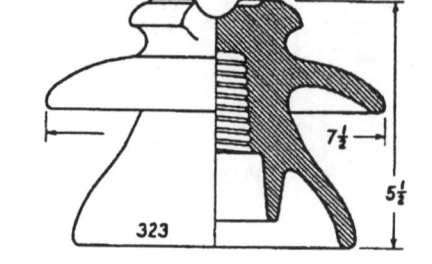

CD 324

PYREX

[010] (Skirt) PYREX T.M. REG. U.S. PAT. OFF. MADE IN U.S.A. 353 {Embossed all the way around the skirt} {1 3/8" pinhole} SB
Straw 20-30
Carnival 100-125

[020] (Skirt) PYREX T.M. REG. U.S. PAT. OFF. MADE IN U.S.A. 353/[Letter] {Embossed all the way around the skirt} {1 3/8" pinhole} SB
Clear, Straw 20-30
Carnival 100-125

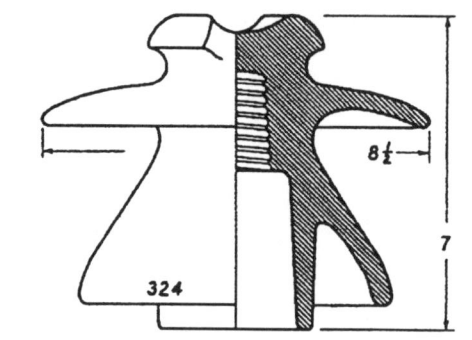

CD 325

PYREX

[010] (Skirt) CORNING PYREX T.M. REG. U.S. PAT. OFF. MADE IN U.S.A. [Letter]/401 {Embossed all the way around the skirt} {1 3/8" or 1 5/8" pinhole} SB
Clear, Green Tint, Light Pink, Straw 20-30
Carnival 100-125

[020] (Skirt) CORNING PYREX T.M. REG. U.S. PAT. OFF. MADE IN U.S.A. [Letter]/401 {Embossing allocated to the four quadrants} {1 3/8" or 1 5/8" pinhole} SB
Carnival 100-125

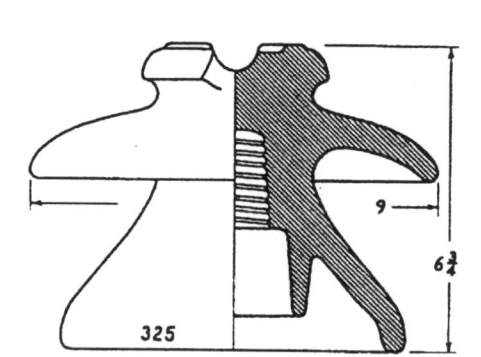

CD 326

PYREX

[010] (Skirt) PYREX T.M. REG. U.S. PAT. OFF. MADE IN U.S.A. 453 [Letter] {1 3/8" pinhole} SB
Clear, Straw 20-30
Yellow 50-75
Carnival 100-125

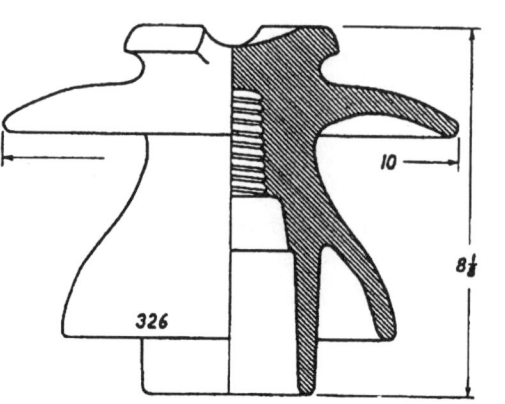

170

CD 327

PYREX

[005] (Crown) 461 (F-Skirt) CORNING PYREX T.M. REG. U.S. PAT. OFF. (R-Skirt) MADE IN U.S.A. 441 SB
Straw 20-30

[010] (Crown) 461 (F-Skirt) PYREX T.M. REG. U.S. PAT. OFF. (R-Skirt) MADE IN U.S.A. 441 SB
Straw 20-30
Green Tint 40-50
Carnival 125-150

[012] (Crown) [Number]/461 (F-Skirt) CORNING PYREX T.M. REG. U.S. PAT. OFF. (R-Skirt) MADE IN U.S.A. 441 SB
Green Tint 40-50
Carnival 125-150

[014] (F-Skirt) CORNING PYREX T.M. REG. U.S. PAT. OFF. (R-Skirt) MADE IN U.S.A. 441 {1 3/8" pinhole} SB
Straw 20-30

[016] (F-Skirt) PYREX T.M. REG. U.S. PAT. OFF. (R-Skirt) MADE IN U.S.A. 441 SB
Carnival 125-150

[018] (Skirt) CORNING PYREX T.M. REG. U.S. PAT. OFF. MADE IN U.S.A. 441 {Embossed all the way around the skirt; embossing is hand scribed} {1 3/8" or 1 5/8" pinhole} SB
Carnival 125-150

[020] (Skirt) CORNING PYREX T.M. REG. U.S. PAT. OFF. MADE IN U.S.A. 441 {Embossed all the way around the skirt} {1 3/8" or 1 5/8" pinhole} SB
Clear, Green Tint, Straw 20-30
Carnival 100-125

[030] (Skirt) CORNING PYREX T.M. REG. U.S. PAT. OFF. MADE IN U.S.A. 441 {Embossed all the way around the skirt} {Small foreign line top variation} {1 3/8" or 1 5/8" pinhole} SB
Clear, Straw 100-125

[040] (Skirt) PYREX T.M. REG. U.S. PAT. OFF. MADE IN U.S.A. 441 {Embossed all the way around the skirt} {1 3/8" or 1 5/8" pinhole} SB
Straw 20-30

CD 330

CD 328

(Illustration on next page)

PYREX

[010] (F-Skirt) PYREX T.M. REG. U.S. PAT. OFF. MADE IN U.S.A. (R-Skirt) 553 SB
Clear 2,000-2,500

CD 328.2

(Illustration on next page)

PYREX

[010] (F-Skirt) PYREX T.M. REG. U.S. PAT. OFF. MADE IN U.S.A. 553 (R-Skirt) C SB
Clear 2,000-2,500
Carnival 2,500-3,000

CD 330

(Illustration on next page)

PYREX

[010] (Dome) A (F-Skirt) PYREX T.M. REG. U.S. PAT. OFF. MADE IN U.S.A. (R-Skirt) 663 A SB
Clear, Light Yellow 2,000-2,500

CD 328 - CD 330

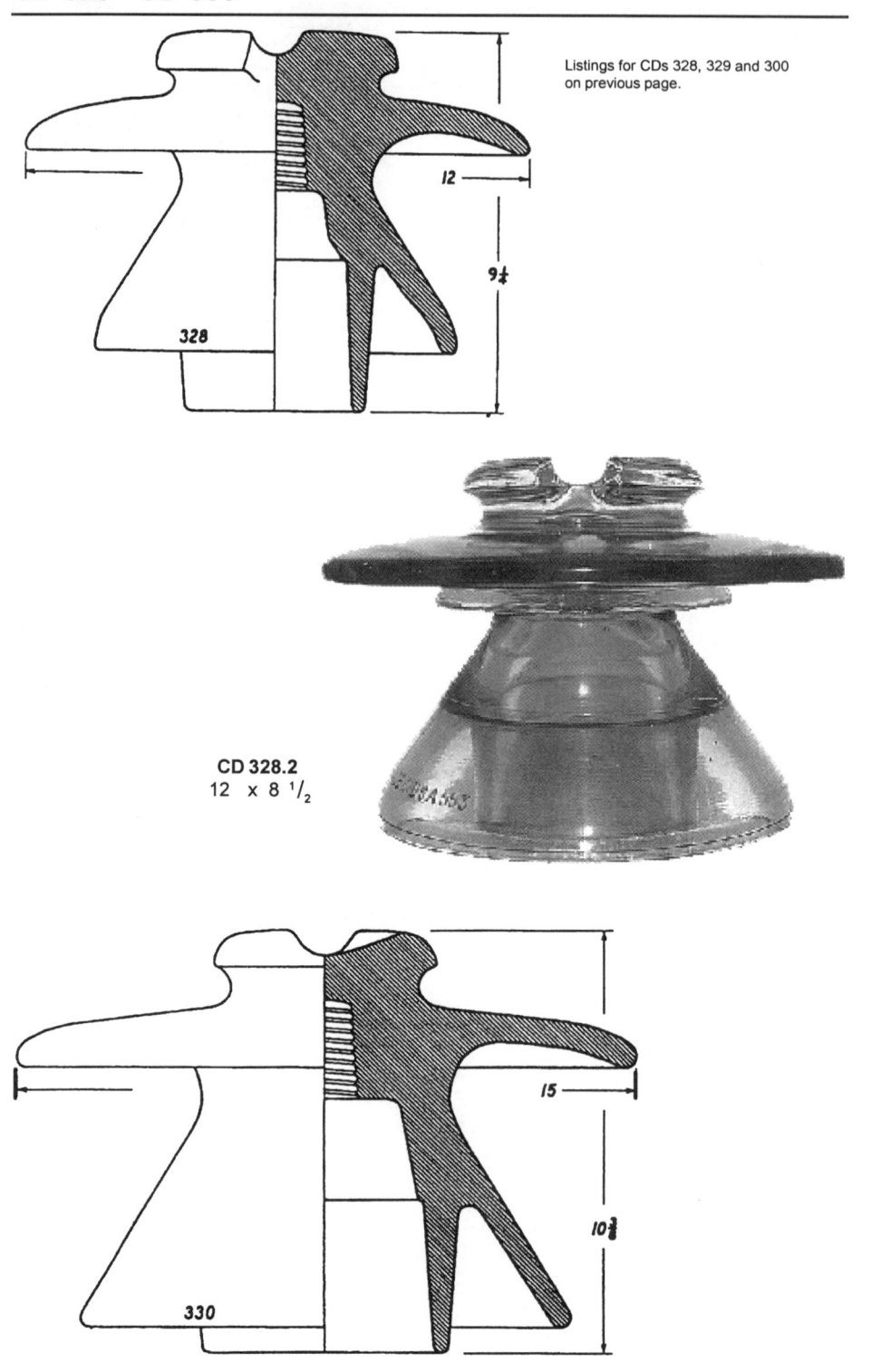

Listings for CDs 328, 329 and 300 on previous page.

328

12

9¼

CD 328.2
12 x 8 ½

330

15

10¾

172

CD 338

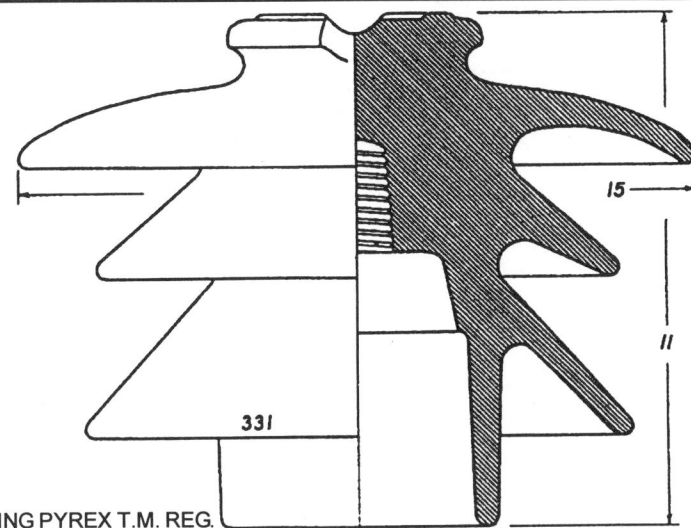

CD 331

PYREX

[010] (Umbrella) CORNING PYREX T.M. REG. U.S. PAT. OFF. MADE IN U.S.A. {Embossed all the way around the umbrella} {2 1/4" pinhole} SB
Clear, Green Tint, Yellow Tint 250-300

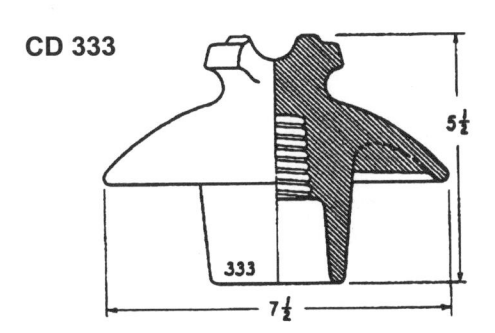

CD 333

PRISM ▣

[010] (F-Umbrella) (Arc)PATENTED/▣/JUNE 17 1890 SB
Blue Aqua, Light Blue 75-100
Blue 100-125
Olive Green 400-500

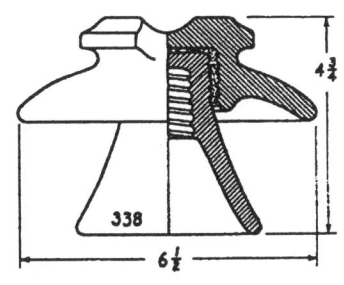

[020] (F-Umbrella) ▣ SB
Aqua, Green Aqua 50-75
Blue 100-125
Green 175-200
Yellow Green 300-350
Olive Green 400-500

CD 338

BROOKFIELD

[010] (Top of ear) BROOKFIELD {Specimen only} SB
Dark Aqua 3,000-3,500

173

CD 342

CD 342

LOCKE

[010] (F-Skirt) LOCKE PATENTS APR. 29.1902 JUNE 7.1898 NOV. 2.1896/SEPT. 28.1897 DEC. 15.1896 APR. 30.1891 {Note 'NOV. 2.1896' and 'APR. 30.1891' dates} (R-Skirt) NO 25 {Sleeve-not a separate CD number} (F-Center) LOCKE PAT/DEC. 15. 96 JUNE 7. 98. SB
Aqua 3,500-4,000

THREADLESS GLASS PINTYPES

CD 700

NO EMBOSSING

CD 700
$2\,^3/_8 \times 3\,^1/_2$

[010] [No embossing] SB
Dark Aqua, Light Aqua,
Light Green Aqua 500-600
Light Sage Green 600-700
Light Purple w/ Blue Cast,
Light Purple w/ Pink Cast 1,000-1,250

CD 700.1

NO EMBOSSING

CD 700.1
$2\,^1/_4 \times 3\,^3/_4$

[010] [No embossing] SB
Aqua, Green Aqua, Light Aqua 500-600
Blue, Light Blue 600-700
Bubbly Aqua 800-1,000
Bubbly Clear, Emerald Green 1,250-1,500
Dark Olive Green, Olive Green 1,500-1,750
Root Beer Amber 2,000-2,500

CD 701

NO EMBOSSING

CD 701
$3\,^1/_8 \times 4\,^1/_8$

[010] [No embossing] SB
Dark Aqua 600-700
Green Blackglass, Light Aqua 800-1,000
Dark 7-up Green,
Dark Yellow Green, Emerald Green ... 1,000-1,250
Teal Blue, Teal Green 1,250-1,500

CD 701.1

NO EMBOSSING

CD 701.1
$3\,^1/_8 \times 4$

[010] [No embossing] SB
Aqua, Blue Aqua, Light Aqua 700-800
Light Green, Light Lime Green 1,250-1,500
7-up Green, Emerald Green 1,500-1,750
Light Straw, Off Clear,
Olive Amber Blackglass 1,750-2,000
Citrine, Light Pink 2,000-2,500

CD 701.2

NO EMBOSSING

CD 701.2
$3\,^1/_8 \times 3\,^7/_8$

[010] [No embossing] SB
Dark Green 2,500-3,000

CD 701.3

NO EMBOSSING

CD 701.3
3×4

[010] [No embossing] SB
Dark Forest Green, Dark Green 2,500-3,000
Dark Orange Amber 5,000-7,500

CD 701.5

CD 701.5

NO EMBOSSING

CD 701.5
$3^1/_4 \times 4$

[010] [No embossing] SB
Dark Green, Dark Teal Green,
Teal Aqua 1,250-1,500

CD 701.6

NO EMBOSSING

CD 701.6
$3^1/_4 \times 4^1/_4$

[010] [No embossing] SB
Dark Emerald Green, Dark Teal Green,
Green Blackglass, Olive Blackglass 350-400
Dark Olive Amber 500-600
Aqua, Green 2,000-2,500
Light Green 3,000-3,500
Dark Cobalt Blue 4,500-5,000

CD 701.8

NO EMBOSSING

CD 701.8
$3^1/_8 \times 3^1/_2$

[010] [No embossing] SB
Blue Aqua, Light Aqua 800-1,000

CD 718

NO EMBOSSING - CANADA

CD 718
$3^1/_8 \times 3^7/_8$

[010] [No embossing] {Both Canada and U.S. manufacture} SB
Olive Amber Blackglass 500-600
Aqua, Blue Aqua, Light Aqua,
Olive Green Blackglass 600-700
Dark Aqua, Light Green, Sky Blue 700-800
Dark Forest Green 800-1,000
Dark Olive Amber, Emerald Green,
Olive Green, Root Beer Amber,
Teal Green 1,000-1,250
Gray 1,500-1,750
Dark Orange Amber,
Dark Yellow Amber, Gingerale,
Jade Aqua Milk 1,750-2,000
Light Cornflower Blue 2,000-2,500
Light Purple 3,000-3,500
Dark Cornflower Blue, Indigo Blue,
Plum, Purple Blackglass 4,000-4,500
Dark Lavender, Light Cobalt Blue 5,000-7,500
Cobalt Blue, Cranberry Red 7,500-10,000

THAMES GLASS WORKS

[010] (F-Crown) THAMES GLASS WORKS
(R-Crown) NEW LONDON, CT SB
Olive Amber Blackglass 10,000-15,000

TILLOTSON & CO.

[010] (F-Crown) (Arc)TILLOTSON & Co/16/
BROADWAY/N.Y. SB
Aqua, Light Aqua, Light Blue,
Light Green 2,000-2,500
Root Beer Amber, Teal Green 2,500-3,000
Dark Green 3,000-3,500
Dark Yellow Olive Green,
Emerald Green, Jade Green Milk 3,500-4,000

CD 721

NO EMBOSSING - CANADA

CD 721
$1^7/_8 \times 3^5/_8$

[010] [No embossing] SB
Aqua, Light Aqua 500-600
Powder Blue 600-700
Cornflower Blue, Jade Aqua Milk 1,000-1,250

CD 724.5

CD 722

NO EMBOSSING - CANADA **CD 722**
$1\,^3/_4$ x $3\,^1/_2$

[010] [No embossing] SB
Aqua 500-600
Green 800-1,000
Jade Aqua Milk, Milky Green 1,750-2,000

CD 723

NO EMBOSSING **CD 723**
$2\,^1/_4$ x $4\,^1/_8$

[010] [No embossing] SB
Aqua, Blue 400-500
Celery Green, Light Green 500-600
Clear 800-1,000
Green 1,500-1,750
Light Olive Green 1,750-2,000

CD 723.3

NO EMBOSSING **CD 723.3**
$2\,^1/_4$ x $3\,^7/_8$

[010] [No embossing] SB
Aqua, Blue Aqua 300-350
Blue, Green Aqua 350-400
Green 1,000-1,250
7-up Green 4,000-4,500
Cobalt Blue 5,000-7,500

CD 723.5

NO EMBOSSING **CD 723.5**
2 x $3\,^3/_4$

[010] [No embossing] SB
Blue, Green Aqua 600-700
Green 800-1,000

CD 723.6

NO EMBOSSING **CD 723.6**
$1\,^1/_2$ x $2\,^3/_4$

[010] [No embossing] SB
Dark Green Aqua 600-700

CD 724

CHESTER **CD 724**
$2\,^3/_8$ x $3\,^5/_8$

[010] (Base) CHESTER N.Y. BE
Clambroth w/ Purple Tint,
Cobalt Blue 5,000-7,500

NO EMBOSSING

[010] [No embossing] SB
Lime Green 1,750-2,000
Cobalt Blue, Pink, Yellow Green 3,000-3,500

CD 724.3

NO EMBOSSING - CANADA **CD 724.3**
$2\,^1/_8$ x 3

[010] [No embossing] SB
Olive Amber Blackglass,
Olive Blackglass 800-1,000
Olive Green 1,000-1,250
Dark Orange Amber 1,500-1,750

CD 724.5

NO EMBOSSING **CD 724.5**
$2\,^1/_8$ x 3

[010] [No embossing] SB
Lime Green 1,250-1,500

177

CD 724.6

CD 724.6

NO EMBOSSING

CD 724.6
3 x 4 1/4

[010] [No embossing] {Specimen only} SB
Dark Teal Green 1,250-1,500

CD 725

NO EMBOSSING

CD 725
2 3/4 x 4 1/2

[010] [No embossing] SB
Aqua, Blue Aqua 1,500-1,750

CD 726

CD 726
3 1/4 x 4

NO EMBOSSING - CANADA

[005] [No embossing] GB
Aqua 500-600

[010] [No embossing] SB
Aqua, Dark Aqua, Light Aqua, Light Blue,
Light Blue Aqua, Sky Blue 600-700
Green Blackglass,
Olive Amber Blackglass,
Olive Green Blackglass 1,500-1,750
Gray 1,750-2,000
Dark Cornflower Blue 3,000-3,500
Dark Puce, Violet Blue 5,000-7,500
Light Purple 7,500-10,000
Purple 10,000-15,000
Cranberry Red 15,000-20,000

CD 727

CD 727
2 7/8 x 4

NO EMBOSSING

[010] [No embossing] SB
Dark Green, Olive Green Blackglass .. 1,500-1,750

CD 728

NO EMBOSSING

CD 728
3 x 4 3/4

[010] [No embossing] SB
Aqua, Ice Aqua, Light Aqua, Light Blue ... 125-150
Ice Green 200-250

CD 728.2

NO NAME

CD 728.2
3 x 3 7/8

[010] (Dome) ['M' or 'W'] SB
Aqua, Blue, Blue Aqua 400-500
Green 700-800

CD 728.4

BROOKFIELD

CD 728.4
2 7/8 x 3 7/8

[010] (F-Crown) (Arc)W. BROOKFIELD/55
FULTON ST./N.Y.
(R-Crown) (Arc)CAUVET'S/PAT/JULY 25
1865/PAT JAN 25 1870/PAT FEB 22 1870
{MLOD} SB
Light Aqua 800-1,000

CD 729.4

[020] (F-Crown) (Arc)W. BROOKFIELD/
[Number]/55 FULTON ST/N.Y.
(R-Crown) (Arc)CAUVET'S/PAT./JULY 25,
1865 {MLOD} SB
Aqua, Light Blue Aqua 700-800

CD 728.5

NO EMBOSSING

CD 728.5
$2 \, ^1/_2$ x $3 \, ^3/_4$

[010] [No embossing] SB
Light Blue Aqua 1,500-1,750

CD 728.7

NO EMBOSSING

CD 728.7
$2 \, ^7/_8$ x $3 \, ^7/_8$

[010] [No embossing] {Oakman pinhole} SB
Blue Aqua, Light Blue 1,250-1,500

[020] [No embossing] {Standard pinhole}
{MLOD} SB
Aqua 1,250-1,500

OAKMAN MFG.CO.

[010] (Base of threads) OAKMAN'S PATENT.
JULY 26.1870. SB
Aqua 2,000-2,500

CD 728.8

BOSTON BOTTLE WORKS

CD 728.8
$3 \, ^1/_8$ x 4

[010] (Base) BOSTON BOTTLE
WORKS-OAKMAN'S PAT JULY 26 1870 BE
Aqua, Light Aqua 4,000-4,500

[020] (Base) BOSTON BOTTLE
WORKS-OAKMAN'S PATENT. JULY 26.1870.
BE
Aqua 4,000-4,500

CD 729

NO EMBOSSING

CD 729
$3 \, ^1/_8$ x $3 \, ^5/_8$

[010] [No embossing] SB
Aqua, Blue Aqua,
Dark Green Aqua, Green 400-500
Dark Forest Green, Dark Green,
Dark Teal Green, Emerald Green,
Green Blackglass, Light Green 500-600
Dark Orange Amber, Dark Red Amber 700-800

CD 729.3

KEELING

CD 729.3
$3 \, ^1/_8$ x $3 \, ^3/_4$

[010] (F-Crown) (Arc)J.S.KEELING (F-Skirt) 16
B. WAY N.Y. SB
Aqua, Light Aqua, Light Green,
Olive Green Blackglass,
Opaque Blackglass, Teal Aqua,
Teal Blue 2,000-2,500

CD 729.4

MULFORD & BIDDLE

CD 729.4
$3 \, ^1/_8$ x $3 \, ^5/_8$

[010] (F-Skirt) MULFORD & BIDDLE/83 JOHN
ST. N.Y. SB
Aqua, Dark Aqua, Dark Green Aqua,
Green Aqua, Light Aqua 1,250-1,500
Blue Aqua 1,500-1,750
Blue, Teal Aqua 1,750-2,000
Green 2,500-3,000
Emerald Green 3,000-3,500

CD 729.6

CD 729.6

NO EMBOSSING

CD 729.6
$2\,^7/_8$ x $3\,^3/_8$

[010] [No embossing] SB
Green Aqua 2,500-3,000

PAT APP FOR

[010] (Base) PT AP^D FOR BE
Light Aqua 2,000-2,500
Green, Light Green 2,500-3,000

CD 731

BROOKFIELD

CD 731
$2\,^7/_8$ x 4

[010] (F-Crown) (Arc)CAUVET'S/PAT/JULY 25 1865 (F-Skirt) W. BROOKFIELD/55 FULTON ST N.Y. (R-Skirt) 6. SB
Aqua, Blue Aqua, Light Aqua 800-1,000

M^cKEE & CO.

[010] (F-Skirt) S.M^cKEE &.Co. SB
Aqua, Blue, Blue Aqua 350-400

NO EMBOSSING

[010] [No embossing] SB
Aqua, Light Aqua, Light Blue 125-150
Green Aqua 150-175
Light Green 250-300
Olive Blackglass, Red Amber Blackglass . 400-500
Green 600-700
Dark Orange Amber 700-800
Dark Honey Amber 800-1,000
White Milk 7,500-10,000

NO NAME

[010] (Dome) ['M' or 'W'] SB
Aqua 350-400

TILLOTSON & CO.

[010] (F-Crown) (Arc)TILLOTSON SB
Aqua, Green Aqua, Ice Aqua,
Light Aqua, Light Blue 300-350
Light Green, Lime Green 400-500
Emerald Green, Light Sapphire Blue 700-800
Teal Blue, Teal Green 1,000-1,250
Cobalt Blue Blackglass,
Dark Brown Amber,
Puce Blackglass, Yellow Green 1,250-1,500
Citrine, Dark Yellow Amber 1,500-1,750
Sapphire Blue 1,750-2,000

CD 731.2

NO EMBOSSING

CD 731.2
3 x $3\,^7/_8$

[010] [No embossing] SB
Aqua, Light Blue 200-250

CD 731.3

NO EMBOSSING

CD 731.3
$2\,^7/_8$ x $3\,^7/_8$

[010] [No embossing] SB
Amber Blackglass, Olive Blackglass 800-1,000
Aqua, Blue, Dark Yellow Amber,
Emerald Green, Light Blue 1,000-1,250

CD 732

CD 732
$2^{7}/_{8}$ x $4^{1}/_{8}$

NO EMBOSSING

[010] [No embossing] SB
Aqua, Dark Aqua, Green Aqua,
Light Aqua 300-350
Lime Green 400-500
Dark Olive Amber, Green,
Green Aqua Slag, Light Green Slag,
Olive Blackglass, Teal Green 700-800
Jade Aqua Milk 800-1,000
Gingerale 2,000-2,500
Light Purple 3,000-3,500

CD 732.2

CD 732.2
3 x $4^{1}/_{8}$

L.G.T.& CO.

[010] (F-Skirt) L.G.T.& CO SB
Light Aqua 1,750-2,000

NO EMBOSSING

[010] [No embossing] SB
Blue Aqua, Celery Green 1,000-1,250
Dark Lime Green, Light Green 1,250-1,500

PATENT - DEC. 19, 1871

[010] (F-Crown) PATENT/DEC. 19. 1871
(R-Crown) 1 SB
Light Sage Green 4,000-4,500
Light Purple 5,000-7,500

CD 734

CD 734
$2^{3}/_{8}$ x $3^{3}/_{4}$

MCMICKING

[010] (F-Skirt) MCMICKING
(R-Skirt) VICTORIA.B.C 75 {MLOD} SB
Aqua, Light Blue Aqua, Light Green Aqua ... 50-75

[020] (F-Skirt) MCMICKING
(R-Skirt) VICTORIA.B.C. 75 {MLOD} SB
Blue Aqua, Light Aqua 50-75
Light Green 150-175

CD 734.1

CD 734.1
$2^{1}/_{4}$ x 3

NO EMBOSSING

[010] [No embossing] SB
Aqua 3,000-3,500

CD 734.3

CD 734.3
$2^{3}/_{8}$ x $3^{3}/_{4}$

NO EMBOSSING

[010] [No embossing] SB
Amber Blackglass 3,000-3,500
Oxblood 10,000-15,000

CD 734.5

CD 734.5
$1^{3}/_{4}$ x $2^{3}/_{8}$

NO EMBOSSING - CANADA

[010] [No embossing] SB
Dark Olive Amber 2,500-3,000

CD 734.8

CD 734.8
$2^{7}/_{8}$ x $2^{3}/_{8}$

NO EMBOSSING - CANADA

[010] [No embossing] SB
Olive Amber Blackglass,
Olive Blackglass 250-300
Aqua, Light Aqua 800-1,000
Light Green 1,000-1,250

181

CD 735

CD 735

CD 735
2 $^3/_4$ x 3 $^3/_8$

CHESTER

[010] (F-Skirt) CHESTER/N.Y. SB
Aqua 500-600
Ice Aqua, Light Green 700-800
Light Yellow, Lime Green 1,000-1,250
Brown Amber Blackglass,
Dark Olive Amber, Olive Amber,
Yellow Amber Blackglass 1,500-1,750
Emerald Green 1,750-2,000

MULFORD & BIDDLE

[010] (F-Skirt) MULFORD & BIDDLE SB
Aqua, Dark Aqua, Light Aqua, Light Blue . 400-500
Green 1,250-1,500

[020] (F-Skirt) MULFORD & BIDDLE
(R-Skirt) U.P.R.R. SB
Aqua, Blue Aqua, Dark Aqua,
Light Aqua 400-500
Light Green, Powder Blue 500-600
Dark Blue, Teal Blue 1,250-1,500
Cobalt Blue, Light Yellow Green 1,500-1,750
Green 2,000-2,500

SO EX CO

[010] (F-Skirt) SO EX CO (R-Skirt) CHESTER/N.Y. SB
Aqua, Light Aqua 1,000-1,250
Amber Blackglass 2,000-2,500
Cobalt Blue 3,500-4,000

TILLOTSON & CO.

[010] (F-Crown) (Arc)TILLOTSON & CO SB
Light Teal Green 800-1,000
Dark Teal Green 1,000-1,250

[020] (F-Crown) (Arc)TILLOTSON & Co
(F-Skirt) (Arc)16 BROADWAY/N.Y. SB
Aqua, Dark Aqua, Green Aqua,
Light Aqua 800-1,000
Blue Aqua, Ice Aqua 1,000-1,250
Blue, Green 1,250-1,500
Lime Green, Teal Blue 1,500-1,750
7-up Green, Emerald Green 2,500-3,000
Yellow Green 3,000-3,500
Olive Green 3,500-4,000
Dark Orange Amber,
Dark Yellow Amber, Olive Amber 5,000-7,500

CD 735.3

U.S.TEL.CO.

CD 735.3
2 $^7/_8$ x 3 $^3/_4$

[010] (F-Skirt) U.S.TEL.CO.
(R-Skirt) CHESTER/N.Y. SB
Dark Aqua, Dark Teal Aqua,
Light Aqua 800-1,000

CD 735.5

CD 735.5
3 $^1/_4$ x 3

LINEA DEL SUPº

[010] (Skirt) LINEA DEL SUPº GOBIERNO SB
Green Aqua 2,500-3,000
Milky Aqua 3,500-4,000

CD 735.6

CD 735.6
3 $^3/_4$ x 4 $^5/_8$

NO EMBOSSING

[010] [No embossing] SB
Aqua, Blue Aqua 4,000-4,500
Green 5,000-7,500

CD 735.7

CD 735.7
3 $^3/_4$ x 4 $^5/_8$

NO EMBOSSING

[010] [No embossing] SB
Aqua, Blue Aqua, Light Aqua 4,500-5,000

CD 736

E.R.W.

[010] (F-Skirt) E.R.W. SB
Light Blue Aqua, Light Green 3,500-4,000

N.Y.& E.R.R.

[010] (F-Skirt) N.Y.& E.R.R. SB
Aqua, Blue Aqua, Light Aqua,
Light Green Aqua 2,500-3,000
Light Cornflower Blue 3,000-3,500
Sapphire Blue 4,500-5,000
Brown Puce, Burgundy Puce 5,000-7,500

[020] (F-Skirt) N.Y.&. (R-Skirt) E.R.R. SB
Light Aqua 2,500-3,000
Puce 5,000-7,500

NO EMBOSSING

[010] [No embossing] SB
Aqua 2,000-2,500
Dark Yellow Amber,
Root Beer Amber 2,500-3,000

TILLOTSON & CO.

[010] (F-Crown) (Arc)TILLOTSON/& Co
(F-Skirt) 16 BROADWAY N.Y. SB
Amber Blackglass, Aqua,
Dark Yellow Amber, Emerald Green,
Green Aqua, Green Blue Aqua,
Light Green 3,500-4,000

[020] (F-Crown) (Arc)TILLOTƧON/& Co
(F-Skirt) 16 BROADWAY N.Y. {Note backwards letter} SB
Blue Aqua, Green 3,500-4,000

CD 736
$3\,^1/_4$ x $3\,^5/_8$

CD 736.1

NO EMBOSSING

[010] [No embossing] SB
Aqua, Green Aqua, Light Aqua 1,750-2,000
Dark Aqua, Dark Teal Green 2,000-2,500
Emerald Green, Green 2,500-3,000

CD 736.1
$3\,^3/_8$ x $4\,^1/_4$

CD 736.3

NO EMBOSSING

[010] [No embossing] SB
Aqua, Dark Aqua, Green Aqua,
Light Aqua, Sky Blue 1,250-1,500

CD 736.3
3 x $3\,^1/_2$

CD 736.4

NO EMBOSSING

[010] [No embossing] SB
Light Green 1,250-1,500
Dark Sapphire Blue, Yellow Green 2,000-2,500
Dark Amber 2,500-3,000
Clear 5,000-7,500

CD 736.4
$2\,^5/_8$ x $3\,^5/_8$

CD 736.5

NO EMBOSSING

[010] [No embossing] SB
Emerald Green, Ice Green 4,000-4,500

CD 736.5
$3\,^5/_8$ x $4\,^3/_4$

CD 736.7

NO EMBOSSING

[010] [No embossing] SB
Dark Emerald Green 4,000-4,500

CD 736.7
4 x $5\,^1/_8$

CD 737

CD 737

LEFFERTS

CD 737
$3\ ^3/_4$ x 4

[010] (F-Skirt) LEFFERTS SB
Blue, Blue Aqua, Dark Teal Aqua,
Light Aqua, Light Green Aqua 2,500-3,000
Light Citrine 3,500-4,000

NO EMBOSSING

[010] [No embossing] SB
Aqua, Green Aqua 1,250-1,500
Green 1,750-2,000
Dark Green, Dark Olive Amber,
Root Beer Amber, Teal Blue 2,500-3,000

TILLOTSON & CO.

[010] (F-Skirt) L.G. TILLOTSON & Co
(R-Skirt) 26 DEY ST. N.Y. SB
Amber Blackglass, Teal Aqua 5,000-7,500

CD 737.5

CD 737.5
$2\ ^3/_8$ x $2\ ^7/_8$

NO EMBOSSING

[010] [No embossing] SB
Apple Green, Green Tint 3,000-3,500
Amber/Apple Green Two Tone 10,000-15,000

[020] [No embossing] {Square pinhole} SB
Light Green 4,000-4,500

CD 737.6

CD 737.6
$2\ ^3/_8$ x $2\ ^7/_8$

NO EMBOSSING

[010] [No embossing] SB
Green Blackglass,
Olive Amber Blackglass, Olive Blackglass 400-500
Amber Blackglass 1,000-1,250
Dark Green 1,500-1,750

[020] [No embossing] {MLOD} SB
Aqua, Light Green 1,500-1,750
7-up Green 2,500-3,000
Light Purple 4,000-4,500

CD 737.9

CD 737.9
$2\ ^1/_4$ x $2\ ^7/_8$

NO EMBOSSING

[010] [No embossing] SB
Amber Blackglass 2,000-2,500

CD 738

CD 738
$3\ ^3/_4$ x $4\ ^1/_8$

CHESTER

[010] (F-Skirt) CHESTER, N.Y. SB
Root Beer Amber 5,000-7,500

NO EMBOSSING

[010] [No embossing] SB
Amber Blackglass, Aqua,
Dark Forest Green, Dark Teal Green,
Green, Olive Amber Blackglass,
Olive Blackglass, Teal Green 1,500-1,750
Teal Blue 1,750-2,000
Root Beer Amber 2,000-2,500

CD 738.5

CD 738.5
3 x $2\ ^7/_8$

NO EMBOSSING

[010] [No embossing] SB
Amber Blackglass 1,750-2,000

CD 738.6

CD 738.6
$2\ ^7/_8$ x 3

NO EMBOSSING

[010] [No embossing] SB
Olive Blackglass 2,000-2,500

CD 740

CD 739

CD 739
3 ³/₄ x 5

NO EMBOSSING

[010] [No embossing] SB
Light Blue, Light Blue Aqua 2,000-2,500
Amber Blackglass 3,500-4,000

CD 739.2

CD 739.2
3 ¹/₂ x 4 ⁷/₈

NO EMBOSSING

[010] [No embossing] SB
Aqua 5,000-7,500

CD 739.5

CD 739.5
3 x 3 ¹/₂

NO EMBOSSING

[010] [No embossing] {Smooth dome} SB
Aqua, Aqua Tint, Blue Aqua 800-1,000
Green Aqua 1,000-1,250
Olive Amber, Teal Green 1,500-1,750

[020] [No embossing] {"Washboard" dome} SB
Aqua, Green Aqua, Light Green 800-1,000
Light Sapphire Blue 2,500-3,000
Light Gray Purple 3,000-3,500
Puce 4,000-4,500

CD 740

CD 740
3 ³/₈ x 3 ⁵/₈

DUPONT

[010] (Base) E. DUPOИT ST. JEAN {Note backwards letter} BE
Aqua 1,000-1,250
Light Green 1,500-1,750
Amber Blackglass, Dark Green 1,750-2,000
Light Yellow Green 2,000-2,500
Milky Green 2,500-3,000

FOSTER

[010] (Base) FOSTER.BROTHERS. ST. JOHИ.C.E.1858 {Note backwards letter} BE
Green Blackglass,
Olive Amber Blackglass,
Olive Blackglass 600-700
Light Green 700-800
Aqua 800-1,000
Root Beer Amber 1,000-1,250
Emerald Green 1,500-1,750
Olive Green 1,750-2,000

NO EMBOSSING - CANADA

[010] [No embossing] {Both Canada and U.S. manufacture} SB
Aqua, Olive Amber Blackglass,
Olive Blackglass,
Orange Amber Blackglass 350-400
Green, Olive Amber 800-1,000
Dark Yellow Green 2,000-2,500
Honey Amber 2,500-3,000

TILLOTSON & CO.

[010] (F-Crown) (Arc)TILLOTSON/& Co (F-Skirt) 16 BROADWAY.N.Y. SB
Aqua 1,750-2,000
Amber Blackglass, Emerald Green,
Olive Amber, Yellow Green 2,000-2,500

[020] (F-Crown) (Arc)TILLOTSON/& Co (F-Skirt) 1ә BROADWAY.N.Y. {Note backwards number} SB
Amber Blackglass, Aqua 1,500-1,750

185

CD 740.1

CD 740.1

CD 740.1
$3\,^3/_8$ x $3\,^1/_8$

NO EMBOSSING - CANADA

[010] [No embossing] SB
Green Blackglass,
Olive Green Blackglass 350-400
Aqua, Blue Aqua, Dark Yellow Amber,
Green Aqua, Root Beer Amber 400-500
Dark Orange Amber, Light Blue 500-600
Light Green 600-700
Green 1,000-1,250
Dark Green 1,250-1,500
Olive Green 1,750-2,000
Light Olive Amber, Olive Amber ... 2,500-3,000

CD 740.3

CD 740.3
$3\,^1/_2$ x $3\,^3/_4$

NO EMBOSSING - CANADA

[010] [No embossing] SB
Amber Blackglass, Green Blackglass,
Olive Amber Blackglass, Olive Blackglass,
Root Beer Amber,
Yellow Amber Blackglass 400-500
Dark Green, Teal Green 700-800
Dark Golden Amber 1,250-1,500
Olive Green 1,500-1,750
Golden Amber, Yellow Amber 2,000-2,500

CD 740.4

CD 740.4
$3\,^3/_8$ x $3\,^1/_2$

NO EMBOSSING

[010] [No embossing] SB
Aqua, Blue Aqua 500-600
Root Beer Amber 800-1,000

CD 740.6

CD 740.6
$3\,^3/_8$ x $3\,^3/_8$

CHESTER

[010] (Base) CHESTER N.Y. BE
Dark Green 5,000-7,500
Dark Cobalt Blue 7,500-10,000

CD 740.7

CD 740.7
$3\,^1/_4$ x $3\,^3/_4$

NO EMBOSSING - CANADA

[010] [No embossing] SB
Amber Blackglass, Dark Green,
Green, Olive Green Blackglass 800-1,000
Ice Green 1,750-2,000
Clear 2,000-2,500

CD 740.8

CD 740.8
$2\,^3/_4$ x $3\,^1/_8$

TELEGRAFICA

[010] (F-Skirt) TELEGRAFICA DE
(R-Skirt) JALISCO. COMPA!. SB
Dark Forest Green 1,000-1,250

TELEGRAFO

[010] (F-Skirt) [Blotted out embossing]
(Base) TELEGRAFO DE JALISCO P BE
Dark Forest Green, Dark Green 1,000-1,250

186

CD 742

D.T.CO.

CD 742
$3 \frac{1}{4} \times 3 \frac{1}{2}$

[010] (Base) D.T.Co. BE
Aqua 2,000-2,500

M.T.CO.

[010] (Base) M.T.Co BE
Aqua, Green Aqua, Light Aqua,
Light Blue 200-250
Light Green 300-350
Blue 500-600
Sky Blue 700-800
Teal Blue 1,000-1,250
Cornflower Blue 1,250-1,500
Dark Cornflower Blue 1,750-2,000
Milky Cornflower Blue,
Milky Teal Blue 2,000-2,500
Cobalt Blue, Purple 7,500-10,000

[020] (Base) M.T.Co. {Embossing is backwards} BE
Aqua w/ Milk Swirls 800-1,000
Sage Green, Teal Green 1,000-1,250
Jade Blue Milk 1,750-2,000

[030] (Base) M.T.C° ['.' under 'o'] BE
Aqua, Green Aqua, Light Blue 200-250
Cobalt Blue 7,500-10,000

NO EMBOSSING - CANADA

[010] [No embossing] SB
Aqua, Light Aqua, Light Green Aqua 150-175
Light Green 250-300
Blue 400-500
Hemingray Blue 500-600
Dark Sky Blue, Dark Teal Blue 800-1,000
Cornflower Blue 1,250-1,500
Light Cobalt Blue 3,000-3,500

CD 742.3

M.T.CO.

CD 742.3
$3 \frac{1}{8} \times 3 \frac{3}{4}$

[010] (Base) M.T.Co BE
Light Blue, Light Teal Blue 500-600
Teal Blue 800-1,000
Teal Green 1,000-1,250
Jade Green Milk 1,500-1,750
Light Cobalt Blue, Sapphire Blue 3,500-4,000

NO EMBOSSING - CANADA

[010] [No embossing] SB
Aqua 300-350
Blue Aqua, Dark Aqua 350-400
Light Teal Blue 500-600
Dark Green, Dark Teal Blue 800-1,000
Electric Blue 2,000-2,500
Light Cobalt Blue 2,500-3,000

CD 743.1

CD 743.1
NO EMBOSSING - CANADA $2 \frac{7}{8} \times 3 \frac{7}{8}$

[010] [No embossing] GB
Aqua 600-700

[020] [No embossing] SB
Aqua, Light Aqua 400-500
Light Green 600-700
Light Yellow Green 1,000-1,250
Dark Green 1,750-2,000
Yellow Green 3,000-3,500
Light Cobalt Blue 3,500-4,000
Cobalt Blue 5,000-7,500

CD 743.2

NO EMBOSSING - CANADA

CD 743.2
$2 \frac{7}{8} \times 3 \frac{3}{4}$

[010] [No embossing] SB
Aqua, Blue Aqua, Light Blue,
Light Blue Aqua, Light Green Aqua 500-600
Light Green 800-1,000

187

CD 743.3

CD 743.3

NO EMBOSSING - CANADA

[010] [No embossing] SB
Aqua, Light Aqua 500-600

CD 743.3
2 $^3/_4$ x 3 $^3/_4$

CD 780

NO EMBOSSING

[010] [No embossing] SB
Light Aqua 4,000-4,500

CD 780
2 $^3/_4$ x 1 $^3/_4$

CD 784

NO EMBOSSING

[010] [No embossing] SB
Aqua 1,000-1,250

CD 784
2 x 2

CD 785

NO EMBOSSING

[010] [No embossing] SB
Amber Blackglass,
Dark Olive Amber,
Dark Olive Green,
Dark Yellow Green Slag 7,500-10,000

CD 785
3 $^3/_8$ x 3 $^7/_8$

CD 786

NO EMBOSSING

[010] [No embossing] SB
Clear 10,000-15,000

CD 786
1 $^7/_8$ x 3 $^1/_2$

CD 788

NO EMBOSSING

[010] [No embossing] SB
Apple Green, Dark Green,
Dark Olive Amber, Light Green,
Olive Amber Blackglass 3,500-4,000
Blue Aqua 4,500-5,000

CD 788
2 $^7/_8$ x 3 $^1/_2$

CD 790

NO EMBOSSING

[010] [No embossing] SB
Bubbly Dark Green, Dark Green,
Dark Teal Green, Green,
Olive Amber Blackglass 7,500-10,000

CD 790
4 $^5/_8$ x 4 $^1/_4$

CD 791

NO EMBOSSING

[010] [No embossing] SB
Dark Green, Root Beer Amber 15,000-20,000

CD 791
4 $^3/_8$ x 3 $^5/_8$

CD 796

BOSTON BOTTLE WORKS

[010] (F-Skirt) BOSTON BOTTLE WORKS/NO SOMERVILLE/MASS. (R-Skirt) OAKMAN'S/ JULY 26 '70/PATENT (Base around pinhole) OAKMAN'S PATENT.JULY 26. 1870. SB
Aqua 15,000-20,000

CD 796
3 x 4

THREADLESS GLASS BLOCKS

CD 1000

NO EMBOSSING — CD 1000 — 2 $1/4$ x 3 $3/8$

[010] [No embossing] SB
Aqua, Blue Aqua, Light Aqua,
Light Green Aqua 500-600
Christmas Tree Green, Lime Green ... 1,000-1,250
7-up Green, Emerald Green,
Jade Green Milk 1,250-1,500
Dark Olive Amber,
Dark Yellow Amber, Olive Green 1,750-2,000
Light Yellow Amber 2,000-2,500

CD 1002

NO EMBOSSING — CD 1002 — 2 $1/8$ x 3 $3/8$

[010] [No embossing] SB
Aqua 1,000-1,250
Blue 1,250-1,500

CD 1003

NO EMBOSSING — CD 1003 — 2 $3/8$ x 3 $1/2$

[010] [No embossing] SB
Jade Aqua Milk 2,000-2,500

CD 1004

NO EMBOSSING — CD 1004 — 2 $1/8$ x 4

[010] [No embossing] SB
Blue 1,500-1,750
Jade Green Vaseline 2,500-3,000

CD 1005

NO EMBOSSING — CD 1005 — 1 $1/2$ x 2 $1/2$

[010] [No embossing] SB
Light Aqua 2,000-2,500

CD 1006

NO EMBOSSING — CD 1006 — 1 $5/8$ x 2 $1/4$

[010] [No embossing] SB
Ice Green, Light Aqua 1,500-1,750

CD 1007

NO EMBOSSING — CD 1007 — 1 $3/4$ x 1 $3/4$

[010] [No embossing] SB
Aqua, Dark Aqua 1,500-1,750

CD 1008

NO EMBOSSING — CD 1008 — 2 x 2

[010] [No embossing] SB
Blue Aqua 2,500-3,000

CD 1010

NO EMBOSSING — CD 1010 — 1 $7/8$ x 2 $3/8$

[010] [No embossing] SB
Aqua 500-600
Dark Olive Green, Emerald Green,
Olive Blackglass,
Olive Green Blackglass 600-700
Dark Orange Amber,
Gray Green Vaseline 800-1,000

CD 1012

CD 1012

NO EMBOSSING

CD 1012
$2 \, ^1/_8 \times 2 \, ^1/_4$

[010] [No embossing] SB
Aqua 1,500-1,750

CD 1014

NO EMBOSSING

CD 1014
$2 \, ^3/_8 \times 2 \, ^1/_2$

[010] [No embossing] SB
Aqua, Blue, Dark Celery Green,
Light Aqua 700-800
Milky Green 800-1,000

CD 1015

NO EMBOSSING

CD 1015
$1 \, ^3/_4 \times 3$

[010] [No embossing] SB
Aqua 2,500-3,000

CD 1016

NO EMBOSSING

CD 1016
$1 \, ^7/_8 \times 3 \, ^1/_8$

[010] [No embossing] SB
Aqua 2,000-2,500

MISCELLANEOUS GLASS

CD 1030

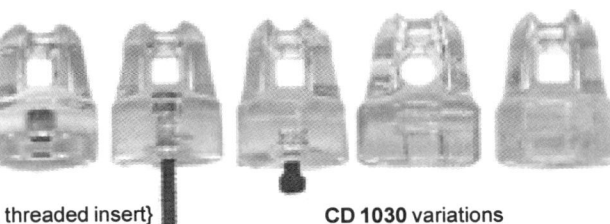

NO EMBOSSING

[010] [No embossing] {Knurled, threaded insert} {Hemingray product} SB
Ice Blue 200-250

[020] [No embossing] {Lag screw in base} {Hemingray product} SB
Blue Tint 200-250

[030] [No embossing] {Machine screw in base} {Hemingray product} SB
Ice Blue 200-250

CD 1030 variations
$2\ 5/8\ \times\ 3\ 3/8$

[040] [No embossing] {Slotted base, round hole variant} {Hemingray product} SB
Green Tint 200-250

[050] [No embossing] {Slotted base} {Hemingray product} SB
Clear 200-250

CD 1038

CD 1038
$4\ 1/4\ \times\ 3\ 1/8$

CUTTER

[010] (Dome) (Circle)CUTTER PAT APRIL 26, 04 {Coffin bottom variant} SB
Aqua 250-300

[020] (Dome) (Circle)CUTTER PAT APRIL 26, 04 {Reverse coffin bottom variant} SB
Aqua 350-400

[030] (Dome) (Circle)CUTTER PAT APRIL 26, 04 {Ribbed base variant} SB
Aqua, Dark Aqua 175-200
Green 400-500
Emerald Green 500-600
Yellow Green 700-800
Olive Green 1,000-1,250

CD 1040

CD 1040
$3\ \times\ 4\ 1/4$

NO EMBOSSING

[010] [No embossing] {Garity patent} SB
Light Aqua 5,000-7,500

GLASS SPOOLS and NAIL KNOBS

CD 1049

NO NAME

[010] (F-Skirt) [Number] SB
Clear, Straw 1-2

OWENS ILLINOIS

[010] (F-Skirt) [Number] ① [Number and dots] SB
Clear 1-2

[020] (F-Skirt) [Number] [Number and dots] SB
Clear 1-2

CD 1050

HEMINGRAY

[010] (F-Above wire groove) HEMINGRAY-519
(F-Below wire groove) MADE IN U.S.A.
(R-Below wire groove) [Numbers and dots] SB
Clear 5-10

CD 1052

HEMINGRAY

[010] (Bottom) HEMINGRAY-518 MADE IN U.S.A. SB
Clear 1-2
Carnival, Flashed Amber,
Light Flashed Amber 30-40

NO EMBOSSING

[010] [No embossing] SB
Ice Blue, Straw 10-15
Carnival 20-30

NO NAME

[010] (F-Below wire groove) D-518 [Number]
(R-Below wire groove) MADE IN U.S.A.
{Hemingray product} SB
Clear 1-2
Ice Blue 3-5
Dark Amber, Red Amber 20-30

ST. LOUIS MALLEABLE

[010] (F-Above wire groove) [Number]
(Bottom) ST.LOUIS MALLEABLE CASTING CO./ST.LOUIS MO. SB
Clear 30-40

[020] (F-Below wire groove) D-518 [Number]
(R-Below wire groove) MADE IN U.S.A.
(Bottom) ST.LOUIS MALLEABLE CASTING CO./ST.LOUIS MO. SB
Yellow Amber 40-50

CD 1053

NO EMBOSSING

CD 1053 Variation with lag screw

[010] [No embossing] {Hex top; brass lag screw} {Hemingray product} SB
Ice Blue 200-250

NO NAME

[010] (F-Below wire groove) D-519 {Hemingray product} SB
Clear 5-10

[020] (F-Below wire groove) ᗡ-519 {Note backwards letter} {Hemingray product} SB
Clear 10-15

CD 1070

HEMINGRAY

[010] (F-Umbrella) HEMINGRAY
(R-Umbrella) № 109 SB
Aqua 10-15

[020] (F-Umbrella) HEMINGRAY-109
(R-Umbrella) MADE IN U.S.A. SB
Ice Blue, Ice Green 5-10
Aqua 10-15
Orange Amber, Yellow Amber 20-30

CD 1105

[030] (F-Umbrella) HEMINGRAY-109/
[Numbers and dots] (R-Umbrella) MADE IN
U.S.A./[Number] SB
Clear, Ice Green, Light Lemon 5-10

[040] (F-Umbrella) HEMINGRAY/MADE IN
U.S.A. (R-Umbrella) № 109 SB
Hemingray Blue 15-20
Orange Amber, Red Amber 20-30

CD 1071

HEMINGRAY

[010] (F-Skirt) HEMINGRAY-110
(R-Skirt) MADE IN U.S.A./[Numbers and dots]
SB
Clear 5-10

CD 1080

NO EMBOSSING

[010] [No embossing] SB
Aqua, Ice Aqua 3-5

CD 1085

NO EMBOSSING

[010] [No embossing] SB
Aqua 3-5
Ice Aqua, Ice Blue, Ice Green 5-10
Blue, Hemingray Blue 10-15

S.B.T.& T.CO.

[010] (F-Below wire groove) S.B.T.& T.Co. {CD
variant; wire groove is smaller and higher up}
SB
Ice Blue, Light Green Aqua 30-40

CD 1087

B

[010] (F-Below wire groove) B (R-Below wire
groove) No. 7 SB
Aqua 3-5

HEMINGRAY

[010] (F-Below wire groove) HEMINGRAY/107
(R-Below wire groove) MADE IN/U.S.A. SB
Aqua, Hemingray Blue, Ice Blue, Ice Green 3-5
Clear 5-10

NO EMBOSSING

[010] [No embossing] SB
Aqua 3-5

CD 1090

B

[010] (F-Below wire groove) B SB
Dark Aqua 20-30
Green 30-40

NO EMBOSSING

[010] [No embossing] SB
Dark Aqua 20-30
Hemingray Blue 30-40
Clear 50-75

CD 1095

NO EMBOSSING

[010] [No embossing] SB
Light Aqua 20-30

CD 1104

NO EMBOSSING

[010] [No embossing] SB
Aqua Tint, Dark Aqua, Ice Blue,
Ice Green, Light Aqua 3-5
Hemingray Blue 5-10
Olive Green 50-75

CD 1105

B

[010] (F-Below wire groove) B (R-Below wire
groove) No 3 SB
Dark Aqua, Light Aqua 3-5
Green 15-20

CD 1105

HEMINGRAY

[010] (F-Above wire groove) 103 (F-Below wire groove) HEMINGRAY (R-Below wire groove) MADE IN U.S.A. SB
Hemingray Blue, Ice Blue 3-5

NO EMBOSSING

[010] [No embossing] SB
Dark Aqua, Light Aqua 3-5
Hemingray Blue 5-10
Green 15-20
Cornflower Blue 20-30

CD 1110

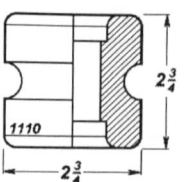

B.E.L.CO.

[010] (F-Below wire groove) B.E.L.Co. SB
Dark Aqua 50-75
Green 100-125

NO EMBOSSING

[010] [No embossing] SB
Dark Aqua 20-30
Green 30-40

CD Unassigned

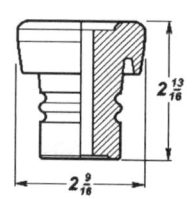

C.& P.TEL.CO.

[010] (F-Skirt) C.& P.TEL.Co. SB
Aqua 40-50
Green 50-75

S.B.T.& T.CO.

[010] (F-Skirt) S.B.T.& T.Co. SB
Aqua 40-50

CD Unassigned

NO EMBOSSING

[010] [No embossing] SB
Light Aqua 20-30

CD Unassigned

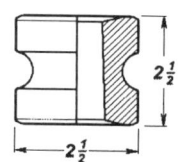

NO EMBOSSING

[010] [No embossing] SB
Light Aqua ....

GUY WIRE STRAINS

CD 1130

CD 1130
2 $^1/_2$ x 1 $^3/_4$

CALIFORNIA

[010] (F-Above wire groove) 10/CALIFORNIA
(F-Below wire groove) PAT APL'D FOR SB
Sage Green 200-250

CD 1131

CD 1131
3 x 2 $^1/_8$

CALIFORNIA

[010] (F-Above wire groove) 15/CALIFORNIA
(F-Below wire groove) PAT APL'D FOR SB
Purple, Sage Green, Smoke 200-250

CD 1138

CD 1138
2 $^1/_2$ x 2 $^1/_2$

NO EMBOSSING

[010] [No embossing] {2 1/2" Johnny Ball} SB
Aqua, Dark Aqua, Light Aqua 15-20
Ice Aqua 20-30

CD 1140

CD 1140
3 x 2 $^1/_2$

NO EMBOSSING

[010] [No embossing] {3" Johnny Ball} SB
Aqua, Dark Aqua, Light Aqua 15-20

CD 1142

CD 1142
3 $^3/_4$ x 3 $^1/_8$

NO EMBOSSING

[010] [No embossing] {3 3/4" Johnny Ball} SB
Aqua 15-20

BATTERY RESTS

CD 10

NO EMBOSSING

$2 \times 1\frac{1}{16}$

[010] [No embossing] SB
Light Aqua 100-125

CD 12

NO EMBOSSING

$2\frac{1}{4} \times 1\frac{3}{8}$

[010] [No embossing] SB
Light Aqua 100-125

CD 20

E.S.B.CO.

$2\frac{3}{16} \times 1\frac{1}{2}$

[010] (F-Skirt) E.S.B.CO. (R-Skirt) PHILA. PA. SB
Ice Green, Light Aqua, Light Blue 10-15

GOULD

[010] (F-Skirt) GOULD BATTERY (R-Skirt) PAT DEC. 1. 1896 SB
Ice Aqua 10-15
Clear, Light Yellow Green 30-40
Yellow Green 40-50
Smoke 50-75
Light Lavender 75-100

CD 22

NO EMBOSSING

$2\frac{3}{16} \times 1\frac{3}{16}$

[010] [No embossing] SB
Aqua 40-50
Emerald Green 50-75
Violet Cobalt Blue 75-100

CD 22.5

NO EMBOSSING

$2\frac{1}{8} \times 1\frac{1}{2}$

[010] [No embossing] SB
Aqua, Ice Aqua 10-15
Dark 7-up Green 50-75

CD 23

U.S.L.

$2\frac{1}{2} \times 1\frac{1}{2}$

[010] (F-Skirt) USL № 2 SB
Aqua 150-175

CD 24

NATIONAL BATTERY CO.

$2\frac{1}{2} \times 1\frac{13}{16}$

[010] (Crown) NATIONAL BATTERY CO. UNIT ACCUMULATOR SB
Aqua 30-40

U.S.L.

[010] (F-Skirt) THE U.S.L.& H.Co. SB
Aqua 150-175

U.S.LIGHT & HEATING CO.

[010] (Crown) THE UNITED STATES LIGHT AND HEATING COMPANY SB
Aqua 75-100

CD 25

GOULD

CD 25
$4\ ^{1}/_{8} \times 2\ ^{1}/_{2}$

[010] (F-Skirt) GOULD BATTERY (R-Skirt) PAT DEC. 1. 1896 SB
Lime Green 200-250

CD 26

NO EMBOSSING

$3\frac{1}{4} \times 1\frac{5}{16}$

[010] [No embossing] SB
Ice Aqua 75-100

CD 28

NO EMBOSSING

$4 \times 2\frac{1}{4}$

[010] [No embossing] SB
Dark Aqua 75-100

CD 29

NO EMBOSSING

$5 \times 2\frac{1}{2}$

[010] [No embossing] SB
Dark Red Amber 400-500

CD 50

NO NAME

[010] (Under crown) B-17307 SB
Aqua, Ice Green, Light Aqua 75-100

CD 29.5

NO EMBOSSING

[010] [No embossing] SB   5 × 2½
Light Green Aqua 75-100

CD 30

E.S.B.CO.

[010] (Base) CHLORIDE ACCUMULATOR/
THE E.S.B.CO BE
Emerald Green 150-175

CD 31

E.S.B.CO.

[010] (Base) CHLORIDE ACCUMULATOR/
THE E.S.B.CO BE
Ice Aqua, Light Aqua 10-15
Emerald Green 50-75

CD 35

E.S.B.CO.

[010] (F-Skirt) THE E.S.B.CO. (R-Skirt) MADE
IN U.S.A. SB
Aqua 40-50

NO EMBOSSING

[010] [No embossing] SB
Aqua 40-50

CD 36

E.S.B.CO.

[010] (F-Skirt) THE E.S.B.CO. (R-Skirt) MADE
IN U.S.A. (Under crown) B-15993 SB
Aqua 40-50

[020] (F-Skirt) THE E.S.B.CO. (R-Skirt) MADE
IN U.S.A. (Under crown) B-15993-3 SB
Aqua 40-50

NO EMBOSSING

[010] [No embossing] SB
Blue 50-75

CD 40

NO EMBOSSING

[010] [No embossing] SB
Dark Blue Aqua 400-500

CD 45

WARE

[010] (Base) THE WARE BATTERY
INSULATOR PATENTED 1900 BE
Honey Amber 400-500

CD 50

GOULD

[010] (F-Skirt) GOULD BATTERY
(R-Skirt) PATENTED DEC. 1. 1896 SB
Aqua 30-40
Yellow Green 200-250

197

CD 50

NO EMBOSSING

[010] [No embossing] SB
Aqua, Light Aqua 30-40

CD 51

CD 51
$5 \frac{1}{2} \times 3 \frac{1}{2}$

GOULD

[010] (F-Skirt) GOULD BATTERY
(R-Skirt) PATENTED DEC. 1. 1896 SB
Aqua 50-75

NO EMBOSSING

[010] [No embossing] SB
Aqua 50-75

CD 53

CD 53
$2 \frac{1}{2} \times 2$

U.S.L.

[010] (F-Skirt) USL № 2 SB
Light Aqua 75-100

CD 55

CD 55
$5 \, ^1/_2 \times 2 \, ^3/_4$

WARE

[010] (Base) THE WARE BATTERY
INSULATOR PATENTED 1900 BE
Honey Amber 400-500

CD 61

CD 61
$5 \, ^1/_2 \times 2 \, ^3/_4$

M.T.CO.

[010] (F-Skirt) M.T.Co. {Embossing is upside down} {Previously shown as CD 782 in the 1995 edition} SB
Aqua 3,000-3,500

COMMEMORATIVES

NATIONAL SHOW

Initiated by Frank Miller following the first national swap meet as a souvenir. Currently produced annually commemorating the National Insulator Association Convention by John and Carol McDougald.

{1970} (Dome) (Outer circle)NEW CASTLE, INDIANA/NAT'L MEET (Inner circle)JUNE 20-21, 1970/FIRST SB
Apple Green, Blackglass 175-200
Straw 500-600

{1970} (Dome) (Outer circle)NEW CASTLE, INDIANA/NAT'L MEET (Inner circle)JUNE 20-21, 1970/FIRST {Penny imprint in dome} SB
Straw 800-1,000

{1970} (Dome) (Outer circle)NEW CASTLE, INDIANA/NAT'L MEET (Inner circle)JUNE 20-21, 1970/FIRST {Solid pour} SB
Straw 800-1,000

{1971} (Dome) (Outer circle)COLORADO SPRINGS, COLO./NAT'L MEET (Inner circle)JULY 10-11, 1971/SECOND SB
Apple Green, Blackglass 175-200
Straw 700-800

{1971} (Dome) (Outer circle)COLORADO SPRINGS, COLO./NAT'L MEET (Inner circle)JULY 10-11, 1971/SECOND {Penny imprint in dome} SB
Blackglass 350-400
Straw 1,000-1,250

{1971} (Dome) (Outer circle)COLORADO SPRINGS, COLO./NAT'L MEET (Inner circle)JULY 10-11, 1971/SECOND {Solid pour} SB
Straw 1,000-1,250

{1971} (Dome) (Outer circle)COLORADO SPRINGS, COLO./NAT'L MEET (Inner circle)JULY 11-12, 1971/SECOND {Note 'JULY 11-12, 1971' date} SB
Blackglass 250-300

{1971} (Dome) (Outer circle)COLORADO SPRINGS, COLO./NAT'L MEET (Inner circle)JULY 11-12, 1971/SECOND {Note 'JULY 11-12, 1971' date} {Solid pour} SB
Blackglass 600-700

{1972} (Base) KANSAS CITY, MO. JULY 1-2, 1972/THIRD NAT'L MEET BE
Dark Cobalt Blue 100-125
Dark Cobalt Blue w/ Carnival 400-500
Green Speckled Cullet,
Orange Carnival, Red Speckled Cullet .. 800-1,000

{1972} (Dome) JOE MC (Base) KANSAS CITY, MO. JULY 1-2, 1972/THIRD NAT'L MEET BE
Dark Cobalt Blue 400-500

{1972} (Dome) MIKE/MITCHELL (Base) KANSAS CITY, MO. JULY 1-2, 1972/ THIRD NAT'L MEET BE
Dark Cobalt Blue 400-500

{1973} (Base) FOURTH NAT'L MEET/ HUTCHINSON, KANSAS JULY 7-8 1973 BE
Orange Carnival 100-125
Clear 175-200

{1973} [Unembossed] {Solid pour} SB
Clear 800-1,000

{1974} (Base) NIA FIFTH NAT'L SHOW/ HERSHEY, PA. JUNE 29-30 1974 BE
Sapphire Blue 75-100

{1975} (Base) 6TH NIA CONVENTION/SAN DIEGO, CA. JULY 11-13, 1975 BE
Amberina Red 100-125

{1975} (Dome) JOE ST./CLAIR (Base) 6TH NIA CONVENTION/SAN DIEGO, CA. JULY 11-13, 1975 BE
Amberina Red w/ Carnival 400-500

{1976} (Base) 7TH NIA CONVENTION/ BEREA, OHIO AUGUST 20-22 1976 BE
White Opalescent Carnival 50-75

{1977} (Base) 8TH NIA CONVENTION LAKELAND, FLA. JULY 15-17 1977 BE
Golden Amber 20-30

{1978} (Base) 9TH NIA CONVENTION RENO, NEVADA JULY 21-23 1978 BE
Yellow Vaseline 30-40

{1978} [Unembossed] {Solid pour} SB
Yellow Vaseline 400-500

{1979} (Base) 10TH NIA CONVENTION DENVER, COLORADO JULY 20-22 1979 BE
Burgundy 20-30

{1979} [Unembossed] {Ash tray variation} SB
Burgundy 150-175

{1979} [Unembossed] {Solid pour} SB
Burgundy 200-250

{1980} (Base) 11TH NIA CONVENTION HERKIMER N.Y. JULY 11-13 1980 BE
Blue Milk 20-30

{1980} [Unembossed] {Solid pour} SB
Blue Milk 150-175

{1981} (Base) 12TH NIA CONVENTION SACRAMENTO, CALIF. JULY 10-12 1981 BE
Emerald Green 20-30

{1981} [Unembossed] {Solid pour} SB
Emerald Green 150-175

{1982} (Base) 13TH NIA CONVENTION CEDAR RAPIDS, IOWA JULY 16-18 1982 BE
White Milk 20-30

{1982} [Unembossed] {Solid pour} SB
White Milk 150-175

{1983} (Base) 14TH NIA CONVENTION ROCHESTER N.Y. JULY 8-10 1983 BE
Lilac 20-30

{1983} [Unembossed] {Solid pour} SB
Lilac 150-175

{1984} (Base) 15TH NIA CONVENTION TACOMA, WASHINGTON JULY 20-22 1984 BE
White Milk w/ Red Orange Swirls {Run 1} ... 40-50
White Milk w/ Red Orange Slag {Run 2} 50-75
Red Dome/White Base,
White Dome/Red Base 100-125
Cobalt Blue w/ Milk Swirls 400-500

{1984} [Unembossed] {Solid pour} SB
White Milk w/ Red Orange Slag 150-175

{1985} (Base) 16TH NIA CONVENTION ST. CHARLES, ILLINOIS JUNE 7-9, 1985 BE
Teal Blue 20-30

{1985} [Unembossed] {Solid pour} SB
Teal Blue 175-200

{1986} (Base) 17TH NIA CONVENTION SARATOGA SPRINGS, NY JULY 11-13, 1986 BE
Amber, Lime Green w/ Amber Swirls 20-30
Lime Green 30-40

{1986} [Unembossed] {Solid pour} SB
Lime Green w/ Amber Swirls 150-175

{1987} (Base) 18TH NIA CONVENTION FRESNO, CALIFORNIA JULY 24-26 1987 BE
Orange Milk 20-30

{1987} [Unembossed] {Solid pour} SB
Orange Milk 150-175

{1988} (Dome) [Texas outline] (Base) 19TH NIA CONVENTION HOUSTON, TEXAS JULY 22-24, 1988 BE
Texas Bluebonnet 20-30

{1988} (Dome) [Texas outline] {Solid pour} SB
Texas Bluebonnet 150-175

{1989} (Dome) [Raised Keystone 20] (Base) 20TH NIA CONVENTION ALLENTOWN, PA JULY 14-16 1989 BE
Lavender Opal 30-40

{1989} (Dome) [Raised Keystone 20] {Solid pour} SB
Lavender Opal 40-50

{1990} (Dome) [Oregon outline] (Base) 21ST NIA CONVENTION PORTLAND, OREGON JULY 6-8, 1990 BE
Cobalt Blue w/ White Milk Swirls 20-30

{1990} (Dome) [Oregon outline] {Solid pour} SB
Cobalt Blue w/ White Milk Swirls 40-50

{1991} (Dome) [MVIC logo] (Base) 22ND NIA CONVENTION CEDAR RAPIDS, IOWA JULY 26-28, 1991 BE
Depression Glass Green w/ Milk Swirls 20-30

{1991} (Dome) [MVIC logo] {Solid pour} SB
Depression Glass Green w/ Milk Swirls 40-50

{1992} (Dome) [CFIC logo] (Base) 23RD NIA CONVENTION ORLANDO, FLORIDA JUNE 19-21, 1992 BE
Cobalt Blue Carnival, Frosted Lilac, Lilac 20-30

{1992} (Dome) [CFIC logo] {Solid pour} SB
Cobalt Blue Carnival 40-50
Lilac 50-75

{1993} (Dome) [TRIC logo] (Base) 24TH NIA CONVENTION DENVER, COLORADO JULY 2-4, 1993 BE
Royal Purple w/ White Milk Swirls 20-30
Chrome Coated Royal Purple
w/ White Milk S 50-75

{1993} (Dome) [TRIC logo] {Solid pour} SB
Royal Purple w/ White Milk Swirls 40-50

{1994} (Dome) [LSIC logo] (Base) 25TH NIA CONVENTION HOUSTON, TEXAS JULY 1-3, 1994 BE
Silver Coated Clear 20-30
Chrome Coated Clear, Clear 50-75

{1994} (Dome) [LSIC logo] {Solid pour} SB
Silver Coated Clear 40-50
Clear 150-175

{1995} (Dome) [YPCIC logo] (Base) 26TH NIA CONVENTION MARLBOROUGH, MA JULY 14-16, 1995 BE
Light Cornflower Blue 20-30
Silver Coated Light Cornflower Blue 50-75

{1995} (Dome) [YPCIC logo] {Solid pour} SB
Light Cornflower Blue 30-40
Silver Coated Light Cornflower Blue 200-250

{1996} (Dome) [Seahorse] (Base) 27TH NIA CONVENTION LONG BEACH, CALIF JULY 19-21, 1996 BE
Clear w/ Cobalt Blue 20-30

{1996} (Dome) [Seahorse] {Solid pour} SB
Clear, Clear w/ Cobalt Blue, Cobalt Blue 40-50

{1997} (Dome) [Illinois logo] (Base) 28TH NIA CONVENTION CHICAGO, ILLINOIS JULY 25-27, 1997 BE
Amethyst Blackglass 20-30
Amethyst Blackglass Carnival 150-175
Dull Salmon Milk, Salmon Milk 200-250

{1997} (Dome) [Illinois logo] {Solid pour} SB
Amethyst Blackglass 30-40

{1998} (Dome) [CBIC B & O logo] (Base) 29TH NIA CONVENTION WILLIAMSBURG, VA JULY 31 - AUG 2, 1998 BE
Aqua 20-30

{1998} (Dome) [CBIC B & O logo] {Solid pour} SB
Aqua 30-40

{1999} (Dome) [GCSIC logo] (Base) 30TH NIA CONVENTION SCOTTSDALE, ARIZONA JUNE 25-27, 1999 BE
Ruby Red 20-30

{1999} (Dome) [GCSIC logo] {Solid pour} SB
Ruby Red 20-30

TELEPHONE PIONEERS (1971)

The distribution and sale of this commemorative by the Columbus (Ohio) Council of the Telephone Pioneers of America benefited community service projects.

(Dome) № 1 (F-Skirt) [Telephone Pioneers of America logo] (R-Skirt) HUMAN SERVICE {"Pilgrim hat" style} SB
Amber, Blue, Blue Carnival, Lilac,
White Milk 10-15
Azalea, Cobalt Blue, Dark Emerald Green,
Dark Forest Green, Emerald Green,
Orange Carnival, Teal Green 15-20
Clear 100-125

S.N.E.T. - 1978

The distribution and sale of this commemorative by the Southern New England Telephone Council of the Telephone Pioneers of America benefited community service projects.

(F-Skirt) [Telephone Pioneers of America logo] (R-Skirt) 1878 SNET CO. 1978/100 YEARS OF SERVICE {"Hat" style} SB
Cobalt Blue 30-40
Purple 100-125

PRIVATE ISSUE COMMEMORATIVES

(F-Skirt) MCLAUGHLIN № 19 (R-Skirt) 1897-1972 {CD 162 style} SB
White Milk 30-40
Blue Milk w/ Slag, White Milk w/ Slag 250-300
Red Milk,
White Milk w/ Red/Blue Drops 800-1,000
Red Milk w/ Yellow Milk Dome 1,000-1,250
Produced by William McLaughlin.

(Base) CANADA GLASS WORKS .1858 - 1998. {CD 740 style} {MLOD} BE
Blue, Bubbly Burgundy, Bubbly Cobalt Blue,
Bubbly Dark Green,
Bubbly Dark Yellow Amber,
Cornflower Blue w/ Cobalt Swirls,
Ice Aqua w/ Cobalt & White Swirls,
Ice Aqua w/ Cobalt & Yellow Swirls 20-30
Produced by Mark Lauckner.

(Base) TELEGRAPH IN CANADA 1846 - 1996
{CD 718 style} BE
Bubbly Cornflower Blue,
Bubbly Dark Cornflower Blue,
Bubbly Light Forest Green,
Bubbly Light Honey Amber,
Bubbly Light Yellow Olive w/ Milk Swirls,
Cobalt Blue, Dark Burgundy,
Electric Blue w/ Amber Swirls,
Light Electric Blue w/ Milk Swirls 20-30
Produced by Mark Lauckner.

(F-Skirt) 1847 - 1997 {Incuse}
(Base) MONTREAL TELEGRAPH - 150 YEARS {CD 742 style} {MLOD} BE
7-up Green w/ Milk Swirls,
Blue w/ Red Milk Swirls,
Bubbly Cornflower Blue w/ Milk Swirls,
Bubbly Electric Blue, Cobalt Blue,
Cranberry Plum, Light 7-up Green,
Light Aqua w/ Red Milk Swirls,
Puce w/ Milk Swirls,
Sapphire/Cobalt Blue Two Tone 20-30
Produced by Mark Lauckner.

(F-Skirt) № 2 COLUMBIA (R-Skirt) L.S.V. {CD 262 style} SB
Blue/Red/Green Swirls,
Dark Cobalt Blue {1988 run}, Light Blue,
Ruby Red, White Opalescent Milk,
White Slag, Yellow Vaseline 20-30
Cobalt Blue {1974 run}, Dark Olive Amber,
Depression Glass Green, Olive Green 50-75
Cobalt Blue w/ Milk Swirls 75-100

(F-Skirt) № 2 COLUMBIA (R-Skirt) L.S.V. {CD 262 style} {Solid pour} SB
Peacock Cobalt Blue 75-100
Produced by Larry S. Veneziano.

(Base) NATIONAL INSULATOR ASSOCIATION/10TH ANNIVERSARY 1983 {CD 257 style} BE
7-up Green, White Milk 15-20
Peacock Blue, Red 20-30
Honey Amber 50-75

[No embossing] {CD 257 style} {Solid pour} SB
Cobalt Blue, Honey Amber, Red,
Slag Glass 400-500
Produced by John and Carol McDougald.

(F-Skirt) "VTS INDUSTRIAL Co"
(R-Skirt) "NUMBER 8" {CD 102 style} SB
Ruby Red 200-250
Peacock Blue 400-500
Produced by VTS Industrial Co., distributors of telephone supplies.

(F-Skirt) WGS - 94 {Tall CD 102 style} SB
Cobalt Blue 10-15

(F-Skirt) WGS - 95 {Tall CD 102 style} SB
Yellow 10-15
Amethyst 100-125

(F-Skirt) WGS - 96 {Tall CD 102 style} SB
Bottle Green 10-15

(F-Skirt) WGS - 97 {Tall CD 102 style} SB
Irridized Clear 10-15

(F-Skirt) WGS - 98 {Tall CD 102 style} SB
Pink 10-15
Produced by William G. Scandariato.

SALESMAN MINIATURE INSULATORS

BROOKFIELD

(F-Crown) WESTERN (F-Skirt) BROOKFIELD (R-Crown) ELEC CO (R-Skirt) STANDARD {CD 162 style} SB
Light Aqua 200-250
Green, Light Green 250-300
Light Yellow Green 400-500

(F-Skirt) BROOKFIELD (R-Skirt) U.S.A. {CD 145 style} SB
Light Aqua 800-1,000

HEMINGRAY

(F-Skirt) HEMINGRAY (R-Skirt) U.S.A. {CD 154 style; original with strong embossing} SDP
Clear, Light Aqua 40-50
Golden Amber, Yellow Green 50-75

(F-Skirt) HEMINGRAY (R-Skirt) U.S.A. {CD 154 style; reproduction with weak embossing} SDP
Amber, Clear 20-30

ARMSTRONG

(F-Skirt) Ⓐ (R-Skirt) Ⓐ (Base of pedestal) *Armstrong's* WHITALL TATUM COMMUNICATION INSULATORS {CD 155 style on a glass pedestal} SB
Clear 300-350

(F-Skirt) Ⓐ (R-Skirt) Ⓐ (Base of pedestal) *Armstrong's* WHITALL TATUM DISTRIBUTION INSULATORS {CD 216 style on a glass pedestal} SB
Clear 300-350
Amber 500-600

WHITALL TATUM

(F-Skirt) WHITALL TATUM Cº (R-Skirt) ▽ SB
Clear 100-125

PRIVATE ISSUE - MINIATURE INSULATORS

(F-Crown) EC&M Co. SF {CD 123 style} SB
Amber, Clear 75-100
Produced by John and Carol McDougald.

(F-Skirt) HEMINGRAY {CD 162 style} RDP
7-up Green, Amber, Clear, Cobalt Blue,
Hemingray Blue, Purple 20-30
Silver Carnival 100-125
Produced by John and Carol McDougald.

(F-Skirt) DOMINION {CD 154 style} RDP
Amber, Clear, Ice Blue, Ice Green,
Peacock Blue 5-10
Produced by John and Carol McDougald.

(F-Skirt) HOLLY [Holly leaves] CITY N.J.
(R-Skirt) DP1 71 [Dot] {CD 155 style} SB
7-up Green, Cornflower Blue,
Dark Yellow Green, Light Blue, Neodymium,
Orange Amber, Peacock Blue, Plum,
Royal Cobalt Blue,
Sage Green w/ Titanium Iridizing 5-10
Olive Green 10-15
Green Aqua 15-20
Blue Aqua 20-30
Apple Green 30-40
Clear w/ Titanium Iridizing 40-50
Olive Amber 50-75
Honey Amber w/ Bubbles 75-100

(F-Skirt) HOLLY [Holly leaves] CITY N.J.
(R-Skirt) DP1 71 [Dot] {CD 155 style} {Solid pour} SB
Apple Green, Peacock Blue 75-100

(F-Skirt) HOLLY [Holly leaves] CITY N.J.
(R-Skirt) DP1 71 {CD 155 style} SB
Amber, Clear, Cobalt Blue 50-75
Blue, Green, Light Green,
Neodymium, Opalescent 75-100
Clear w/ Swirls, Honey Amber,
Lime Green, Royal Purple Blackglass 100-125
Produced by Don Wentzel and Calvin Cobb.

(F-Skirt) WENTZEL-COBB TW 1971 [Dot] {CD 203 style} SB
Dark Yellow Green, Green Aqua,
Honey Amber, Light Brooke's Blue,
Light Cornflower Blue, Neodymium,
Peacock Blue, Royal Cobalt Blue,
Sage Green w/ Titanium Iridizing 5-10
Clear w/ Titanium Iridizing 50-75
Cobalt Blue w/ Opal Swirls,
Green w/ Black Swirls 75-100
Teal Blue 100-125

(F-Skirt) WENTZEL-COBB TW 1971 [Dot] {CD 203 style} {Solid pour} SB
Apple Green, Peacock Blue 75-100

(F-Skirt) WENTZEL-COBB TW 1971 {CD 203 style} SB
Bubbly Green Aqua, Clear,
Emerald Green, Royal Purple Blackglass ... 75-100
Candle Blue, Cobalt Blue, Neodymium,
Olive Green, Rose 100-125
Opalescent 200-250

(F-Skirt) WENTZEL-COBB TW 1971 {CD 203 style} {Ground base} SB
Bubbly Green Aqua 100-125

(F-Skirt) WENTZEL-COBB TW 1971 {CD 203 style}{Solid pour} SB
Clear 75-100
Produced by Don Wentzel and Calvin Cobb.

(F-Skirt) HOLLY [Holly leaves] CITY, N.J.
(R-Skirt) 1986 WC {CD 297 style} SB
Cornflower Blue, Honey Amber,
Neodymium, Royal Cobalt Blue 5-10
7-up Green, Clear Rainbow 10-15
Produced by Don Wentzel and Calvin Cobb.

(F-Skirt) HOLLY [Holly leaves] CITY, N.J.
(R-Skirt) 1997 WCJ {CD 285 style} SB
7-up Green, Clear Rainbow, Cobalt Blue,
Light Cornflower Blue,
Light Pink Rainbow, Peach 5-10
Light Pink Rainbow w/ Frit, Peach w/ Frit ... 10-15

(F-Skirt) HOLLY [Holly leaves] CITY, N.J.
(R-Skirt) 1997 WCJ {CD 297 style} SB
7-up Green, Clear Rainbow 5-10
Produced by Don Wentzel and Calvin Cobb.

(F-Skirt) PAT. NOV. 1870 {CD 154 style} SB
Clear 10-15
Production is unattributed.

APPENDIX I
CROSS REFERENCE OF MANUFACTURER'S STYLE NUMBER TO CD NUMBER

COMPANY Style No.	CD No.	COMPANY Style No.	CD No.
ARMSTRONG Ⓐ		**CABLE**	
2	122	2 Cable	252
3	115	3 Cable	254
4	163	4 Cable	267
5	164		
9	107	**CALIFORNIA**	
10	214	10	1130
13	113	15	1131
14	160	303	178
51-2U	216	A007	121
51-C1	167	A011	166
51-C1A	230.2	A021	190/191
51-C2	228.5		
51-C3	217	**CHAMBERS**	
511A	272	4	124.5
CSC	128		
DP1	155	**COLUMBIA**	
TS	129	2	262
TW	203 & 203.2		
		CRIMSA	
B		45	155
3	1105		
7	1087	**CRISA**	
32	160	9	107
44	145		
B-1	185	**CRISOL TEXCOCO** ▽	
		9	106.5
BIRMINGHAM		45	155
10	106		
		DERFLINGHER	
BROOKFIELD		TN-1	154.5
3 Trans	205		
6	731	**DOMINION** ◇	
7	133	6 (error)	108
9	101	9	108
17	300	10	115
20	133	16	122
31	112	42	154 & 155
36	162	614	164
36 (error)	134		
38	164	**DRY SPOT INSULATOR**	
41	134	10	182
48	152		
53	205	**E.S.B.CO.**	
		B-15993	36
		B-15993-3	36

205

APPENDIX I
CROSS REFERENCE OF MANUFACTURER'S STYLE NUMBER TO CD NUMBER

COMPANY Style No.	CD No.	COMPANY Style No.	CD No.
E.S.S.CO.		42	154
401	252	WU 5	125
		8	112.4
ELECTRICAL SUPPLY CO.		9	106 & 107
905	134	9 (error)	113
		10	115
ERICSSON		11	114
LD-1	250.2	12	113
		13	124
GAYNER		14	102, 160 & 202
6	103	15	125
36-190	162	16	121 & 122
38-20	164	16 (error)	115
44	154	17	122
48-400	153	18	134
90	106	19	162 & 163
91	107	19	186, 186.1 & 186.2
140	160	20	133, 164 & 165
160	121	21	145
530	205	23	241
620	252	24	241.2
		25	175
GOOD		38	157
16	121	40	152
		42	154, 155 & 233
H.G.CO.		43	213 & 214
2 Transposition	201	44	208
7	133	45	155
20	133	50	190/191
40	160	51	196
51	196	53	197 & 202
		53S	197
HEMINGRAY		54-A	195
0 Provo	249	54-B	194
1 Cable	251	55	205
1 High Voltage	280, 281 & 291	56	203
1 Provo	283	60	257
2 Cable	252	60-A	257
2 High Voltage	281	61	251
2 Provo	282	61 Cable	251
2 Transposition	201	62	252
3 Cable	254	62 Cable	252
3 High Voltage	291	63	254
4	124	63 (error)	257
4 High Voltage	295	63 Cable	254
4 Provo	303.5	63-A (error)	257

APPENDIX I
CROSS REFERENCE OF MANUFACTURER'S STYLE NUMBER TO CD NUMBER

COMPANY Style No.	CD No.	COMPANY Style No.	CD No.
HEMINGRAY (continued)		Section 2	313
66	219 & 242	TS	129
67	220	TS-2	142
68	221	TYPE 1	169
71	183 & 281		
71 High Voltage	281	**K.C.G.W.**	
72	237 & 295	5	120
72 High Voltage	295		
72A	237 & 295	**KERR**	
75	302	2	122
76	303	22 (error)	122
79	248	CSC	128
88	243	DP1	155
95	185	TS	129
100 H.V.	280	TW	203
103	1105		
107	1087	**KIMBLE**	
109	1070	820	231 & 231.2
110	1071	830	239
511	233.2	850-1	239.2
512	230		
513	232	**KNOWLES**	
514	238	2 Cable	252
515	301.5	5 1/2	282
518	1052	6	292.5
519	1050		
660	218 & 219	**LOCKE**	
661	216	14	202
670	220	15	287
680	221	16	297
720	237	17	300
820	231	18	288
830	239	19	318
CSA	128	20	296
CSC	128	21	275
CSO	128	23	319
D-510	168	25	342
D-512	230 & 230.1	135	287
D-513	232		
D-514	238 & 238.1	**LOWEX**	
D-990	137	512	230
E-1	100.6 & 128	D-514	299.7
E-14B	128		
E-2	122.4	**LYNCHBURG** Ⓛ	
E-3	233	1	251
SB	128.4	2 Cable	252

207

APPENDIX I
CROSS REFERENCE OF MANUFACTURER'S STYLE NUMBER TO CD NUMBER

COMPANY Style No.	CD No.	COMPANY Style No.	CD No.
LYNCHBURG Ⓛ (continued)		**NO NAME**	
10	106	2	132.2
30	121 & 122	2 Transposition	200
31	112	5	291
32	160	7	126
36	162	9	106
36-190	162	14	202
38	164	15	287
38-20	164	16	297
43	145	19	162
44	154	20	133 & 296
53	205	21	275
62 Cable (error)	252	36-19	162
180	281	38-20	164
181	306	40	152
530	205	44	154
		48	152
M. & E. CO.		48-40	153
58	254	51	202
401	252	68	221
		90	106
MAYDWELL		100	308
9	106	115	226
10	115	135	280 & 287
14	160	180	280
16	122	512	230
16W	121	1025	240.2
19	162	9200	245
20	164	B-17307	29
42	154	D-518	1052
62	252	D-519	1053
		RD149959	126
McLAUGHLIN		RD154745	133
9	106	TS 3	142.4
10	115	TS-2	142
14	160	W1	127
16	121 & 122		
19	162	**NO NAME - CANADA**	
20	164	1673	162.4
40	152	1678	162 & 162.4
40 (error)	154		
42	154	**NO NAME - MEXICO**	
62	252	3	155
USLD	139.9	C45	155
		SM-1	106
		SM-2	162

208

APPENDIX I
CROSS REFERENCE OF MANUFACTURER'S STYLE NUMBER TO CD NUMBER

COMPANY Style No.		CD No.	COMPANY Style No.		CD No.
O.V.G.CO.				441	327
	11	112		453	326
				553	328 & 328.2
OWENS ILLINOIS				660	139.6
	51-C1	167		661	233
	E-14B	128		662	235
				663	330
P.S.S.A.				70005	194.5
	9	106, 106.2 & 107		CSA	128
	19	162.7			
	42	155	R.Y T.		
				45	155
PAT APP FOR					
	2	132	S.S.& CO.		
	3	133.1		19	162
	5	125		40	160
PAT'D			SANTA ANA		
	2	132		303	178
PATENT - DEC. 19, 1871			ST.LOUIS MALLEABLE		
	1	131.4 & 732.2		D-518	1052
	2	132			
	3	133, 133.1 & 133.4	T-H.		
	4	104, 124, 124.2 & 124.3		9200	245
	5	120	U.S.L.		
				2	23 & 53
PATENT - OTHER					
	Section 1	317.8	W.F.G.CO.		
				16	121
PRISM					
	2	252	W.U.		
	4 1/2	299.1		5	125
PYREX			WESTINGHOUSE		
	61	233		2	162
	62	235		3	102
	63	234		4	113
	131	240		6	134
	161	240.5		TYPE B	309.5
	171	320		TYPE C	286.9
	233	322			
	271	323			
	353	324			
	401	325			

APPENDIX I
CROSS REFERENCE OF MANUFACTURER'S STYLE NUMBER TO CD NUMBER

COMPANY Style No.	CD No.	COMPANY Style No.	CD No.

WHITALL TATUM ▽

Style	CD No.
1	153, 154 & 155
2	122
3	115
4	162, 163, 163.4 & 169
5	164 & 165.1
9	107 & 108
10	214
11	168
12	176
13	113
14	160
15	197
62 (error)	272
511A	272
512U	216
514	221
CSA	128
CSC	128

APPENDIX II
CROSS REFERENCE OF PRIMARY EMBOSSING TO CD NUMBER

A.A. (AA)
154

A.T.& T.CO.
121   190/191

A.U.
116   121.4   126.3

AM.INSULATOR CO.
105   126.4   134   143.4   145
156   156.1   160.4   160.7

AM.TEL.& TEL.CO.
106   121   160.6   190/191
192/193   192.1/193.1

ARMSTRONG (A)
107   113   115   122   128
129   155   160   163   164
167   203   203.2   214   216
217   228.5   230.2   238.2   272

AYALA
106

B
102   103   112   116.5   145
152   160   164   185   188
190/191   196   207   226.3   1087
1090   1105

B & O
136

B.C.DRIP
164

B.E.L.CO.
185.2   1110

B.F.G.CO.
133.2

B.G.M.CO.
102   133   134   145   162
164

B.T.C.
102   112.4   121

BARCLAY
150

BEAL'S
309

BIRMINGHAM
106

BOSTON BOTTLE WORKS
136.5   143.6   145.6   158   158.2
158.9   728.8   796

BROOKE'S
120   127   133.1

BROOKFIELD
101   102   104   110   112
115   116   119   121   126
126.1   126.3   127   131   133
134   134.6   136   138   143.4
145   149   150   151   152
153   157   160   161.2   162
162.1   162.3   164   169.5   173
174   205   208   211   225
228   236   287.2   299   300
316   338   728.4   731

BUZBY
141.8

C.& P.TEL.CO.
121

C.D.& P.TEL.CO.
121

C.E.L.CO.
134

C.E.W.
120

C.G.I.CO.
102

C.N.R.
143   154

C.P.R.
143

CABLE
252   254   258   259   260
267

CAL.ELEC.WORKS
130   130.1

CALIFORNIA
102   112   121   133   134
145   152   160   161   162
166   178   187   190/191   200
201   208   260   1130   1131

CANADA
121

CANADIAN PACIFIC RY.CO.
143

CHAMBERS
124.5   317   317.5

CHESTER
123.2   158.1   724   735   738
740.6

CHICAGO INSULATING CO.
109   135

211

APPENDIX II
CROSS REFERENCE OF PRIMARY EMBOSSING TO CD NUMBER

CIA COMERCIAL
106

CIA TELEFONICA
154

CITY FIRE ALARM
133

CLIMAX
184

COLUMBIA
262   263   264

COMBINATION SAFETY
139

CONVERSE
283

CRIMSA
155

CRISA
107

CRISOL TEXCOCO ▽
106.5   154   155

CROWN
268

CUTTER
1038

D.T.CO.
742

DERF
162.7

DERFLINGHER
154.5

DIAMOND ◇
102   106   108   112.4   112.5
115   121   122   152   154
155   164   190/191

DIAMOND P ◇P
134

DOMINION ◇D
108   115   122   154   155
164

DRY SPOT INSULATOR
182

DUPONT
740

DUQUESNE
106.1   106.3   113.2

DWIGHT
143

E.C.& M.CO.
123

E.D.R.
145

E.L.CO.
166

E.R.
133

E.R.W.
135.5   736

E.S.B.CO.
20   30   31   35   36

E.S.S.CO.
252

ELECTRICAL SUPPLY CO.
133.1   134

EMMINGER'S
141.9

ERICSSON
106   250.2

F.F.C.C.N.DE M.
162   162.7

FALL RIVER
134

FISHER
265

FLOY
284

FLUID INSULATOR
180.5

FOSTER
740

FT.W.E.CO.
162

G.E.CO.
134

G.N.R.
143

G.N.W.TEL.CO.
143   145

G.P.R.
143

G.T.P.TEL.CO.
145

212

APPENDIX II
CROSS REFERENCE OF PRIMARY EMBOSSING TO CD NUMBER

GAYNER
103 106 107 121 153
154 160 162 164 205
252

GOOD
106 121 134 162

GOULD
20 25 50 51

GREAT NORTHWESTERN
143

GREELEY
187 244.5

GREGORY
159

H.B.R.
145

H.G.CO.
133 145 151 160 162
164 196 196.2 196.5 201
317.7/314/314

H.G.W.
102

HAMILTON
162

HARLOE ⌬
109.5 206.5

HAWLEY ⌬
102 112 121 145 160
162 164 260

HEMINGRAY
100.6 102 106 107 112.4
113 114 115 121 122
122.4 124 125 128 128.4
129 133 134 137 142
145 147 152 154 155
155.6 157 160 162 163
164 165 168 169 175
178 183 185 186 186.1
186.2 190/191 194/195 196 197
201 202 203 205 208
213 214 216 218 219
220 221 230 230.1 231
232 232.1 233 233.2 237
238 238.1 239 241 241.2
242 243 248/311/311 249
251 252 254 257 263
271 280 281 282 283
291 295 301.5 302
303/310 303.5 304/310 307 313

HEMINGRAY (continued)
1050 1052 1070 1071 1087
1105

JEFFREY MFG.CO.
185

JOHNSON & WATSON
109.7

JUMBO
140 269

K
190/191 202

K.C.G.W.
102.3 120 134 145 160
<CD 162 164

KEELING
729.3

KERR
122 128 129 155 203

KIMBLE
231 231.2 239 239.2

KNOWLES
252 253 282 292.5

L.G.T.& CO.
131.4 732.2

LAWRENCE GRAY
138.2

LEFFERTS
737

LINEA DEL SUP<u>o</u>
735.5

LIQUID INSULATOR
180

LOCKE
178.5 202 204 241 275
286 287 287.1 288 289
293 293.1 296 297 300
318 319 342

LOWEX
216 230 299.7

LYNCHBURG Ⓛ
106 112 121 122 145
154 160 162 164 205
251 252 281 306

M.& E.CO.
252 254

M.T.CO.
61 742 742.3

213

APPENDIX II
CROSS REFERENCE OF PRIMARY EMBOSSING TO CD NUMBER

MANHATTAN
256

MAYDWELL
| 106 | 115 | 121 | 122 | 154 |
| 160 | 162 | 164 | 252 | |

McKEE & CO.
731

McLAUGHLIN
106	115	121	122	139.9
152	154	160	162	164
252				

McMICKING
734

MERSHON
288

MINE INSULATOR
185

MONTREAL
| 102 | 143 |

MULFORD & BIDDLE
| 729.4 | 735 |

N DE M
162

N.E.G.M.CO.
| 102 | 145 | 162 | 250 | 251 |
| 267 | 267.5 | 294 | | |

N.W.& B.I.T.CO.
102

N.Y.& E.R.R.
736

NATIONAL BATTERY CO.
24

NATIONAL INSULATOR CO.
| 104 | 110.5 | 110.6 | 138.2 |

NEW ENG.TEL.& TEL.CO.
| 102.4 | 104 | 110.5 | 112 | 121 |

NO EMBOSSING
10	12	22	22.5	26
28	29	29.5	35	36
40	50	51	100.6	102
103	104	105	106	106.3
112	118	120	121	121.4
122	125	126	126.3	127
127.4	128	131	131.4	131.8
132	132.2	133	133.2	133.3
133.4	134	134.4	136.4	136.7
137.5	138.2	139.8	140	141
141.5	142.4	144	145	145.6

NO EMBOSSING (continued)
146.4	146.5	147	149	151
152	154	155	156.1	156.2
158	158.1	158.2	162	162.3
164	164.4	166.2	170	170.1
175.5	178	181	181.5	185
188	190/191	192/193	192.1/193.1	
196	204	205.5	206	210
219	229.6	233	240.2	244
247	248/311/311		252	254
260	263	265	267	270
272	273	285	288	289
292	297	298	299.2	301
301.2	304/310	308	310	311
312	313.1	314	317.9	700
700.1	701	701.1	701.2	701.3
701.5	701.6	701.8	723	723.3
723.5	723.6	724	724.5	724.6
725	727	728	728.5	728.7
729	729.6	731	731.2	731.3
732	732.2	734.1	734.3	735.6
735.7	736	736.1	736.3	736.4
736.5	736.7	737	737.5	737.6
737.9	738	738.5	738.6	739
739.2	739.5	740.4	780	784
785	786	788	790	791
1000	1001	1002	1003	1004
1005	1006	1007	1008	1010
1012	1013	1014	1015	1016
1030	1040	1052	1053	1080
1085	1087	1090	1095	1104
1105	1110	1138	1140	1142

NO EMBOSSING - CANADA
102	112.5	115.1	120	121
133	133.2	134	143	144
145	162	162.4	718	721
722	724.3	726	734.5	734.8
740	740.1	740.3	740.7	742
742.3	743.1	743.2	743.3	

NO EMBOSSING - MEXICO
| 103.4 | 106.4 | 107.2 | 154.5 | 155 |
| 162.7 | | | | |

NO NAME
29	102	102.3	104	106
113	113.2	117	121	124
126	126.3	127	128	132.2
133	133.2	133.3	134	136.4
140.5	142	142.4	144.5	145
146.5	151	152	153	154
160	162	163.2	164	168.5
168.6	187	200	202	221
226	230	240.2	245	275
280	287	291	296	297

214

APPENDIX II
CROSS REFERENCE OF PRIMARY EMBOSSING TO CD NUMBER

NO NAME (continued)
  308   728.2   731   1049   1052
  1053

NO NAME - CANADA
  102   112.5   113   121   162
  162.4

NO NAME - MEXICO
  106   155   162

O.V.G.CO.
  106   112   121   133   145
  160   162   196

OAKMAN MFG.CO.
  134   140   259   263   269
  728.7

OWENS ILLINOIS
  128   167   1049

P & W
  133.2

P.L.W.
  162.5

P.R.R.
  162.5   190/191

P.S.S.A.
  106   106.2   107   155   162.7

PAISLEY
  132.2

PAT APP FOR
  109.9   120.2   125   132   133.1
  138.9   147   729.6

PAT'D
  102   126   132   133.4   134
  187

PATENT - DEC. 19, 1871
  104   116   120   124   124.2
  124.3   127.4   131.4   132   132.4
  133   133.1   133.4   134   732.2

PATENT - OTHER
  102   106   113   116   121
  125   126   134   136.7   141.6
  144   144.5   147   156.1   156.2
  158   162   178.5   187   188
  257   259   263   266   289.9
  292   317.8/313/313/313.1   321

PETTINGELL ANDREWS
  134

PONY
  120.2

POSTAL
  145   156   156.1   210

PRISM
  190/191   199   252   280   282
  283   292   296   299.1   315
  321   333

PYREX
  100.5   111   122.4   128   139.6
  194.5/195.5   233   234   235
  240   240.5   248/311/311   311
  320   322   323   324   325
  326   327   328   328.2   330
  331

R.Y T.
  155

S.B.T.& T.CO.
  112   1085

S.F.
  102   112

S.S.& CO.
  160   162

SANTA ANA
  178

SEILERS
  130.2

SO EX CO
  735

SO.MASS.TEL.CO.
  102

ST.LOUIS MALLEABLE
  1052

STANDARD
  143

STANDARD GLASS INSULATOR CO.
  104   114.2   121   138.2   145
  157.5   268.5

STAR
  102   104   106   112   113
  133   134   145   160   161
  162   162.1   162.3   164   185
  200   260

STERLING £
  102   112   160   164

SURGE
  100   100.2

T-H.
  245

215

APPENDIX II
CROSS REFERENCE OF PRIMARY EMBOSSING TO CD NUMBER

T-H.E.CO.
134   143.5   162

T.C.R.
145

TEL.FED.MEX.
133   133.1   133.5

TELEGRAFICA
740.8

TELEGRAFO
740.8

TELEGRAFOS NACIONALES
214

THAMES GLASS WORKS
718

TILLOTSON & CO.
131   718   731   735   736
737   740

TWIGGS
141.7

TWO PIECE TRANSPOSITION
190/191

U.S.L.
23   24   53

U.S.LIGHT & HEATING CO.
24

U.S.TEL.CO.
121   735.3

V.B.
106

W.E.MFG.CO.
126.4

W.F.G.CO.
106   121   134   162

W.G.M.CO.
106   121   134   145   162

W.U.
125   127

WARE
45   55

WESTINGHOUSE
102   113   134   162   286.9
309.5

WHITALL TATUM ▽
107   108   113   115   122
128   153   154   155   160
162   163   163.4   164   165.1
168   169   176   197   214
216   221   272

APPENDIX III
CROSS REFERENCE OF PATENT DATES TO CD NUMBER

JUL 25, 1865
B.F.G.CO.
133.2
BROOKFIELD
126   126.3   131   133   728.4
731
P & W
133.2
PAISLEY
132.2
W.U.
127

JAN 14, 1870 {Should be JAN 14, 1879}
BROOKFIELD
126   126.3

JAN 25, 1870
B & O
136
BROOKE'S
127
BROOKFIELD
126   126.3   127   131   133
134   136   138   145   728.4
CHESTER
123.2
L.G.T.& CO.
131.4
PATENT - OTHER
102   125   126
TILLOTSON & CO.
131

FEB 20, 1870 {Should be FEB 22, 1870}
BROOKFIELD
126

FEB 22, 1870
BROOKFIELD
126   126.1   126.3   127   131
133   728.4
L.G.T.& CO.
131.4
PATENT - OTHER
102
TILLOTSON & CO.
131

MAR 20, 1870 {Should be MAR 20, 1877}
BROOKFIELD
126

MAR 25, 1870 {Unattributed date}
PATENT - OTHER
102

APR 4, 1870 {Should be APR 4, 1871}
BROOKFIELD
127

JUL 26, 1870
BOSTON BOTTLE WORKS
728.8   796
OAKMAN MFG.CO.
728.7

APR 4, 1871
BROOKFIELD
127

DEC 19, 1871
A.U.
116   121.4   126.3
B & O
136
CHAMBERS
124.5   317   317.5
HEMINGRAY
124   125
PATENT - DEC. 19, 1871
104   116   120   124   124.2
124.3   127.4   131.4   132   132.4
133   133.1   133.4   134   732.2
SEILERS
130.2
T-H.E.CO.
134
W.E.MFG.CO.
126.4
W.U.
125

EC 19, 1871 {Should be DEC 19, 1871}
PATENT - DEC. 19, 1871
133.1

PEC 19, 1871 {Should be DEC 19, 1871}
B & O
136

FEB 20, 1872
EMMINGER'S
141.9

OCT 15, 1872
BOSTON BOTTLE WORKS
136.5   143.6   145.6   158.2
PATENT - OTHER
136.7   156.2   158

FEB 6, 1877
SEILERS
130.2

217

APPENDIX III
CROSS REFERENCE OF PATENT DATES TO CD NUMBER

MAR 20, 1877
BROOKFIELD
126
PATENT - OTHER
102

MAR 22, 1877 {Should be MAR 20, 1877}
BROOKFIELD
126

AUG 14, 1877
CHAMBERS
124.5   317   317.5

JAN 14, 1879
BROOKFIELD
126   126.3
PATENT - OTHER
102

JAN 25, 1879 {Should be JAN 25, 1870}
BROOKFIELD
126

FEB 22, 1879 {Should be FEB 22, 1870}
BROOKFIELD
126

FEB 12, 1881 {Should be FEB 12, 1884}
BROOKFIELD
145

SEP 13, 1881
AM.INSULATOR CO.
105   126.4   134   143.4   145
156   156.1   160.4   160.7
OAKMAN MFG.CO.
134
PATENT - OTHER
134   156.1
POSTAL
156   156.1

JUL 1, 1882
A.U.
116
PATENT - DEC. 19, 1871
116
PATENT - OTHER
116

MAY 1, 1883
NATIONAL INSULATOR CO.
110.5   110.6
NEW ENG.TEL.& TEL.CO.
110.5

AUG 14, 1883
BROOKE'S
120   133.1

OCT 16, 1883
CHICAGO INSULATING CO.
109   135

NOV 1883 {Should be NOV 13, 1883}
AM.INSULATOR CO.
145

NOV 13, 1883
AM.INSULATOR CO.
143.4   145   156.1   160.4   160.7
AM.TEL.& TEL.CO.
160.6
BROOKFIELD
143.4   145   160   162   164
FT.W.E.CO.
162
PATENT - OTHER
162
T-H.E.CO.
162

DEC 25, 1883
NATIONAL INSULATOR CO.
110.5   110.6
NEW ENG.TEL.& TEL.CO.
110.5

JAN 1, 1884
NATIONAL INSULATOR CO.
104   110.5   110.6   138.2
NEW ENG.TEL.& TEL.CO.
104   110.5

JAN 29, 1884 {Unattributed date}
AM.INSULATOR CO.
145

EEB 12, 1884 {Should be FEB 12, 1884}
BROOKFIELD
145

FAB 12, 1884 {Should be FEB 12, 1884}
BROOKFIELD
145

FEB 12, 1884
AM.INSULATOR CO.
143.4   145   160.7
BROOKFIELD
143.4   145   162

FEB 15, 1884 {Should be FEB 12, 1884}
AM.INSULATOR CO.
160.7

OCT 7, 1884
NATIONAL INSULATOR CO.
104   110.5   110.6   138.2
NEW ENG.TEL.& TEL.CO.
104   110.5

APPENDIX III
CROSS REFERENCE OF PATENT DATES TO CD NUMBER

APR 28, 1885
 BROOKFIELD
  119

AUG 11, 1885
 DIAMOND P ◇P◇
  134

NOV 23, 1886
 GREELEY
  187
 PATENT - OTHER
  187   188

MAY 7, 1889
 PATENT - OTHER
  289.9

MAY 6, 1890
 BUZBY
  141.8

JUN 17, 1890
 HEMINGRAY
  257
 KNOWLES
  252   253   292.5
 LYNCHBURG Ⓛ
  251
 MANHATTAN
  256
 N.E.G.M.CO.
  251   267
 OAKMAN MFG.CO.
  140   259   263   269
 PATENT - OTHER
  257   259   263   266   292
  321
 PRISM ▱
  315   321   333
 STAR ✶
  260

AUG 19, 1890
 OAKMAN MFG.CO.
  140   259   263   269
 PATENT - OTHER
  144.5   259

DEC 23, 1890
 PATENT - OTHER
  144

APR 30, 1891 {Unattributed date}
 LOCKE
  342

MAY 12, 1891
 COLUMBIA
  262   263
 HEMINGRAY
  263
 OAKMAN MFG.CO.
  263

MAY 1893 {Should be MAY 2, 1893}
 HEMINGRAY
  106

MAY 2, 1893
 CONVERSE
  283
 H.G.CO.
  133   151   160   162   164
  196   196.2   196.5   201
  317.7/314/314
 HEMINGRAY
  106   113   114   115   121
  124   125   133   134   152
  157   160   162   164   178
  201   202   251   252   257
  280   281   282   283   291
  295   302   303/310   303.5
  304/310   307   313
 PATENT - DEC. 19, 1871
  116   120   134
 PATENT - OTHER
  106   116   257
 W.U.
  125

JUN 17, 1893 {Should be JUN 17, 1890}
 HEMINGRAY
  257

AUG 8, 1893
 STANDARD GLASS INSULATOR CO.
  268.5

AUG 29, 1893
 PATENT - OTHER
  141.6

MAY 22, 1894
 LOCKE
  202   204   286   287   287.1
  288   289   293   293.1   297
  300

SEP 25, 1894
 JEFFREY MFG.CO.
  185

APR 7, 1896
 BROOKFIELD
  134.6

APPENDIX III
CROSS REFERENCE OF PATENT DATES TO CD NUMBER

MAY 22, 1896 {Should be MAY 22, 1894}
  LOCKE
   286   300   318   319

MAY 24, 1896 {Should be NOV 24, 1896}
  LOCKE
   300

NOV 2, 1896 {Should be NOV 24, 1896}
  LOCKE
   342

NOV 24, 1896
  LOCKE
   286   288   293   293.1   297
   300   318   319

DEC 1, 1896
  GOULD
   20   25   50   51

DEC 15, 1896
  LOCKE
   286   293   293.1   300   318
   319   342

SEP 28, 1897
  LOCKE
   286   288   293   293.1   297
   300   318   319   342

JUN 7, 1898
  LOCKE
   286   288   300   318   319
   342

DEC 15, 1898 {Should be DEC 15, 1896}
  LOCKE
   297

MAR 21, 1899
  HARLOE ⌗
   109.5   206.5

APR 25, 1899
  HEMINGRAY
   249   282   283

SEP 19, 1899
  PATENT - OTHER
   121

OCT 10, 1899
  BEAL'S
   309

MAR 12, 1901
  HARLOE ⌗
   109.5   206.5

APR 29, 1902
  LOCKE
   342

JUN 10, 1902
  PATENT - OTHER
   317.8/313/313/313.1

DEC 9, 1902
  HARLOE ⌗
   109.5   206.5

APR 7, 1903
  GREGORY
   159

APR 26, 1904
  CUTTER
   1038

AUG 29, 1905
  TWIGGS
   141.7

OCT 8, 1907
  BARCLAY
   150
  BROOKFIELD
   110   150
  HEMINGRAY
   147
  PATENT - OTHER
   147

MAR 31, 1914
  LOCKE
   178.5   241

FEB 2, 1915
  LOCKE
   178.5   241
  PATENT - OTHER
   178.5

OCT 12, 1915
  LOCKE
   241

MAY 27, 1919
  PYREX
   139.6   248/311/311   311

APPENDIX IV
ABBREVIATED PRIMARY EMBOSSING CROSS REFERENCE

PRIMARY EMBOSSING TO ABBREVIATION

A.A. (AA)	AA
A.T.& T.CO.	ATT
A.U.	AU
AM.INSULATOR CO.	AMINS
AM.TEL.& TEL.CO.	AMTEL
ARMSTRONG (A)	ARM
AYALA	AYALA
B	B
B & O	B&O
B.C.DRIP	BCD
B.E.L.CO.	BEL
B.F.G.CO.	BFG
B.G.M.CO.	BGM
B.T.C.	BTC
BARCLAY	BARC
BEAL'S	BEALS
BIRMINGHAM	BIRM
BOSTON BOTTLE WORKS	BOS
BROOKE'S	BRKS
BROOKFIELD	BF
BUZBY	BUZBY
C.& P.TEL.CO.	CP
C.D.& P.TEL.CO.	CDP
C.E.L.CO.	CEL
C.E.W.	CEW
C.G.I.CO.	CGI
C.N.R.	CNR
C.P.R.	CPR
CABLE	CABLE
CAL.ELEC.WORKS	CALEW
CALIFORNIA	CAL
CANADA	CAN
CANADIAN PACIFIC RY.CO.	CANP
CHAMBERS	CHAMB
CHESTER	CHEST
CHICAGO INSULATING CO.	CHI
CIA COMERCIAL	CIAC
CIA TELEFONICA	CIAT
CITY FIRE ALARM	CITY
CLIMAX	CLIM
COLUMBIA	COL
COMBINATION SAFETY	COMB
CONVERSE	CONV
CRIMSA	CRIM
CRISA	CRISA
CRISOL TEXCOCO	CRISO
CROWN	CROWN
CUTTER	CUT
D.T.CO.	DT
DERF	DERF
DERFLINGHER	DERFL
DIAMOND ◇	DIAM
DIAMOND P	DIAMP
DOMINION	DOM
DRY SPOT INSULATOR	DRY
DUPONT	DUP
DUQUESNE	DUQ
DWIGHT	DWI
E.C.& M.CO.	ECM
E.D.R.	EDR
E.L.CO.	EL
E.R.	ER
E.R.W.	ERW
E.S.B.CO.	ESB
E.S.S.CO.	ESS
ELECTRICAL SUPPLY CO.	ELECT
EMMINGER'S	EMMIN
ERICSSON	ERIC
F.F.C.C.N.DE M.	FFCCN
FALL RIVER	FALL
FISHER	FISH
FLOY	FLOY
FLUID INSULATOR	FLUID
FOSTER	FOS
FT.W.E.CO.	FTWE
G.E.CO.	GE
G.N.R.	GNR
G.N.W.TEL.CO.	GNW
G.P.R.	GPR
G.T.P.TEL.CO.	GTP
GAYNER	GAY
GOOD	GOOD
GOULD	GOULD
GREAT NORTHWESTERN	GNORW
GREELEY	GREEL
GREGORY	GREG
H.B.R.	HBR
H.G.CO.	HG
H.G.W.	HGW
HAMILTON	HAM
HARLOE	HAR
HAWLEY	HAW
HEMINGRAY	HEMI
JEFFREY MFG.CO.	JEFF
JOHNSON & WATSON	J&W
JUMBO	JUMBO
K	K
K.C.G.W.	KCGW
KEELING	KEEL
KERR	KERR
KIMBLE	KIM
KNOWLES	KNOWL
L.G.T.& CO.	LGT
LAWRENCE GRAY	LGRAY
LEFFERTS	LEFF
LINEA DEL SUP<u>o</u>	LINEA

221

APPENDIX IV
ABBREVIATED PRIMARY EMBOSSING CROSS REFERENCE

PRIMARY EMBOSSING TO ABBREVIATION

LIQUID INSULATOR	LIQ
LOCKE	LOCKE
LOWEX	LOWEX
LYNCHBURG Ⓛ	LYNCH
M.& E.CO.	M&E
M.T.CO.	MT
MANHATTAN	MAN
MAYDWELL	MAYD
McKEE & CO.	MCKEE
McLAUGHLIN	MCL
McMICKING	MCMIC
MERSHON	MERSH
MINE INSULATOR	MINE
MONTREAL	MONT
MULFORD & BIDDLE	M&B
N ᴰᴱ M	NDEM
N.E.G.M.CO.	NEGM
N.W.& B.I.T.CO.	NWBIT
N.Y.& E.R.R.	NYERR
NATIONAL BATTERY CO.	NB
NATIONAL INSULATOR CO.	NAT
NEW ENG.TEL.& TEL.CO.	NET
NO EMBOSSING	NE
NO EMBOSSING - CANADA	NECAN
NO EMBOSSING - MEXICO	NEMEX
NO NAME	NN
NO NAME - CANADA	NNCAN
NO NAME - MEXICO	NNMEX
O.V.G.CO.	OVG
OAKMAN MFG.CO.	OAK
OWENS ILLINOIS ⊙	OI
P & W	P&W
P.L.W.	PLW
P.R.R.	PRR
P.S.S.A. ⓢ	PSSA
PAISLEY	PAIS
PAT APP FOR	PATAP
PAT'D	PATD
PATENT - DEC. 19, 1871	PTDEC
PATENT - OTHER	PTOTH
PETTINGELL ANDREWS	PETAN
PONY	PONY
POSTAL	POST
PRISM ▭	PRISM
PYREX	PYREX
R.Y T. ⓡ	RYT
S.B.T.& T.CO.	SBT
S.F.	SF
S.S.& CO.	SS
SANTA ANA	SANTA
SEILERS	SEIL
SO EX CO	SOEX
SO.MASS.TEL.CO.	SOMA
ST.LOUIS MALLEABLE	STLM
STANDARD	STAN
STANDARD GLASS INSULATOR CO.	STANG
STAR ✶	STAR
STERLING £	STERL
SURGE	SURGE
T-H.	TH
T-H.E.CO.	THE
T.C.R.	TCR
TEL.FED.MEX.	TELFM
TELEGRAFICA	TELGA
TELEGRAFO	TELGO
TELEGRAFOS NACIONALES	TELNA
THAMES GLASS WORKS	TGW
TILLOTSON & CO.	TILL
TWIGGS	TWIGG
TWO PIECE TRANSPOSITION	TWO
U.S.L.	USL
U.S.LIGHT & HEATING CO.	USLH
U.S.TEL.CO.	USTEL
V.B.	VB
W.E.MFG.CO.	WEMFG
W.F.G.CO.	WFG
W.G.M.CO.	WGM
W.U.	WU
WARE	WARE
WESTINGHOUSE	WEST
WHITALL TATUM ▽	WT

APPENDIX IV
ABBREVIATED PRIMARY EMBOSSING CROSS REFERENCE

ABBREVIATION TO PRIMARY EMBOSSING

AA	A.A. (AA)	DRY	DRY SPOT INSULATOR
AMINS	AM.INSULATOR CO.	DT	D.T.CO.
AMTEL	AM.TEL.& TEL.CO.	DUP	DUPONT
ARM	ARMSTRONG (A)	DUQ	DUQUESNE
ATT	A.T.& T.CO.	DWI	DWIGHT
AU	A.U.	ECM	E.C.& M.CO.
AYALA	AYALA	EDR	E.D.R.
B	B	EL	E.L.CO.
B&O	B & O	ELECT	ELECTRICAL SUPPLY CO.
BARC	BARCLAY	EMMIN	EMMINGER'S
BCD	B.C.DRIP	ER	E.R.
BEALS	BEAL'S	ERIC	ERICSSON
BEL	B.E.L.CO.	ERW	E.R.W.
BF	BROOKFIELD	ESB	E.S.B.CO.
BFG	B.F.G.CO.	ESS	E.S.S.CO.
BGM	B.G.M.CO.	FALL	FALL RIVER
BIRM	BIRMINGHAM	FFCCN	F.F.C.C.N.DE M.
BOS	BOSTON BOTTLE WORKS	FISH	FISHER
BRKS	BROOKE'S	FLOY	FLOY
BTC	B.T.C.	FLUID	FLUID INSULATOR
BUZBY	BUZBY	FOS	FOSTER
CABLE	CABLE	FTWE	FT.W.E.CO.
CAL	CALIFORNIA	GAY	GAYNER
CALEW	CAL.ELEC.WORKS	GE	G.E.CO.
CAN	CANADA	GNORW	GREAT NORTHWESTERN
CANP	CANADIAN PACIFIC RY.CO.	GNR	G.N.R.
CDP	C.D.& P.TEL.CO.	GNW	G.N.W.TEL.CO.
CEL	C.E.L.CO.	GOOD	GOOD
CEW	C.E.W.	GOULD	GOULD
CGI	C.G.I.CO.	GPR	G.P.R.
CHAMB	CHAMBERS	GREEL	GREELEY
CHEST	CHESTER	GREG	GREGORY
CHI	CHICAGO INSULATING CO.	GTP	G.T.P.TEL.CO.
CIAC	CIA COMERCIAL	HAM	HAMILTON
CIAT	CIA TELEFONICA	HAR	HARLOE (H)
CITY	CITY FIRE ALARM	HAW	HAWLEY (H)
CLIM	CLIMAX	HBR	H.B.R.
CNR	C.N.R.	HEMI	HEMINGRAY
COL	COLUMBIA	HG	H.G.CO.
COMB	COMBINATION SAFETY	HGW	H.G.W.
CONV	CONVERSE	J&W	JOHNSON & WATSON
CP	C.& P.TEL.CO.	JEFF	JEFFREY MFG.CO.
CPR	C.P.R.	JUMBO	JUMBO
CRIM	CRIMSA	K	K
CRISA	CRISA	KCGW	K.C.G.W.
CRISO	CRISOL TEXCOCO (T)	KEEL	KEELING
CROWN	CROWN	KERR	KERR
CUT	CUTTER	KIM	KIMBLE
DERF	DERF	KNOWL	KNOWLES
DERFL	DERFLINGHER	LEFF	LEFFERTS
DIAM	DIAMOND ◇	LGRAY	LAWRENCE GRAY
DIAMP	DIAMOND P (P)	LGT	L.G.T.& CO.
DOM	DOMINION ◇	LINEA	LINEA DEL SUP.º

223

APPENDIX IV
ABBREVIATED PRIMARY EMBOSSING CROSS REFERENCE

ABBREVIATION TO PRIMARY EMBOSSING

LIQ	LIQUID INSULATOR
LOCKE	LOCKE
LOWEX	LOWEX
LYNCH	LYNCHBURG Ⓛ
M&B	MULFORD & BIDDLE
M&E	M.& E.CO.
MAN	MANHATTAN
MAYD	MAYDWELL
MCKEE	McKEE & CO.
MCL	McLAUGHLIN
MCMIC	McMICKING
MERSH	MERSHON
MINE	MINE INSULATOR
MONT	MONTREAL
MT	M.T.CO.
NAT	NATIONAL INSULATOR CO.
NB	NATIONAL BATTERY CO.
NDEM	N DE M
NE	NO EMBOSSING
NECAN	NO EMBOSSING - CANADA
NEGM	N.E.G.M.CO.
NEMEX	NO EMBOSSING - MEXICO
NET	NEW ENG.TEL.& TEL.CO.
NN	NO NAME
NNCAN	NO NAME - CANADA
NNMEX	NO NAME - MEXICO
NWBIT	N.W.& B.I.T.CO.
NYERR	N.Y.& E.R.R.
OAK	OAKMAN MFG.CO.
OI	OWENS ILLINOIS
OVG	O.V.G.CO.
P&W	P & W
PAIS	PAISLEY
PATAP	PAT APP FOR
PATD	PAT'D
PETAN	PETTINGELL ANDREWS
PLW	P.L.W.
PONY	PONY
POST	POSTAL
PRISM	PRISM
PRR	P.R.R.
PSSA	P.S.S.A.
PTDEC	PATENT - DEC. 19, 1871
PTOTH	PATENT - OTHER
PYREX	PYREX
RYT	R.Y T.
SANTA	SANTA ANA
SBT	S.B.T.& T.CO.
SEIL	SEILERS
SF	S.F.
SOEX	SO EX CO
SOMA	SO.MASS.TEL.CO.
SS	S.S.& CO.
STAN	STANDARD
STANG	STANDARD GLASS INSULATOR CO.
STAR	STAR ✶
STERL	STERLING £
STLM	ST.LOUIS MALLEABLE
SURGE	SURGE
TCR	T.C.R.
TELFM	TEL.FED.MEX.
TELGA	TELEGRAFICA
TELGO	TELEGRAFO
TELNA	TELEGRAFOS NACIONALES
TGW	THAMES GLASS WORKS
TH	T-H.
THE	T-H.E.CO.
TILL	TILLOTSON & CO.
TWIGG	TWIGGS
TWO	TWO PIECE TRANSPOSITION
USL	U.S.L.
USLH	U.S.LIGHT & HEATING CO.
USTEL	U.S.TEL.CO.
VB	V.B.
WARE	WARE
WEMFG	W.E.MFG.CO.
WEST	WESTINGHOUSE
WFG	W.F.G.CO.
WGM	W.G.M.CO.
WT	WHITALL TATUM
WU	W.U.

INSULATOR REFERENCE MATERIAL

Please contact the authors for current publication pricing.

Albers, Marilyn, 14715 Oak Bend Drive, Houston TX 77079
Glass Insulators From Outside North America, 2nd Revision, Albers/Woodward, 1993
Price Guide for G.I.F.O.N.A., 2nd Revision, Albers/Woodward, 1996
Worldwide Porcelain Insulators, Albers/Tod, 1982
Worldwide Porcelain Insulators, 1986 Supplement, Albers/Tod, 1986

Burnet, Robert G., P.O. Box 40526, 5230 Dundas St. West, Etobicoke, Ontario M9B 6K8 Canada
Canadian Railway Telegraph History, Burnet, 1996
Autographs and Memoirs of the Telegraph by Jeff W. Hayes, reprint by Burnet, 1999

Cranfill, Gary, 6353 Buckeye Lane, Granite Bay, CA 95746-9681
The Glass Insulator - A Comphrehensive Reference, Cranfill, 1969, Price List updated 1996

Gish, Elton, P.O. Box 1317, Buna, TX 77612
Multipart Porcelain Insulators, Gish, 1988
Fred M. Locke, A Biography, Gish, 1994
Value Guide for Unipart and Multipart Porcelain Insulators, Gish, 1995
Porcelain Insulators: Guide Book for Collectors, Tod, 1988, reprint by Gish, 1995
Insulator Patents, 1880-1960, Tod, 1985, reprint by Gish, 1995
1990 Fred M. Locke Catalog #5, reprint by Gish
1902 Fred M. Locke Catalog #6, reprint by Gish
1904 Lima Insulator Co. Catalog #1, reprint by Gish
1908 Pittsburg Catalog #1, reprint by Gish

Howard, Dan, 2940 S.E. 118th Avenue, Portland, OR 97266-1602
Old Familiar Strains, bimonthly magazine

Hyve, H.G. "Bea", 3269 N. Mtn. View Dr., San Diego, CA 92116-1736
The Hemingray Glass Company-A Most Colorful History, Hyve, 1998

Lauckner, Mark, Mayne Island, B.C., Canada V0N 2J0
Spec-True, Collectible Glass Standard Color Reference, Lauckner, 1991
Canadian Railway Communications Insulators 1880-1920, Lauckner, 1995

McDougald, John & Carol, Box 1003, St. Charles, IL 60174
Crown Jewels of the Wire, monthly magazine
INSULATORS - A History and Guide to North American Glass Pintype Insulators, Volumes 1& 2, McDougald, 1990
Price Guide for the McDougald books, McDougald, 1999
What Are Insulators?, brochure, McDougald/Hyve, 1997

Padgett, Fred, P.O. Box 1122, Livermore, CA 94551
Dreams of Glass, The Story of William McLaughlin and His Glass Company, Padgett, 1996

Tucker, Mike, 4527 Middle Road, Cherokee, IA 51012
In Search of Threadless-A Source Guide for Threadless Diggers, Tucker, 1990

Walsh, C. D., 3289 Carriage Court, Clinton, WA 98236
Most About Glass Insulators, Milholland, 1976

Woodward, N.R., Box 171, Houston, TX 77001
Searching for Railway Telegraph Insulators, Neal, 1982 (England)
Railway and Other Rare Insulators, Neal, 1987 (England)
The Glass Insulator In America-1988 Report, Woodward, 1988

GO-WITHS

CROWN POINT, P.O. Box 23, Winfield, IL 60190
The Crown Point, lightning rods/balls magazine, 4 issues annually

Krupka, Rod, 2615 Echo Lane, Ortonville, MI 48462
The Complete Book of Lightning Rod Balls with Prices, Krupka & Bruner

Linscott, Jacqueline, 3557 Nicklaus Drive, Titusville, FL 32780
Blue Bell Paperweights, 1992 Revised, Linscott, 1992 and *Addendum,* Linscott, 1995

INTERNET RESOURCES

www.insulators.com
Bill and Jill Meier, designers and producers of the website.

On line since April 1995, this is the oldest and largest website for all on-line insulator and go-with collectors. Although the site's primary focus is on pintype insulators, there is also information on LRI's, LRB's, spools, radio strains, water bottles, fruit jars, ashtrays, and related stamps. The site is packed with information on books, magazines, the Insulator Research Service, historical photographs and other reference material. There are hundreds of color photographs of glass, porcelain and composition, U.S. and foreign insulators. Locating insulator and bottle shows around the country is easy with their unique "Show Finder!". The web surfer will also enjoy the weekly updates of photographs, articles, news, show reports and other material. The site also features "for-sale list" posting by anyone, and the popular "Insulator Finder!" which matches people's wants with insulators currently for sale and notifies them via E-mail. There are also pages for people to tell other collectors about themselves, as well as information about all the insulator clubs and how to join them. These services are offered at no cost. To supplement the website, they also provide a mailing list of over 700 Insulator Collectors On the Net (ICON), where on-line collectors can ask questions and exchange information.

www.crownjewelsofthewire.com
John and Carol McDougald, *Crown Jewels of the Wire*

This site will soon feature a searchable library of all of the stories from the monthly magazine *Crown Jewels of the Wire,* which has served the insulator collecting community since 1969. Showcased are winners and nominees of the N.R. Woodward Literary Award given annually by the Lone Star Insulator Club for the best article published in publications and on websites. Also available is information on reference materials for collectors and a calendar of insulator shows and events. Most popular is the on-line version of McDougald's *Price Guide for Insulators* where market value can be researched. "Grampa Mac's Antique Insulator Emporium" is an on-line store stocked with both insulators and lightning rod balls. A color photo, description and price accompanies each item in stock. You can shop 24 hours a day...the emporium never closes. Come visit the site.

www.nia.org
Bob Berry, Webmaster

The National Insulator Association (NIA) is an international organization of collectors and friends interested in communication and electrical insulators, as well as other artifacts connected with insulators, such as telephone, telegraph, power transmission, railroads, and lightning protection devices.

The NIA was established and officially founded on July 7, 1973, at the national show held in Hutchinson, Kansas. In the first year of its operation, more than 800 charter members joined its ranks. Since then, annual paid membership has averaged 900 members. In all, the NIA has had more than 5900 total members.

The site features general information about the organization, information for new collectors, and important articles on these subjects: "Artificially Induced Colors", "History of the National Insulator shows"and "Members' Promotional Activities".

Insulator WebRing

There is also an Insulator WebRing, which provides easy navigation to over two dozen other web sites created by individual collectors and insulators clubs that have insulator content. At the bottom of the home page of any of the sites listed above is the navigation controls for moving around the Insulator WebRing. Click on "next" in the WebRing navigation control, and see what other great sites you can find! Or, jump to a random site!

McDougald's
INSULATORS
A History and Guide to North American Glass Pintype Insulators

VOLUME 1
History - Color - Subject Index
The most extensive history to date on the manufacture, sale and use of glass pintype insulators. New material never published before in an insulator reference book. 90+ full-color photographs assist in identification and serve as a color reference guide. Index aids research of companies, embossings and historical personalities.

VOLUME 2
Pictorial Guide
New, original and full-size black and white photographs of over 320 threaded and 90 threadless CD style are shown for easy identification. Full embossings are included as well as appropriate patent information regarding specific and unusual insulator styles. Cross Reference Table for Manufacturer and User Embossing to CD Number.

-- 500 plus 8 1/2" x 11" pages --

"Has the potential to re-awaken a broad collecting interest in insulators. Vol- ume One presents a detailed history of the development of the glass insulator, de- tailed his- tories of major manufac- turers, and a clear explanation of the various types. Especially helpful are the full color plates show- ing how to iden- tify glass in- sulator colors. T h e McDougald's first volume clearly illus- trates the re- search commit- ment of the collectible's collector and the willingness on the part of the collector to share it. An- tiques collec- tors who rely heavily on museum cura- tors and oth- ers to do their research. take notice. There are few topics in your area covered as well. The second Mc- Dougald vol- ume is de- voted to iden- tifying glass insulators by type. Pho- tographs are large and clear. There is no reason why anyone, whether an insu- lator collector or not, should have trouble identifying any example encountered."

HARRY RINKER, Editor, *Warman's Americana & Collectibles*.

TWO VOLUME SET
$49.⁵⁰ ppd

Orders to:
The McDougalds
P.O. Box 1003
St. Charles, IL 60174-7003
www.crownjewelsofthewire.com

CROWN JEWELS *of the Wire*

Established in 1969, this internationally circulated monthly magazine is devoted exclusively to insulator collecting, telephone/telegraph history and related collectibles. Columns feature porcelain, foreign and glass insulators, insulator go-withs, *Mac's Believe It Or Not, Walking the Lines, Ask Woody, Patent Office, Advertising through the Ages,* articles and letters from collectors, show reports, collector profiles and club news. *Coming Events* with free listings and *CrossRoads* classified ads sections are monthly features. A directory of *Crown Jewels of the Wire* subscribers and National Insulator Association members is available annually in October.

U.S./Canada SUBSCRIPTION RATES:

First Class	$31.50 no directory
First Class	$35.00 with directory
Second Class	$23.50 no directory
Second Class	$27.00 with directory
Overseas Airmail	$55.00 no directory
Overseas Airmail	$58.50 with directory

A current sample copy and informational packet is available for $4.00 from the address below.

CROWN JEWELS *of the Wire*
Box 1003
St. Charles, IL 60174-7003
editor@crownjewelsofthewire.com

www.crownjewelsofthewire.com

Grampa Mac's
Antique
Insulator Emporium

"Old Fashioned Service at Old Fashioned Prices"

Quality insulators are still available at reasonable prices.

Grampa Mac, Prop.

What is Grampa Mac's Antique Insulator Emporium?

It is an on-line stockroom filled with hundreds of insulators for retail sale and, in some cases, at wholesale prices. A complete description of each glass insulator in stock along with a photograph of the actual piece is available. If a photo is not posted on a specific insulator you wish to see, you can request that a picture be taken and made available. You can even haggle on the price with Grampa Mac if you feel the insulator is overpriced. And, if you can't sleep and want to look at insulators in the middle of the night, the emporium never closes its doors.

What kind of insulators are sold?

Currently glass pintype insulators are stocked, but U.S. porcelain and both glass and porcelain from outside of North America will be added shortly.

How expensive are the insulators?

Over 3,000 insulators have been sold annually from the Emporium's shelves. The insulators are priced to sell. Items listed range from $1.00 to $1,500. And, Grampa Mac guarantees your satisfaction on goods received if you are not entirely pleased.

Is this an auction?

No, but auctions have been new and exciting additions to the collector's marketplace. A word of caution. Before you purchase an insulator at auction, it might be wise to check on whether a similar item can be purchased outright for less. Recently, an item sold for over $100 at auction. Five identical pieces were available at the Emporium for $25 each. The same day the auction closed, all five sold for 75% less than the auction price. Become an educated buyer! You can always E-mail the Emporium to see if an item will be available soon. It may save you some money.

Want to sell your collection?

Grampa Mac is actively purchasing collections of all sizes, so crank the wall phone and give him a call. (630) 513-1544

CLICK ON THE "EMPORIUM" BUTTON AT:
www.crownjewelsofthewire.com

Notes

Notes

Notes

Notes